The Costs of Climate Change Mitigation Innovations

The notion that humanity may be too late to alter climate change could potentially lead to fear and therefore the advocacy of implementing radical strategies and/or hastening the execution of certain measures to the extreme. There is evidence that extensive and intensive implementation of some climate change solutions can significantly alter the environment and ecosystems in unintended ways. For example, the microclimate of a field in the proximity and downstream of a closely packed array of wind turbines can be noticeably altered by the modified lower atmospheric fluxes caused by the turbines, which can then negatively affect crop yields. Additionally, some studies have found that large-scale solar fields can result in the modulation of atmospheric circulation, leading to changes in regional precipitation. *The Costs of Climate Change Mitigation Innovations: A Pragmatic Outlook* provides a forum for discussion on the long-term consequences of various climate strategies. It promotes our striving toward minimizing the potential negative impact of new interventions by performing objective, holistic analyses. The bottom line is that we do not want today's solutions to become tomorrow's problems.

David S-K. Ting studied Combustion and Turbulence, followed by Convection Heat Transfer and Fluid-Structure Interactions, prior to joining the University of Windsor. Dr. Ting is the founder of the Turbulence and Energy Laboratory and a professor in the Department of Mechanical, Automotive and Materials Engineering. Professor Ting supervises students on a wide range of research topics including Energy Conservation and Renewable Energy. To date, he has co-supervised over 90 graduate students, co-authored more than 170 journal papers, authored five textbooks, and co-edited more than 25 volumes.

Jacqueline A. Stagner is the Undergraduate Programs Coordinator in the Faculty of Engineering at the University of Windsor, and an adjunct faculty member in the Department of Mechanical, Automotive and Materials Engineering. Dr. Stagner co-advises students primarily in sustainable energy in the Turbulence and Energy Laboratory. Prior to working at the University of Windsor, she attained a PhD in Materials Science and Engineering, a Master of Business Administration, and a bachelor's degree in Mechanical Engineering. She also worked as a release engineer in the automotive industry for six years. She has co-edited ten volumes.

The Costs of Climate Change Mitigation Innovations

A Pragmatic Outlook

Edited by
David S-K. Ting and
Jacqueline A. Stagner

CRC Press is an imprint of the
Taylor & Francis Group, an informa business

Cover image: © William Wang

First edition published 2024
by CRC Press
6000 Broken Sound Parkway NW, Suite 300, Boca Raton, FL 33487-2742

and by CRC Press
4 Park Square, Milton Park, Abingdon, Oxon, OX14 4RN

CRC Press is an imprint of Taylor & Francis Group, LLC

ISBN: 978-1-032-51681-3 (hbk)
ISBN: 978-1-032-51683-7 (pbk)
ISBN: 978-1-003-40345-6 (ebk)

DOI: 10.1201/9781003403456

Typeset in Times
by codeMantra

To all those who strive to exercise objectivity and sensitivity in mitigating climate change.

Contents

Editors

David S-K. Ting studied Combustion and Turbulence, followed by Convection Heat Transfer and Fluid-Structure Interactions, prior to joining the University of Windsor. Dr. Ting is the founder of the Turbulence and Energy Laboratory and a professor in the Department of Mechanical, Automotive and Materials Engineering. Professor Ting supervises students on a wide range of research topics including Energy Conservation and Renewable Energy. To date, he has co-supervised over 90 graduate students, co-authored more than 170 journal papers, authored five textbooks, and co-edited 25 volumes.

Jacqueline A. Stagner is the Undergraduate Programs Coordinator in the Faculty of Engineering at the University of Windsor, and an adjunct faculty member in the Department of Mechanical, Automotive and Materials Engineering. Dr. Stagner co-advises students primarily on sustainable energy in the Turbulence and Energy Laboratory. Prior to working at the University of Windsor, she attained a PhD in Materials Science and Engineering, a Master of Business Administration, and a bachelor's degree in Mechanical Engineering. She also worked as a release engineer in the automotive industry for six years. She has co-edited ten volumes.

Preface

A multitude of ideas have been put forward in recent years to tackle the challenges faced by humanity. The notion that we may be too late inevitably led to fear and the advocacy of implementing radical strategies and/or hastening the execution of some measures to the extreme. There is evidence that extensive and intensive implementation of some climate change solutions can significantly alter the environment and ecosystem. Case in point, the microclimate of a field in the proximity and downstream of a closely packed array of wind turbines can be noticeably altered by the modified lower atmospheric fluxes caused by the turbines. For a crop field, the yields are thus accordingly affected. Some studies have found that large-scale solar panels can result in the modulation of atmospheric circulation, leading to changes in regional precipitation. This volume provides a forum for discussion of the pros and cons of various strategies. This is explained by **Ting** and **Stagner** in the opening chapter, "The Pros and Cons of Climate Change Mitigation Innovations," Chapter 1. This book aims to promote our striving toward understanding the overall impact of new inventions by performing objective analyses similar to "cradle to grave" or "circular economics." The bottom line is that we do not want our solutions to be tomorrow's problems. **Gökgöz** and **Başbilen** enlighten us with "Investigating the R&D and Innovation Economic Efficiencies of the Renewable Energy Sectors in EU" in Chapter 2. The key to success is efficiency, specifically, research and development efficiencies and innovation economic efficiencies. For greening buildings, robust building information modeling is a potent strategy to be applied at an early stage. This is expounded in Chapter 3, "BIM-Powered Energy Efficiency and Life-Cycle Cost Analysis for Greener Design," by **Balo**, **Sua** and **Boydak**. Tall buildings are unique, including being known for excessive energy demand, and must be treated so. **Armstrong** and **Al-Kodmany** provide an up-to-date review on "Improving Energy Efficiency of Tall Buildings Using Innovative Environmental Systems," in Chapter 4. This chapter conveys the use of modern curtain wall systems, atriums, sky gardens, etc. to environmentally furnish thermal comfort, minimize energy consumption, and enhance livability. Supertall buildings can mitigate the rising demand for housing and commercial space. In Chapter 5, **Ilgın** discloses "Efficiency of Space Utilization in Supertall Towers with Free Forms." Based on 35 selected case studies, their analysis finds that, among other things, the average space utilization efficiency is below 75%, that is, over one quarter of the space is not being utilized. They recommend using structural logic to improve space efficiency, with emphasis on the service core and structural elements. From a regional perspective, the energy security of the national economy is paramount. As such, **Chygryn** and **Khomenko** convey "Linking between Renewables Development and Energy Security: A Scoping Review," in Chapter 6. They surveyed more than 3000 publications and identified seven clusters devoted to renewables development and energy security. Politics dictates almost everything and, hence, a good energy policy is essential. **Patel**, **Sarkar**, **Singh** and **Tyagi** present Chapter 7, "Energy Policy: Formulation, Monitoring and Adaptation for Moving Towards a Low Carbon Economy." They provide a comprehensive methodology for developing a sustainable energy policy. They highlight the necessity to develop an international, cross-border energy plan. **Patel** and **Patel** furnish us with a detailed disclosure on furthering solar thermal systems in Chapter 8, "Recent Developments in Large-Scale Solar Flat Plate Reflecting Systems: Optical Analysis using Specialized Numerical and Analytical Tools." Among other elements, this comprehensive chapter expounds on cutting-edge methods for analyzing flat plate-type solar radiation reflector-based solar thermal systems. The progress in renewable energy is not always beneficial to those who are in remote, less developed and/or low-income regions. **Suvarna**, **Ting** and **Clement**, in Chapter 9, "Energy Poverty and the Sustainable Development of Renewable Energy Systems," report that well-developed, large-scale renewable energy systems faired less well than the smaller ones of less than 20 MW. Using case studies from varied regions, the factors affecting the success of a renewable project are revealed. This volume concludes with Chapter 10, "The Challenges of Stakeholders Engagement in Climate Change Adaptation in Nigeria," by **Oruonye** and **Ayuba**. This is a case study based on Nigeria, where progress toward climate change mitigation has not been timely. It is found that corrupt leadership emerges at the top of the list of hindrances. An illiterate population, inadequate tools, and working in silos form the threats to effective stakeholder engagement. The solution includes involving all stakeholders. After all, every living soul can contribute to a brighter tomorrow.

Contributors

Kheir Al-Kodmany
Department of Urban Planning
University of Illinois at Chicago
Chicago, Illinois

Paul Armstrong
School of Architecture
University of Illinois at Urbana-Champaign
Champaign, Illinois

H. K. Ayuba
Department of Geography
Nasarawa State University
Keffi, Nigeria

Figen Balo
Department of Industrial Engineering
Firat University
Elazığ, Turkey

Gaye Demirhan Başbilen
Faculty of Political Sciences, Quantitative Methods Division, Department of Management
Ankara University
Ankara, Türkiye

Hazal Boydak
Department of Architecture
Dicle University
Diyarbakır, Turkey

Olena Chygryn
Department of Marketing
Sumy State University
Sumy, Ukraine

Shibu Clement
Department of Mechanical Engineering
Birla Institute of Technology and Science
Pilani, India

Fazıl Gökgöz
Faculty of Political Sciences, Quantitative Methods Division, Department of Management
Ankara University
Ankara, Türkiye

Hüseyin Emre Ilgın
School of Architecture
Tampere University
Tampere, Finland

Liliia Khomenko
Department of Marketing and Management of Innovative Activities
Sumy State University
Sumy, Ukraine

E. D. Oruonye
Department of Geography
Taraba State University
Jalingo, Nigeria

Amit R. Patel
Faculty of Technology & Engineering, Department of Mechanical Engineering
The Maharaja Sayajirao University of Baroda
Vadodara, Gujarat

Jay Patel
Faculty of Technology & Engineering, Department of Mechanical Engineering
The Maharaja Sayajirao University of Baroda
Vadodara, Gujarat

Prabir Sarkar
Indian Institute of Technology Ropar
Rupnagar, Punjab, India

Harpreet Singh
Indian Institute of Technology Ropar
Rupnagar, Punjab, India

Jacqueline A. Stagner
Faculty of Engineering, Turbulence & Energy Laboratory
University of Windsor
Windsor, ON, Canada

Lütfü S. Sua
Department of Decision Sciences and Supply Chain
Southern University
Baton Rouge, Louisiana

Ishan Suvarna
Department of Mechanical Engineering
Birla Institute of Technology and Science
Pilani, India
and
Turbulence & Energy Laboratory
University of Windsor
Windsor, ON, Canada

David S-K. Ting
MAME, Turbulence & Energy Laboratory
University of Windsor
Windsor, ON, Canada

Himanshu Tyagi
Indian Institute of Technology Ropar
Rupnagar, Punjab, India

1 The Pros and Cons of Climate Change Mitigation Innovations

David S-K. Ting and Jacqueline A. Stagner
University of Windsor

"Every solution breeds new problems." – Arthur Bloch.

1.1 INTRODUCTION

It is humbling to recognize the laws of thermodynamics concerning fighting against entropy, that is,

1. You cannot win, you can only break even.
2. You can only break even at absolute zero.
3. You cannot reach absolute zero.

In the context of mitigating climate change, every solution we execute comes with a cost, that is, entropy generation. A little foresight and objectivity would prevent us from creating larger problems from the solution that we attempt to implement. While we cannot beat entropy, there are substantial differences among various solutions for a given problem in terms of the underlying costs. Formulating an accurate entropy, or entropy-like, equation to calculate the amount of entropy generated in a well-defined engineering process is challenging but attainable on most occasions. This undertaking becomes substantially more formidable when we additionally consider the ecological, sociological, and other effects. The bottom line is renewable energy tampers with the natural environment and ecology. After reviewing 132 publications, Pratiwi and Juerges (2020) summarized that the negative impact is manifested in environmental pollution, biodiversity loss, habitat fragmentation, and wildlife extinction. In this article, we provide an overview of the somewhat unpopular, latest studies detailing the effects of renewable energy systems on their environment. Steps to minimize the adverse effects are postulated, assisting in forging a less costly renewable energy path forward.

DOI: 10.1201/9781003403456-1

1.2 ECOLOGY

1.2.1 Biodiversity Is Part of the Equation

Recently, Gorman et al. (2023) advocated that climate mitigation should be implemented in a "Right Action, Right Place" framework. To put it another way, climate change mitigation strategies that appropriately integrate biodiversity are the only sound solutions. Unfortunately, not many studies quantify the costs in terms of the deterioration of our environment, such as biodiversity, among comparative strategies. Furthermore, there are significant discrepancies from one study to another when it comes to the calculations of the costs associated with renewable energy implementations. For some approaches, researchers do not agree even qualitatively if the solution approach is helping or worsening the overall problem at hand. For example, some studies resolve to enhance building-integrated solar energy as a win–win solution while others found this marriage ecologically damaging. Subjectivity of the researchers is likely a contributing factor. Part of the discrepancies may also be attributed to the particular details, such as the local environment, type, and size of the involved renewable energy systems and operating conditions. These disagreements call for the formulation of objective, systematic, quantifiable criteria, especially those explicitly expressed in mathematical equations. The general, overarching equation may be expressed as

$$y = f(x_1, x_2, x_3, \ldots, x_n). \tag{1.1}$$

Here, y is an overall or primary objective function or merit factor. This primary objective function is composed of secondary objective functions, x_i, such as carbon footprint, economics, and ecological impact. Table 1.1 provides a depiction of the hierarchy of involved factors. The deduction of some secondary objective functions, such as ecological impact, can be very challenging because many factors affect the objective function in a convoluted manner. Because clean energy for tomorrow is a pressing issue, it has drawn a multitude of researchers from all over the world to invest much effort into better understanding the problem and devising working solutions. As such, there exist many pieces of valuable information scattered in recent literature, each dealing with one or more secondary objective functions from particular perspectives. Significant scrutiny is called for to gather these pieces into cohesive

TABLE 1.1
A Hierarchy of Factors Involved in Climate Change Innovations

Primary Objective Function, y = overall effect caused by the implementation				
Objective function x_1 = carbon footprint	Objective function x_2 = economics	Objective function x_3 = ecological impact	...	Secondary Objective Function x_n
Materials, installation, operation	Life cycle	Air, water, plants, living organisms	...	...

reports that ultimately can lead to the "universal equation." In all likelihood, there is probably more than one possible universal equation. To put it another way, the primary objective function may be expressed in different forms. An appropriate primary objective function, once formulated, can guide the way forward and be progressively improved as we learn.

1.3 PHOTOVOLTAIC

1.3.1 Photovoltaic Plants and Local Environment

Jiang et al. (2021) measured the actual amount of warming and cooling of local air by a utility-scale photovoltaic (PV) plant in a natural barren field in Wujiaqu in Xinjiang, China. This was possible because a neighboring field was available to serve as an in-situ reference. Findings like this can be synergized into a useful map, guiding the formulation of the equations for approximate calculations of differing climate mitigation innovations.

Zhang et al. (2023) conducted a field experiment to evaluate the effects of PV arrays on greenhouse gas emissions. Their results show that PV arrays can alter soil temperature and moisture, leading to approximately 8% of global warming potential compared to reference open grassland. This increase in greenhouse gas emissions signifies 25% to over 50% overestimation of typical greenhouse reduction calculations without accounting for the side effects of microclimates created by PV arrays.

Not all is a loss with the establishment of PV arrays. Luo et al. (2023) found that PV power plants can promote biological soil crusts and improve vegetation growth. It should be highlighted that this study was conducted in a desert environment and, thus, shading and other changes such as ground-level wind speed and humidity can promote vegetation. Negatively, the concentrations of calcium, sulfur, and chlorine also increase along with soil salinity. In spite of the potential for some natural benefits, the construction of solar parks degrades soil and harms existing native vegetation. Therefore, ecological restoration is necessary. Lambert et al. (2022) monitored a solar park in south-eastern France for four years. They found solar parks hampered restoration and reduced plant species richness. Moscatelli et al. (2022) compared the soil properties under solar panels with those between rows of panels seven years after the installation. Their measurements show a clear decrease in microbial activity under the panels. On the other hand, Wang and Gao (2023) found that on the Chinese Loess Plateau slopes, the installation of PV panels led to 27%–63% less sediment flux. In other words, PV panels can preserve topsoil. What about greening deserts with the help of solar PV arrays? Xia et al. (2022) presented the expansion of vegetation from 0 to 103 km^2 in China's twelve biggest deserts from 2011 to 2018 resulting from the deployment of PV arrays.

What about offshore solar? Li et al. (2022) studied fishery complementary PV panels. They found a PV panel that causes the lake to become a heat sink. With a solar PV panel, the albedo decreased by almost 20% with respect to the free water surface. Thankfully, due to the high heat capacity of water, the effect diminishes rapidly with water depth.

1.3.2 Solar Panels and Global Climate

On the global scale, Hu et al. (2016) argued that solar panels alone induce local cooling by converting incoming solar energy into electricity instead of warming up the ground. According to thermodynamics, heat is the dustbin for all energy including high-quality electricity. That being the case, the harnessed solar-into-electric energy ultimately raises the temperature upon usage of the generated electricity, canceling the cooling due to the initial solar-to-electric energy conversion. This leads to a net balance of global solar thermal energy equal to that without widespread installation and applications of solar panels. The atmospheric circulation, on the other hand, is altered due to the initial cooling and subsequent heating that tends to take place at different localities and possibly at different times. This may contribute to atmospheric turbulence including those studied by researchers such as Lee et al. (2023), Smith et al. (2023), and Nerushev et al. (2020).

1.3.3 Agrophotovoltaics or AgriVoltaics

Ramos-Fuentes et al. (2023) explored AgriVoltaic systems for shielding maize from excessive radiation and reducing water demand. They found that the leaf area index decreased significantly at the cost of irrigation reduction, by up to 47% compared to unshaded plots. Dynamic AgriVoltaic systems can be employed to optimize gain, reduce water consumption, lower cost, and mitigate crop yield decrease.

Williams et al. (2023) looked at AgriVoltaics from a different angle, that is, utilizing agriculture to cool the PV panels and, thus, improving the energy conversion efficiency of the solar panels. They showed that mounting PV panels at a height of 4 m while growing soybeans underneath can reduce a PV module temperature by up to 10°C compared to those mounted at 0.5 m above bare soil. Waghmare et al. (2023) showed a theoretical 18% gain in energy generation with the cooling effectuated by a crop's natural transpiration.

1.4 WIND TURBINES

1.4.1 Turbine Wake and Microclimate

Talking about creating maps to maximize renewable energy while minimizing local climate alteration, Abraham and Hong (2021), based on certain two turbine operational-dependent pathways, created a surface momentum flux map across the United States. Among others, Rajewski et al. (2016) measured higher nighttime surface temperatures in a wind farm. Be mindful of the fact that the impacts of wind turbines are operational dependent in addition to weather dependent. A change in local meteorology, topography, etc. can significantly alter the integrated outcomes. Additional elements can be added to this kind of analysis. Once the maps are constructed, further measurements and analyses can be performed to improve their accuracy.

1.4.2 Wind and Vegetation

Among others, Wang et al. (2023) illustrated that the operation of wind turbines resulted in drying of soil. Interestingly, downwind drying is relatively greater in summer and autumn, while upwind drying dominates in spring. Aksoy et al. (2023) investigated changes in vegetation in Turkey between 2002 and 2017 after the installation of 239 wind power plants. The largest negative impact was found for broad-leaf forests, coniferous forests, and agricultural land.

1.5 INTEGRATED WIND AND SOLAR

Integrating solar and wind is a promising option from certain perspectives such as optimizing land use. Among others, Armstrong et al. (2014) hypothesized that PV panels and wind turbines can notably alter local ground-level climate, affecting the fundamental plant-soil processes that govern carbon dynamics. Agha et al. (2020) asked the right question, do wind and solar energy development "short-circuit" conservation of wildlife? They reviewed the trade-offs between renewable energy development, land use, humans, and wildlife under the context of continuous wind and solar development in the western United States. Desert tortoises and sage-grouse were selected as representative endangered species. Among other measures, they suggested the designing of green energy facilities that allow for safe wildlife passage. Another measure is to identify offsite habitats for providing compensatory mitigation for both volant and non-volant wildlife.

1.6 NUCLEAR

Nuclear energy is known for its waste and safety challenges and less known for its benefits. Zhumadilov (2022) found that local soybean yield can increase by 2% in the presence of a nuclear power plant. The enhanced crop yields are attributed to the increased moisture from evaporative cooling systems. Miernicki et al. (2020) studied the use of nuclear waste heat in agriculture. Utilizing a large amount of low-grade waste heat for agricultural needs boosts reciprocal efficiencies of a nuclear power plant and simultaneously mitigates nuclear power plant environmental impacts.

1.7 OTHER HIDDEN COSTS OF GREEN ENERGY TECHNOLOGIES

Matatiele and Gulumian (2016) cautioned the negative effects of green technologies and practices on the environment. These adverse outcomes can result from energy and raw material consumption and waste generation. There are also employee health and safety issues stemming from transitioning into green jobs. The risk factors are higher in poor, developing countries; see, for example, Mariita (2002). To paraphrase Serpell et al. (2021), wind, solar, hydropower, and geothermal technologies are not free from the limitations of resource availability.

With electric vehicles emerging as a cleaner alternative than their internal combustion engine-powered counterparts, the push for higher capacity batteries logically follows. Among others, Agusdinata et al. (2018) presented the serious issues related to lithium mineral extraction for the making of lithium-ion batteries. They highlighted the urgent need to address the adverse impacts on local mining communities. Lèbre et al. (2020) disclosed the augmented stress on people and the environment in extractive locations. They showed that 84% of platinum resources and 70% of cobalt resources are located in high-risk contexts.

1.8 CONCLUDING REMARKS

In conclusion, we cannot win; the more we do, the more entropy we generate. Moving energy development to remote areas does not stop entropy generation. Aycrigg et al. (2023) disclosed that energy development around wilderness areas threatens wildernesses. Within the constraint that all development results in entropy generation, we can devise means that produce less entropy. Our responsibilities include choosing and devising means to supply energy along the path that results in the least amount of entropy generation.

REFERENCES

A. Abraham, J. Hong, "Operational-dependent wind turbine wake impact on surface momentum flux," *Renewable and Sustainable Energy Reviews*, 144: 111021, 2021.

M. Agha, J. E. Lovich, J. R. Ennen, B. D. Todd, "Wind, sun, and wildlife: do wind and solar energy development 'short-circuit' conservation in the western United States?" *Environmental Research Letters*, 15: 075004, 2020.

D. B. Agusdinata, W. Liu, H. Eakin, H. Romero, "Socio-environmental impacts of lithium mineral extraction: towards a research agenda," *Environmental Research Letters*, 13: 123001, 2018.

T. Aksoy, M. Cetin, S. N. Cabuk, M. A. S. Kurkcuoglu, G. B. Ozturk, A. Cabuk, "Impact of wind turbines on vegetation and soil cover: a case study of Urla, Cesme, and Karaburun Peninsulas, Turkey," *Clean Technologies and Environmental Policy*, 25: 51–68, 2023.

A. Armstrong, S. Waldron, J. Whitaker, N. J. Ostle, "Wind farm and solar park effects on plant-soil carbon cycling: uncertain impacts of changes in ground-level microclimate," *Global Change Biology*, 20: 1699–1706, 2014.

J. L. Aycrigg, T. R. McCarley, S. Martinuzzi, R. T. Belote, M. Bosher, C. Bailey, M. Reeves, "A spatial and temporal assessment of energy development around wilderness areas," *Biological Conservation*, 279: 109907, 2023.

C. E. Gorman, A. Torsney, A. Gaughran, C. M. McKeon, C. A. Farrell, C. White, I. Donohue, J. C. Stout, Y. M. Buckley, "Reconciling climate action with the need for biodiversity protection, restoration and rehabilitation," *Science of the Total Environment*, 857(20): 159316, 2023.

A. Hu, S. Levis, G. A. Meehl, W. Han, W. M. Washington, K. W. Oleson, B. J. Van Ruijven, M. He, W. G. Strand, "Impact of solar panels on global climate," *Nature Climate Change*, 6: 290–294, 2016.

J. Jiang, X. Gao, Q. Lv, Z. Li, P. Li, "Observed impacts of utility-scale photovoltaic plant on local air temperature and energy partitioning in the barren areas," *Renewable Energy*, 174: 157–169, 2021.

Q. Lambert, R. Gros, A. Bischoff, "Ecological restoration of solar park plant communities and the effect of solar panels," *Ecological Engineering*, 182: 106722, 2022.

É. Lèbre, M. Stringer, K. Svobodova, J. R. Owen, D. Kemp, C. Côte, A. Arratia-Solar, R. K. Valenta, "The social and environmental complexities of extracting energy transition metals," *Nature Communications*, 11: 4823, 2020.

J. H. Lee, J.-H. Kim, R. D. Sharman, J. Kim, S.-W. Son, "Climatology of clear-air turbulence in upper troposphere and lower stratosphere in the northern hemisphere using ERA3 reanalysis data," *Journal of Geophysical Research: Atmospheres*, 128(1): e2022JD037679, 2023.

P. Li, X. Gao, Z. Li, X. Zhou, "Physical analysis of the environmental impacts of fishery complementary photovoltaic power plant," *Environmental Science and Pollution Research*, 29(30): 46108–46117, 2022.

L. Luo, Y. Zhuang, H. Liu, W. Zhao, J. Chen, W. Du, X. Gao, "Environmental impacts of photovoltaic power plants in northwest China," *Sustainable Energy Technologies and Assessments*, 56: 103120, 2023.

N. O. Mariita, "The impact of large-scale renewable energy development on the poor: environmental and socio-economic impact of a geothermal power plant on a poor rural community in Kenya," *Energy Policy*, 30(11–12): 1119–1128, 2002.

P. Matatiele, M. Gulumian, "A cautionary approach in transitioning to 'green' energy technologies and practices is required," *Reviews on Environmental Health*, 31(2): 211–223, 2016.

E. A. Miernicki, A. L. Heald, K. D. Huff, C. S. Brooks, A. J. Margenot, "Nuclear waste heat use in agriculture: history and opportunities in the United States," *Journal of Cleaner Production*, 267: 121918, 2020.

M. C. Moscatelli, R. Marabottini, L. Massaccesi, S. Marinari, "Soil properties changes after seven years of ground mounted photovoltaic panels in Central Italy coastal area," *Geoderma Regional*, 29, e00500, 2022.

A. F. Nerushev, K. N. Visheratin, R. V. Ivangorodsry, "Turbulence in the upper troposphere according to long-term satellite measurements and its relationship with climatic parameters," *Sovremennye Problemy Distantslonnogo Zondirovaniya Zemli Iz Kosmosa*, 17(6): 82–86, 2020.

S. Pratiwi, N. Juerges, "Review of the impact of renewable energy development on the environment and nature conservation in Southeast Asia," *Energy, Ecology and Environment*, 5(4): 221–239, 2020.

D. A. Rajewski, E. S. Takle, J. H. Prueger, R. K. Doorenbos, "Toward understanding the physical link between turbines and microclimate impacts from in situ measurements in a large wind farm," *Journal of Geophysical Research: Atmospheres*, 121: 13392–13414, 2016.

I. A. Ramos-Fuentes, Y. Elamri, B. Cheviron, C. Dejean, G. Belaud, D. Fumey, "Effects of shade and deficit irrigation on maize growth and development in fixed and dynamic AgriVoltaic systems," *Agriculture Water Management*, 280: 108187, 2023.

O. Serpell, B. Paren, W-Y. Chu, "*Rare Earth Elements: A Resource Constraints of the Energy Transition*," Kleinman Center for Energy Policy, University of Pennsylvania, Philadelphia, PA, 2021.

I. H. Smith, P. D. Williams, R. Schiemann, "Clear-air turbulence trends over the north Atlantic in high-resolution climate models," *Climate Dynamics*, https://doi.org/10.1007/s00382-023-06694-x, 2023.

R. Waghmare, R. Jilte, S. Joshi, P. Tete, "Review on agrophotovoltaic systems with a premise on thermal management of photovoltaic modules therein," *Environmental Science and Pollution Research*, 30(10): 25591–25612, 2023.

F. Wang, J. Gao, "How a photovoltaic panel impacts rainfall-runoff and soil erosion processes on slopes at the plot scale," *Journal of Hydrology*, 620(B): 129522, 2023.

G. Wang, G. Li, Z. Liu, "Wind farms dry surface soil in temporal and spatial variation," *Science of The Total Environment*, 857(1): 159283, 2023.

H. J. Williams, K. Hashad, H. Wang, K. M. Zhang, "The potential for agrivoltaics to enhance solar farm cooling," *Applied Energy*, 332: 120478, 2023.

Z. Xia, Y. Li, W. Zhang, R. Chen, S. Guo, P. Zhang, P. Du, "Solar photovoltaic program helps turn deserts green in China: evidence from satellite monitoring," *Journal of Environmental Management*, 324: 116338, 2022.

B. Zhang, R. Zhang, Y. Li, S. Wang, F. Xing, "Ignoring the effects of photovoltaic array deployment on greenhouse gas emissions may lead to overestimation of the contribution of photovoltaic power generation to greenhouse gas reduction," *Environmental Science and Technology*, 57(10): 4241–4252, 2023.

D. Zhumadilov, "Effect of nuclear power plants on local crop yields," *Journal of Agricultural and Applied Economics*, 54(1): 114–136, 2022.

2 Investigating the R&D and Innovation Economic Efficiencies of the Renewable Energy Sectors in EU

Fazıl Gökgöz and Gaye Demirhan Başbilen
Ankara University

NOMENCLATURE

BCC: Banker Charnes Cooper
CCR: Charnes Cooper Rhodes
CPIA: Czech Power Industry Alliance
CRS: Constant Returns to Scale
DEA: Data Envelopment Analysis
DMUs: Decision-Making Units
EPO: European Patent Office
EC: European Commission
EU: European Union
ESPVIA: European Solar PV Industry Alliance
GHG: Greenhouse gases
HR S&T: Human Resources in Science and Technology
IEA: International Energy Agency
IPCC: Intergovernmental Panel on Climate Change
IRENA: International Renewable Energy Agency
NDC: Nationally Determined Contribution
OECD: Organization for Economic Co-operation and Development
PV: Photovoltaics
R&D: Research and Development
RES: Renewable Energy Sources
SE: Super-Efficiency
VRS: Variable Returns to Scale

DOI: 10.1201/9781003403456-2

2.1 INTRODUCTION

With Paris Climate Agreement, 196 parties accepted the risks associated with the rising global temperatures and agreed to stabilize them to below 2°C from the pre-industrial levels. For this target, socio-economic transformation built on the best available scientific data is a must. Since 2020, countries have been submitting nationally determined contributions (NDCs).

In December 2020, European Union (EU) updated its NDC, and they will be reducing the emissions by a minimum of 55% until 2030 compared to 1990 levels. The 2050 target is becoming an emission-free continent.

As 75% of GHG emissions was produced by the energy sector in Europe, increasing the renewable energy share in the overall energy has become a priority for most countries. Since 1997, Europe has been setting renewable energy source (RES) targets for its members. The first RES target set was 12% by 2010. Later the Directive 2001/77/EC, 2003/30/EC, 2003/54/EC, 2003/87/EC and 2009/28/EC came into force. With the latter one, three targets were introduced which were abbreviated as "20/20/20" by 2020: decrease the GHG emissions by 20% from 1990 levels; improve energy efficiency by 20% from 2005 levels; and increase the percentage of RE to 20% came into force. Considering EU was on track, the RES deployment target was revised again in 2018, with the new Directive (EU) 2018/2001 briefly known as RED II, it was increased to 32% by 2030. Nowadays EU has provisionally agreed to raise RES share by at least 42.5% by 2030 (European Commission, 2020a).

The innovations are drivers of growth in many sectors, it is so in RE, too. There is a growing interest to analyze the RES within a lifetime in a broader perspective with its effect on the environment through its production and operation stages.

Li et al. (2019b) argue that the pollution of the environment which is a result of the production process of RES is often ignored in many studies. However, especially for photovoltaic cell production processes, there is a growing concern for environmental issues (Hou et al., 2016; Rabaia et al., 2021).

Green technology innovation became very important, too. Innovation is a process composed of two phases. In the first R&D phase, the human, financial and material resources are processed and intermediate outputs, which are scientific and technological accomplishments, and new products were emerged. In the second step, these inputs were turned into the final outputs, which are technological commercial products (Wang et al., 2016).

Studying the innovation efficiencies in RESs can serve the improvement of the inefficient enterprises (Jiang et al., 2021).

For this reason, within the renewable energy industry, saving resources and energy, reducing pollution during the production stage and achieving long-term sustainable development became a research area (Rabaia et al., 2021; Campos-Guzmán et al., 2019).

The studies on the R&D spendings, innovation, exports and economic growth were quite large (Sandu & Ciocanel, 2014; D'Angelo, 2012; Altomonte et al., 2016; Segarra-Blasco et al., 2020).

The purposes of our analysis are as follows:

i. To examine the R&D budgets and number of researchers' efficiency in terms of patent filings for hydro, solar and wind sectors in selected EU countries;
ii. To examine the innovation economic efficiency of hydro, solar and wind sectors in terms of transforming patent inputs to an economic output which can be seen in export figures;
iii. To determine the best-performing countries, to take lessons from their policies.

Throughout the study, the following questions will be addressed:

i. How do EU countries perform in terms of patent fillings when R&D personnel numbers and R&D budgets were taken as inputs?
ii. Do patent fillings lead to an increase in exports from these sectors?
iii. Is there any significant difference between R&D efficiencies between hydro, solar and wind sectors?
iv. Is there any significant difference between innovation economic efficiencies among countries' renewable energy sectors?

The original contribution of this empirical study is its micro-focus on renewable energy sub-sectors. Per our knowledge, there is not any innovation efficiency study comparing the hydro, wind and solar energy sectors of EU. The three most common RESs are studied with the most recent available data from international agencies such as OECD, IRENA and EU. This study can be a tool for governments and countries to evaluate the results of R&D budgets and canalize the monetary resources to those which are the most efficient ones. It would give feedback on over-utilized resources as well.

Our study is structured under seven parts starting with this introduction. The following part elaborates the literature on R&D and innovation efficiency studies. Under the methodology part, the main data envelopment analysis (DEA) models and two-stage super-efficiency (SE) DEA techniques are investigated. In the fourth part, data and variables were defined in detail. In the same section, some descriptive statistics would be clarifying the countries' current situations. The following part is reserved for the results of our empirical study. The conclusion gives some reflections on the studied countries' respective sectors and policy implications.

2.2 LITERATURE REVIEW

Research and innovation takes an important part in the governments' agendas on their goal of achieving economic growth. European Commission is no exception. With the Treaty of Lisbon in 2007, EU took action in the research field for creating a European research area. In 2010, Innovation Union presented a comprehensive innovation strategy. It focused on climate change, energy efficiency and healthy living (EC, 2020b). Later in 2015, European Commission set three main policy targets; open innovation, open science and open to the world (EC, 2015).

The latest strategy of European Commission covers the period between 2020 and 2024, and it states that research and innovation will be the main contributors for achieving six goals towards the environment and climate, digital future, jobs and economy, protecting citizens and their values, strengthening EU's position in the world and protecting democracy and the rights (EC, 2020a)

Countries' performance in terms of innovation is regularly evaluated by European Union. As of today, the latest scoreboard of 2022 lists the countries under four classes; innovation leaders, strong innovators, moderate innovators and emerging innovators (European Union, 2022a).

- Sweden has the highest performance (like the previous year) in the EU followed by Finland, Denmark, the Netherlands and Belgium.
- Ireland, Luxembourg, Austria, Germany, Cyprus and France are regarded as strong innovators. It means that their performance is higher than the EU average.
- Estonia, Slovenia, Czechia, Italy, Spain, Portugal, Malta, Lithuania and Greece are accepted as moderate innovators.
- Hungary, Croatia, Slovakia, Poland, Latvia, Bulgaria and Romania are emerging innovators.

These performances reflect the general status of the country; however, for specific renewable technologies, these rankings can be changed as our study shows where Germany took the best performer position in many areas.

Oslo Manual of OECD points out the important role of both innovation and its measurement. It states that generating, exploiting and diffusing the knowledge are key to the growth of economy and the welfare of nations. Measuring innovation is central to this need (OECD & SOEC, 2005).

The two mainstream methodologies used for innovation efficiency measurements are Stochastic Frontier Analysis and DEA (Wang et al., 2016; Bresciani et al., 2021). As DEA integrates multiple outputs and multiple inputs without setting a particular form at the beginning, it has gained more popularity in efficiency analysis.

DEA is utilized in the analysis of growth and exports. Bilbao-Osorio and Rodriguez-Pose (2004) used a two-step analysis: R&D investments' effects on innovation (measured by patent applications) as the first step, innovation effect on economic growth as the second step.

Dong et al. (2022) studied the innovation impact on firm-level exports in China. The other studies that can be mentioned are the research of Ayllon and Radicic (2019), the study of Cassiman and Golovko (2011) using Spanish companies' data and those of Silva et al. (2017) using Portuguese companies' data.

For industry, regional, or national scale, numerous DEA innovation efficiency studies are conducted, some of which can be summarized as follows:

Chen et al. (2006) used four inputs for measuring technical efficiency; the number of workers, operational capital, R&D spendings and land. The outputs were yearly sales and total patents.

Hashimotoa and Hanedab (2008) measured total factor R&D efficiency by employing R&D expenditures as input whereas patents, sales and operating profits were used as outputs.

In Zhong et al. (2011), R&D spendings and full-time R&D workers were the primary inputs; patent applications were the primary output and intermediate input; and finally, revenue from the sales of new products together with the profit of the main business was taken as the final output.

Guan and Chen (2012) analyzed the efficiency of OECD countries' innovations with two-stage DEA.

Chen et al. (2014) analyzed R&D efficiency on an international level via DEA with the inputs of R&D workers, R&D expenditure and outputs of patents, scientific journal articles, royalty and licensing fees.

Wang et al. (2016) studied innovation efficiency at the enterprise level through two-stage DEA. The first stage named as the R&D stage included fixed assets, staff wages, R&D costs as inputs and software assets and revenues as outputs. The outputs of the first part were inputs of the marketing stage. The following part measured the efficiency in terms of total profits and market value outputs.

Liu et al. (2020) utilized R&D expenditures and R&D workers as inputs and the patent applications and patents in force as output and intermediate output for revenue from the sales of new products and principal business as final outputs.

Bresciani et al. (2021) studied the innovation efficiency of Southern Italian and Spanish regions where private and public expenditures were the inputs and patents together with trademark applications were the outputs.

Jiang et al. (2021) studied green technology innovation in Chinese RE companies. They used DEA with the inputs of R&D workers, R&D investment funds, operational costs, outputs of the number of patents, tax payable income, yearly operational revenue, net profit and undesirable output with sewage charges or environmental taxes.

Zuo et al. (2022) used a two-stage DEA for measuring the mining sector in China. They analyzed the technological innovation efficiency and eco-efficiency using 10-year data divided under nine sub-periods each composed of 3 consecutive years to overcome the innovation lag.

Gökgöz and Yalçın (2022) measured eco-innovation efficiency with R&D expenditures as input and export of high-technological products, electricity generation and patents as outputs.

Li et al. (2019a) analyzed the Chinese semiconductor industry's innovation efficiency with three-stage DEA. The input variables were labor defined as the total workers involved in R&D, total R&D capital expenses and total patent management investment. The outputs were sales[1] and the number of patents.

2.3 METHODOLOGY: TWO-STAGE SUPER-EFFICIENCY DEA

DEA is a technique for performance assessment against the best performer (Cook et al., 2014). Since its first introduction by Charnes, Cooper and Rhodes in 1978, DEA is quickly adopted by many researchers from many fields. It is accepted as a tool that is easy to use for modeling processes where the measurement of the performance is needed. It measures the efficiencies of decision-making units (DMUs) in a

[1] Measured as the annual additional rise in sales.

set of similar units. These are called decision-making units and they convert numerous inputs into a number of outputs (Cooper et al., 2011).

In conventional DEA, the DMUs are regarded as black boxes. The internal processes and the structures are ignored (Lewis & Sexton, 2004). Especially after Seiford and Zhu's article (1999) on US commercial banks, two-stage DEAs started to be implemented. In two-stage model DEA, DMUs have a two-stage network structure with intermediate products used in two stages. The first stage, which can be a process such as profitability as in Seiford and Zhu's study (1999), uses inputs to produce outputs. The first-phase output(s) are later employed as intermediate input resources of the second phase.[2]

Wang et al. (2016) argue that innovation processes are not suitable for one-stage analysis. Typically, two stages were included especially in high technology sectors: development of upstream technology and transformation to economic downstream (Moon & Lee, 2005; Sharma & Thomas, 2008).

According to Zoltan et al. (2002), measures of technological change have one of the three main parts of the innovative process. The inputs such as R&D spendings can be measured, intermediate outputs such as the patented inventions can be measured, or a direct measure can be employed. Our study with the advantage of two-stage DEA measures all the aspects of the innovation process.

Per the advantages of two-stage DEA, in our study, following many scholars (Guan & Chen, 2010b; Bilbao-Osorio & Rodriguez-Pose, 2004; Zuo et al., 2022; Zhong et al., 2011; Liu et al., 2020; Wang et al., 2016; Aytekin et al., 2022), a two-stage Super Radial BCC-Output oriented DEA model with variable returns to scale (VRS) will be employed for evaluating innovation efficiency.

Before presenting the mathematical model of two-stage DEA, it is beneficial to start with the conventional models and their approach.

The first classification DEA can be done under orientation types; whether it is input-oriented (BCC) or output-oriented (CCR). The input-oriented model uses VRS whereas the output-oriented model uses constant returns to scale.

DEA formulation of CCR is given in Eqs (2.1)–(2.3) (Cook et al., 2014):

$$\max \sum_{r=1}^{s} \mu_j y_{rjo} \tag{2.1}$$

Subject to

$$\sum_{r=1}^{s} \mu_r y_{rj} - \sum_{i=1}^{p} \omega_i x_{ij} \leq 0 \tag{2.2}$$

$$\sum_{i=1}^{p} \omega_i x_{ijo} = 1 \tag{2.3}$$

[2] In Seiford and Zhu's study (1999) the inputs at the first stage were labors and assets, they used to generate profits and revenue. In the second part, these outputs become inputs and used to produce market value, returns and earnings per share.

The DEA formulation of BCC can be given as Eqs. (2.4)–(2.7):

$$\min \theta \tag{2.4}$$

Subject to

$$\sum_{j=1}^{n} \lambda_j x_{ij} \leq \theta_{x_{ij0}} \quad i = 1,\ldots,p \tag{2.5}$$

$$\sum_{j=1}^{n} \lambda_j y_{rj} \leq y_{r0} \quad r = 1,\ldots,s \tag{2.6}$$

$$\lambda_j \geq 0 \quad j = 1,\ldots,n \tag{2.7}$$

Classical DEA models usually give multiple efficient DMUs as stated by Tone (2001). SE models present a way to rank these efficient DMUs so that the comparison can be more accurate. The classical model attributes to 1 to every efficient DMU; however, in SE the DMUs can have values larger than 1. The DMU with the highest score ranks the first and the efficiency levels are determined according to the SE scores ranking lower than the previous DMU.

The enveloping surfaces can be shown in Figure 2.1 following Cooper et al. (2006). On CRS envelopment surface, the amount of increase in inputs is equal to the amount of increase in outputs. However, VRS links the DMUs D, C and B. But A is an efficient DMU and it needs to move to A' in order to be efficient.

Efficiency input orientation is shown in Eqs. (2.8)–(2.11). Output orientation can be formulated in Eqs. (2.12)–(2.15) (Andersen and Petersen, 1993).

[Super Radial-Input Oriented CRS]

$$\min \phi^{\text{Super}} \tag{2.8}$$

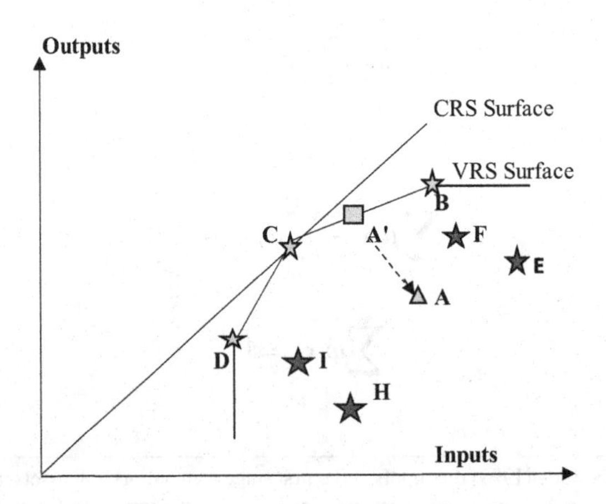

FIGURE 2.1 Enveloping surfaces for the DMUs. (Cooper et al., 2006.)

Subject to

$$\sum_{j=1,\neq 0}^{n} \lambda_j x_j \leq \phi^{\text{Super}} x_0 \tag{2.9}$$

$$\sum_{j=1,\neq 0}^{n} \lambda_j y_j \geq y_0 \tag{2.10}$$

where

$$\phi^{\text{Super}}, \lambda \geq 0 \text{ and } j \neq 0 \tag{2.11}$$

[Super Radial-Output Oriented CRS]

$$\max \theta^{\text{Super}} \tag{2.12}$$

Subject to

$$\sum_{j=1,\neq 0}^{n} \lambda_j x_j \leq x_0 \tag{2.13}$$

$$\sum_{j=1,\neq 0}^{n} \lambda_j y_j \geq \theta^{\text{Super}} \; y_0 \tag{2.14}$$

where

$$\lambda \geq 0 \text{ and } j \neq 0 \tag{2.15}$$

The two-stage approach which was mentioned above was first proposed by Seiford and Zhu (1999). In their article on commercial banks in US, they have analyzed the banks' profit in the first level and the market value in the second level. They employed the output-oriented BCC DEA VRS model. The formulation of the model can be given as below:

$$\max y_0^l + \varepsilon \left(\sum_{i=1}^{m} s_i^- + \sum_{r=1}^{s} s_r^+ \right) \quad l = 1,2 \tag{2.16}$$

Subject to

$$\sum_{j=1}^{n} \lambda_j x_{ij} + s_i^- = x_{i0} \quad i = 1,2,\ldots\ldots,m \tag{2.17}$$

$$\sum_{j=1}^{n} \lambda_j y_{rj} - s_r^+ = \gamma_0^l y_{r0} \quad r = 1,2,\ldots\ldots,s \tag{2.18}$$

where

$$\sum_{i=j}^{n} \lambda_j = 1 \text{ and } \lambda_j s_i^- s_r^+ \geq 0 \tag{2.19}$$

where γ_0^1 and γ_0^2 are the optimal values for Level 1 and Level 2, the scale efficiency is measured by

$$\pi_0^l = \frac{\phi_0^{l*}}{\gamma_0^{l*}} \text{ for } l = 1,2 \tag{2.20}$$

2.4 DATA ANALYSIS

2.4.1 Data and Variables

In this empirical work, we measured the efficiencies of EU countries under two stages. In the first stage, the R&D budget together with the number of R&D personnel of each EU country's hydro, solar and wind sectors was taken as input with the patents filled as output in line with several studies conducted on similar research (Chen et al., 2006; Zhong et al., 2011; Chen et al., 2014; Liu et al., 2020; Li et al., 2019a).

As argued by Wang and Huang (2007), Guan and Chen (2010a) and Wang et al. (2013), patents are the crucial and the most explicit sign of innovation output.

In the second stage, patent fillings are the input whereas the exports from each RES are taken as the output.

A frequently discussed subject in innovation research is the time lag to be determined between the input and the output. There is not any clear consensus on whether there should be a time lag and if there is a need, how much should be the time required to see the results of innovation activities.

Lan et al. (2022) used a 1-year difference between innovation input and output in their study of photovoltaic industry. Guan and Chen (2012) used a 3-year gap for the knowledge production process and a 1-year gap for the knowledge commercialization.

For some high-tech industries, 2–3 years were used (Guan and Chen, 2010a; Wang et al., 2013). There are longer terms as well; Hashimotoa and Hanedab (2008) used an 8-year time gap between the expenditures to economic values for the pharmaceutical industry.

On the other side, some argue that the time lag is not significant in estimating the results (Griliches, 1998; Hollanders and Celikel-Esser, 2007).

In this empirical analysis, we will be using 1-year time lag between the time of expenditures spent and the patent fillings based on the shorter time lag used in similar industries (Table 2.1).

The inputs and outputs of our study are given in Table 2.2.

The research can be visualized in Figure 2.2, which shows two stages clearly.

TABLE 2.1
Data Description

Name		Description	Unit	Source
R&D Budget	*Hydro Sector* *PV Sector* *Budget Wind*	2018–2019 research and development budget of each country for each sector	in 1,000 Euro (2022 prices and exch. rates)	OECD (OECD ilibrary, 2023)
R&D personnel in	*Hydro sector* *PV Sector* *Wind Sector*	2018–2019 research and development personnel of each country for each sector	Number/year	OECD (OECD, 2023), IRENA, Eurobserver (2022–2023)
Patent Fillings in	*Hydro Sector* *PV Sector* *Wind Sector*	2019–2020 total patent fillings of each country for each sector	Number/year	IRENA (IRENA, 2022)
Export from	*PV Sector*[a] *Hydro Sector* *Wind Sector*	2019–2020 total export of each country for each sector	1,000 USD	World Integrated Trade Solution-World Bank (World Integrated Trade Solution, 2023)

[a] For the sectors, World Customs' Organization six digit codes were used. The product code for PV sector is 854140, for Hydro Sector 841013 and for Wind Sector 850230.

TABLE 2.2
Inputs and Outputs

Stage 1 R&D Efficiency	
Input 1	R&D Budget for each energy source
Input 2	Number of R&D personnel
Output	Patent fillings for each energy source
Stage-2 Innovation Economic Efficiency	
Input	Patent fillings for each energy source
Output	Exports from each energy source

The limitations and main assumptions of our study can be summarized as follows:

1. The solar thermal data were not available for all countries; for this reason, only PV values are used.
2. The time lag between RD efforts and the filed patents was taken as 1 year. In the first phase, the R&D Budget and R&D personnel data of 2018 were taken as inputs, whereas filed patents of 2019 were taken as the outputs of the model. In the second phase, filed patents of 2019 were taken as inputs and 2019 exports were the outputs.

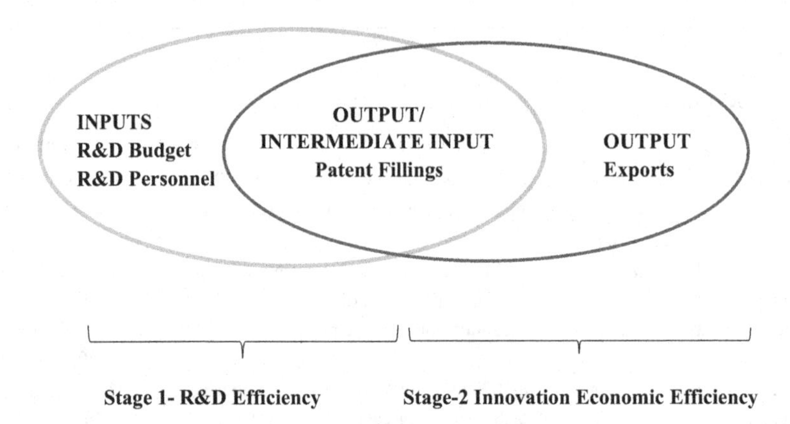

FIGURE 2.2 Illustration of the two-stage DEA analysis.

3. As with DEA no negative values can be used, we had to focus on export values.
4. The R&D personnel data was not directly available. It needs to be calculated by the authors with an approximation of IRENA total employment data for each RES with the OECD data of total R&D personnel per thousand of employment.
5. For the export values, the export of the products with the World Customs' Organization six-digit codes was used.
6. The results of our calculations are dependent on the selected input and output variables and the available data for the DMUs. Furthermore, efficiency results might vary depending on using different input and output variables.

2.4.2 Descriptive Statistics

Hydro, solar and PV are the most developed and largely employed RESs in Europe. As with other RESs, these are highly dependent on natural and climatic conditions. France, Italy, Spain, Austria, Germany and Sweden are the traditional leaders of hydro power plants and have the highest installed capacities during the analyzed years (Figure 2.3). The total installed capacity of hydro did not increase from 2019 to 2020, even though there is a slight decrease in capacity in France, Italy and Spain. The aging of the hydro plants is a common problem in EU and as argued by Eurobserver (2022), the hydro growth potential is depending largely on modernization of the power plants and realization of very few planned projects.

However, with regard to annual electricity generation from hydro, the rank changes. Sweden has the highest electricity generation. France, Italy and Austria follow Sweden. The reason of Sweden being the leader in the generation can be attributed to the dependency of generation on natural and climatic conditions. This is also the case for Finland and Romania where with a relatively low installed capacity a higher generation can be achieved (Figure 2.4).

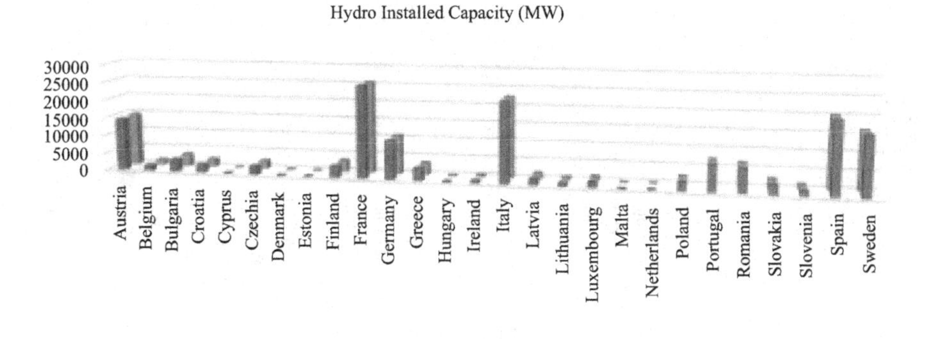

FIGURE 2.3 Installed hydro capacity. (Eurobserver, 2022.)

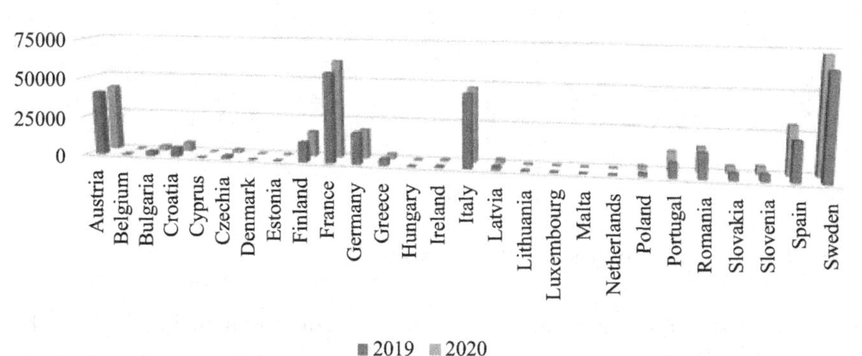

FIGURE 2.4 Electricity generation from hydro. (Eurobserver, 2022.)

The second studied sub-sector is solar (PV). PV is the new star of renewable energy sector. From 2010 to 2020, PV experienced an exponential growth from 0.8% share in cumulative power capacity to 9.4% capacity (matching that of wind) for year 2020. It continued its expansion in 2022 and reached a share of 12.8%, passing the wind sector share by 2% (IEA, 2023). According to Eurobserver (2022), the solar market increase in 2020 dropped in the first month of the pandemic but made a strong comeback in the second term of the year. Nearly all EU countries experienced additional capacities of solar from 2019 to 2020.

EU countries that dominate PV sector in terms of installed capacity are Germany, Italy and the Netherlands (Figure 2.5). Mediterranean countries such as Italy, France and Spain follow these countries. As will be discussed later, Belgium as a strong candidate for being a hub for PV has a considerable capacity, too.

PV electricity generation values of EU countries generally resemble the capacity amounts. Germany, Italy, France, Spain and the Netherlands are the leaders

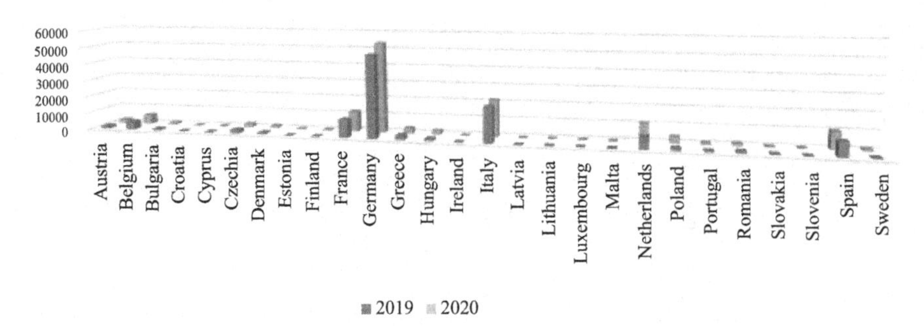

FIGURE 2.5 PV installed capacity. (Eurobserver, 2022.)

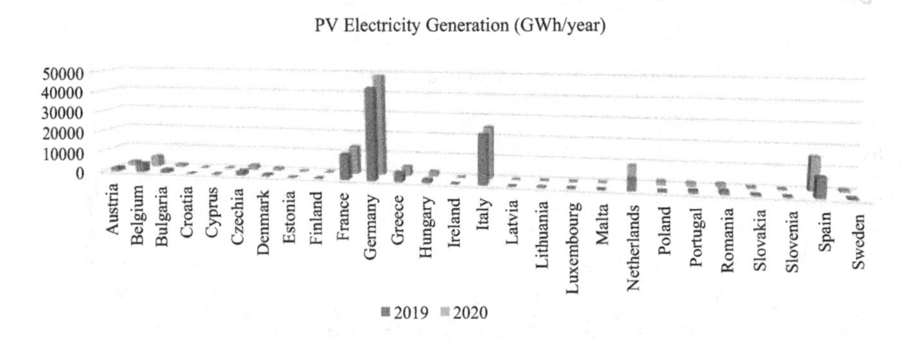

FIGURE 2.6 PV electricity generation. (Eurobserver, 2022.)

in generation values (Figure 2.6). 2020 was the hottest year in Europe and the number of sunshine hours was the highest in the same year. Not only the additional capacity but also the climatic conditions affected the high output from PV (Eurobserver, 2022).

The last sector analyzed in our study is wind energy. The wind electricity capacity leaders in EU are Germany, Spain, France, Italy and Sweden (Figure 2.7). Overall, the increase from 2019 to 2020 was 9.8 GW. According to Eurobserver (2022), the disruptions in supply chains and delays in commissioning of the wind turbines have led to a relatively lower level of annual additions.
With regard to electricity generation, Germany, France, Spain, Italy and Sweden have the highest ranks (Figure 2.8).

Our study focuses on 2019 and 2020 data. It is an important time period, as it coincides with the Covid-19 Pandemic. From the first official report of the virus in December 2019, 2020 had experienced many closures and shutdown of the businesses. The contraction of GDP in Europe for 2020 is calculated as 5.9%. Spain and Italy had experienced the sharpest decrease in GDP as their income is mainly depending on tourism and international travel (Smit et al., 2023).

Covid-19 has brought some positive news to environment and renewable energy targets. In 2020, all EU members, except France, have either achieved or exceeded

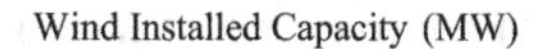

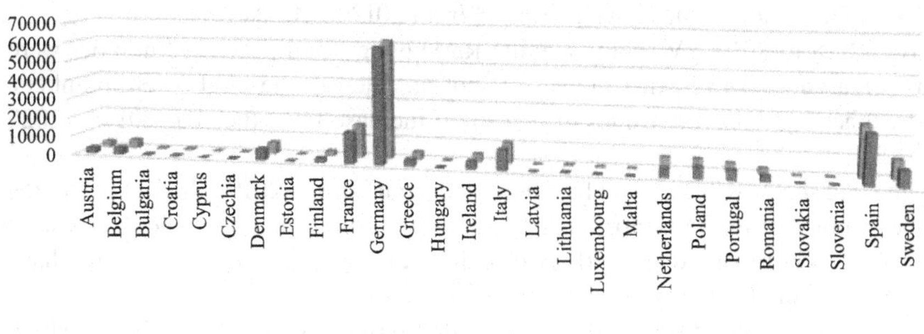

FIGURE 2.7 Wind installed capacity. (Eurobserver, 2022.)

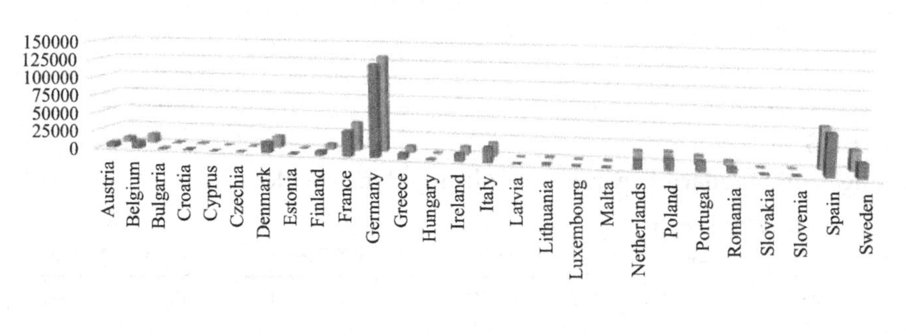

FIGURE 2.8 Electricity generation from wind. (Eurobserver, 2022.)

their RES targets as the total energy consumption was decreased by 8% in 2020 compared to 2019 levels (Euractiv, 2022).

However, there were negative effects as well to renewables in terms of cost increases and delays. IEA (2021) estimated the increase in investment costs of wind as 15%–25%, whereas solar PV had an increase of 10%–20%. The reasons for these were the rise in commodity prices, high transportation costs and supply chain disruptions.

The R&D personnel, budgets, filed patents and export values that are the main inputs and outputs of our study can be described in terms of average and SD values.

Among the three renewable energy sub-sectors, the total R&D budget is highest in the PV industry. Wind and hydro industries follow PV. The highest R&D employment is in the wind industry followed by PV and hydro. For the filled patents, PV ranks the top, followed by wind sector and hydro comes last with a value far below those in wind and PV.

The total R&D budget reserved for hydro energy is increased from 2018 to 2019; however, the R&D personnel is decreased which is also reflected in the filed patent numbers. The export values have increased from 2019 to 2020 (Table 2.3).

From 2018 to 2019, PV sector-specific R&D budget and personnel number were both increased. However, this increase is not reflected in the filed patent numbers. On the other hand, export values have a small increase from the year 2019 to 2020 (Table 2.4).

The wind sector R&D budget has increased from 2018 to 2019. The RD personnel number employed in the sector has decreased by a small number. The decrease in filed patents was recorded as nearly half of the previous year. The export values have experienced a small improvement (Table 2.5).

In all sectors, the patent fillings were decreased from 2019 to 2020, which can be partially attributed to Covid-19 effects on overall economic activities. According to the European Patent Office, the total patent applications from 2019 to 2020 have dropped by 0.7%. However, this decrease could have been higher if the growth in pharmaceuticals (more than 10%), biotechnology (more than 6%) and medical technology (around 3%) patent fillings could be ignored (EPO, 2021a).

TABLE 2.3
Average and Standard Deviation Results for Hydro Sector

Year	Total RD Budget (in 1,000 Euro)		Total RD Personnel		Filed Patent		Export (in 1,000 USD)	
	AV	SD	AV	SD	AV	SD	AV	SD
2018	838.3	986.4	80	93				
2019	1,066.2	1,094.2	42	42.3	10.5	15.8	3,618.6	6,593.7
2020					6.3	9.6	4,589.3	9,796.8

Source: Authors' calculations.

TABLE 2.4
Average and Standard Deviation Results for PV Sector

Year	Total RD Budget (in 1,000 Euro)		Total RD Personnel		Filed Patent		Export (in 1,000 USD)	
	AV	SD	AV	SD	AV	SD	AV	SD
2018	19,963.7	30,300.7	101.8	168.7				
2019	21,207.4	34,944.4	165.5	227.4	67.8	105.2	308,976.1	592,468.1
2020					38.3	66.2	363,080.6	645,317

Source: Authors' calculations.

TABLE 2.5
Average and Standard Deviation Results for Wind Sector

Year	Total RD Budget (in 1,000 Euro)		Total RD Personnel		Filed Patent		Export (in 1,000 USD)	
	AV	SD	AV	SD	AV	SD	AV	SD
2018	17,023.7	21,971	295.5	458.3				
2019	19,687.8	25,190.3	251.2	363.3	46.2	66.9	455,371.4	891,748.3
2020					24.6	30.3	486,728.5	855,207.3

Source: Authors' calculations.

2.4.3 Empirical Analysis and Discussions

In the first stage, R&D efficiency was calculated for the hydro, solar and wind sectors.

Generally speaking, compared to other RESs, it should be noted that the available latest statistical data in terms of hydro sector is very limited. This can be attributed to relatively low levels of installed hydro power plants among individual EU countries. According to Eurobserver (2022), over 70% of the total installed hydro capacity in Europe is in only six countries, namely France, Italy, Spain, Sweden, Austria and Germany and this leads to relatively few data available in other countries analyzed in the report. From 2020 to 2021, only 612 MW hydroelectrical power plants commissioned in all Europe whereas the PV power plants' additions were 25,703 MW and wind additions were 11,312 MW. The installed capacity of wind power plants is around 188 GW in the end of 2021, whereas hydro is 151 GW and solar PV is 161 GW (Eurobserver, 2022).

The average SE scores for each RES are given in Table 2.6 for two consecutive years.

EU countries' hydro sectors' R&D efficiency is higher than the others. Additionally, from 2019 to 2020, the SE scores of hydro have increased considerably. When stage 2 is considered, where knowledge is transferred to economic gains (from patent filings to exports) again hydro sector has the highest efficiency compared to the others.

Among the three sub-sectors, hydro R&D efficiency is highest. Hydro is the most mature source among the renewable energies; its first projects date back to the 1800s (IRENA, 2023). It is also the largest electricity supplier among the renewable energies. However, investments in hydropower were surpassed by solar PV and wind technologies in the last decade. According to S&P Global (2022), the average age of operational hydro plants is around 40 years and the already retired hydro power plants had an average lifetime of 60 years. The aging of the hydro is an important problem and IRENA believes that modernizing hydropower plants, improving their overall flexibility and operational improvements can be key to solve this problem. This can be also achieved through innovations. Considering this need and the possibility of immediate improvements in the existing fleet can be a factor for its high R&D efficiency.

TABLE 2.6
Average Super Efficiency Scores for Hydro, Solar and Wind Sectors (2019–2020)

Stages	Sectors	2019	2020
Stage 1	Hydro	0.955	1.431
R&D Efficiency	Solar	0.625	0.792
	Wind	0.910	0.909
		2019	**2020**
Stage 2	Hydro	1.009	1.056
Innovation Economic	Solar	0.517	0.952
Efficiency	Wind	0.416	0.422

Source: Authors' calculations.

With regard to innovation economic efficiency, as "the pioneering hydropower engineers and manufacturers largely originated from Central and Northern Europe", European manufacturers continue to enjoy the privileges of being the first entrants to the market. Europe hydro equipment manufacturers together with 50 key players of the supply value represent 2/3rd of the global industry (Andritz Hydro, 2023)

The solar sectors' both R&D and innovation economic efficiency scores are also improved during the analyzed years. This can be attributed to the record installations in the market and manufacturing capacity.

ESPVIA which is initiated by EC aims 30 GW yearly PV production capacity by 2025. The data reveals that already 20 GW manufacturing capacity is in the pipeline and this projection is expected to be realized (Solar Alliance EU, 2023).

The SE score of wind in terms of R&D efficiency is close to 1 and stable in the analyzed years. Nevertheless, innovation economic efficiency of wind energy is far from 1 which can be interpreted as there is difficulty in turning the innovation knowledge to exports. According to Eurostat (2021), the export value of Europe for years 2019 and 2020 was dropped from 3 billion Euro to just over 2 billion Euro (Eurostat, 2021).

The country scores calculated with the inputs and outputs summarized in the above sections are provided in Table 2.7. These results are based on the data available for each country. The utmost result of our calculations is the level of efficiency achieved by Germany nearly in all aspects analyzed. Without any exception, it is the leader in R&D efficiency in hydro, solar and wind sectors with values ranging from 2.5 to 5.7. Poland performance follows Germany. The Stip Compass of OECD provides data for science, technology and innovation. When Poland is considered for the recent years, it can be observed that since 1990s, 48 of 264 policies were started in 2019 or 2020 (Stip Compass, 2021). Two programs worth mentioning are the Priority Program for New Energy of 2020, which aims to support new innovation in modern technologies, and the Clean Air Program of 2018, which with a budget of 103 billion Polish Zloty (PLN) aims to increase new low emission heating sources in cities (Stip Compass, 2021; IEA, 2022).

TABLE 2.7
Super Efficiency Scores of EU Countries for R&D (Stage 1)

	Hydro		Solar		Wind	
Countries	**2019**	**2020**	**2019**	**2020**	**2019**	**2020**
Austria	0.071	0.044	0.217	0.251	0.870	0.250
Belgium	0.062	-	0.154	0.053	0.029	0.202
Czech Rep	0.106	0.145	0.491	0.246	0.344	0.130
Denmark	1	-	0.500	0.394	0.654	0.669
Estonia	-	-	0.092	-	-	-
Finland	0.121	0.390	0.343	0.247	-	0.369
France	0.125	0.029	0.500	0.549	0.756	0.568
Germany	4.308	5.667	2.475	3.597	3.218	4.154
Ireland	-	-	-	1	-	-
Italy	-	-	0.238	-	0.590	-
Netherlands	1	1	0.396	0.400	0.710	0.139
Poland	3.008	1.335	1.985	0.726	1.725	2.017
Slovak Republic	-	-	-	-	-	1
Spain	0.622	0.266	0.721	1.530	0.727	0.498
Sweden	0.080	-	0.217	0.297	0.292	-
Average	0.955	1.431	0.625	0.792	0.910	0.909

Source: Authors' calculations.

For year 2020, Ireland and Spain follow Germany. Spain's rank is understandable as it is traditionally a strong PV investor. For Ireland, it can be argued that its SE score might be related to the announcement of a new PV installment grant scheme by Sustainable Energy Authority of Ireland (Solar Generation, 2020). This might have increased the attractiveness of R&D efforts in the sector.

For hydro, the super-efficient ones are Denmark, Netherlands and Poland. When the solar sector is considered, the countries with the highest efficiency scores are Poland, Spain and Ireland.

The last sector analyzed is wind where it is not easy to determine the best performers because of the fluctuations between 2 years' performances (*except Poland which ranked 2nd place after Germany*). However, for 2019, Austria, France, Netherlands and Spain did well, whereas for year 2020 Slovak Republic, Denmark, France and Spain were relatively good performers. Poland's success in three RESs should be scrutinized by other countries' policymakers.

Sweden, Finland and Belgium which were determined as best performers by EU (European Union, 2022a) are not the most efficient ones in our study. However, Denmark and the Netherlands from the same list have a relatively efficient place for hydro and wind. The reason is mainly the focus of our study which is on the subrenewable sectors rather than a general perspective of EU studies.

The second stage of the analysis is innovation economic efficiency where again Germany is the top performer (Table 2.8).

TABLE 2.8
Super Efficiency Scores of EU Countries for Innovation Economic Efficiency (Stage 2)

	Hydro		Solar		Wind	
Countries	**2019**	**2020**	**2019**	**2020**	**2019**	**2020**
Austria	0.292	0.339	0.330	0.257	0.020	0.336
Belgium	-	-	0.577	8.504	1	0.006
Czech Rep	2.054	2.473	0.382	0.242	0.269	1
Denmark	-	-	0.086	0.056	2.706	1.526
Estonia	-	-	0.098	-	-	-
Finland	0.002	0.001	0.015	0.019	-	0.009
France	0.358	0.404	0.370	0.449	0.031	0.018
Germany	6.143	6.816	2.125	1.628	0.805	1.279
Ireland	-	-	1	0.117	-	-
Italy	0.565	-	0.520	-	0.060	-
Netherlands	0.487	0.163	2.345	3.355	0.731	1.600
Norway	-	-	0.011	0.003	0.001	0.0003
Poland	-	0.197	0.023	0.025	0.088	0.066
Republic of Türkiye	0.001	0.0001	0.713	0.128	0.233	0.004
Slovak Republic	-	-	-	-	0.001	0.0001
Spain	0.030	0.001	0.049	0.113	0.318	0.358
Sweden	-	-	0.162	0.089	0.284	-
Switzerland	0.139	0.163	0.384	0.157	0.087	
United Kingdom	-	-	0.125	0.095	0.016	0.014
Average	1.007	1.056	0.517	0.952	0.416	0.422

Source: Authors' calculations.

In the hydro sector, Czech Republic is the second and the one of two with a score above 1. Czech Republic has a strong manufacturer background. CPIA points out that historically 90% of the equipment was manufactured domestically, including key components for nuclear power plants. The alliance members can also give an idea about the Czechian leaders in hydro; Doosan Skoda (a global leader in steam turbines) Sigma the pumping manufacturer, etc. (CPIA, 2023).

In the solar sector, Belgium and the Netherlands (as in the EU innovation scoreboard) are the best performers. For 2019, two countries Ireland and the Republic of Türkiye performed relatively well. As mentioned above, Ireland's special incentives for PV installations can be stimulating this score (Solar Generation, 2020). Republic of Türkiye's relatively higher score can be attributed to its manufacturing capacity. It is the biggest solar panel manufacturer in Europe in 2022 (Balkan Green Energy News, 2022). Strong domestic market coupled with the rising electricity demand created a boom in PV factories which raised to 5.6 GW/year in 2021 (Pv Magazine,

2019). In the coming years, Republic of Türkiye could be among the top achievers for innovation economic efficiency. The manufacturing volume is increased to 7.96 GW by mid of 2022 and the goal is to be in the global top three manufacturing countries (Balkan Green Energy News, 2022).

For the wind sector, Denmark has a higher efficiency level than Germany.[3] The Netherlands should also be mentioned as it has increased its super-efficiency level above 1.

Belgium's extremely high score for 2020 needs further attention. Belgium is a hub for PV sector as well; The European Solar Manufacturing Council and European Solar PV Industry Alliance centers are located in Belgium. The alliance concentrates on the innovation-led expansion of European industrial solar value chain, particularly in the PV manufacturing sector. The knowledge dissemination network together with the improving domestic market can be the reasons for these high SE values, which need to be followed up in new research in the future (Pv-Magazine, 2021; Euronews, 2023).

R&D efficiencies and innovation economic efficiencies of countries can be compared to determine the overall process of R&D inputs to economic outputs. With regard to the hydro sector, the EU countries on average have achieved a substantial improvement from 2019 to 2020 in R&D efficiency. Nevertheless, the improvement in innovation economic efficiency is negligible on average. For the solar sector, both R&D efficiency and innovation economic efficiency on average are improved from 2019 to 2020. However, for innovation economic efficiency improvement is much higher, especially with the success of Belgium becoming a hub for PV and Netherlands inter-industry alliance which will be further discussed below. For the wind sector, it can be argued that the R&D efforts and the economic results from these efforts are stabilized as the averages of R&D and innovation economic efficiencies between the two analyzed years are very similar.

To better visualize the countries' relative super-efficiency levels, the appearance times of the countries as super-efficient in the reference sets can be counted.

Aytekin et al. (2022) used appearance times to evaluate the innovation efficiencies of EU members and candidate countries. They have performed their analysis with DEA package and the appearance times is one of the built-in results of the software. It measures the number of times where the efficiency scores are equal to 1. Benitez et al. (2021) described the number of appearance times as the number of times DMUs appear efficient in the set of inefficient DMUs. As our study is measuring SE levels, the countries with a SE level over 1 are taken as efficient and the total number of being efficient is summed up for calculating the number of appearance times. The study of Bodendorf et al. (2022) on carbon footprints and Georgescu et al. (2022) study on digitalization efficiency have also employed the appearance times in their analysis.

Considering the R&D Efficiency Appearance times, Germany leads the list with a value of six times, followed by Poland with a value of five times. Netherlands has

[3] This result can be understandable. Denmark was a Pioneer country in developing wind turbines and Vestas, a Danish turbine manufacturer, is a leader in global turbine supply and it supplied the highest number of turbines in both analyzed years; 2019 and 2020 (Reglobal, 2021).

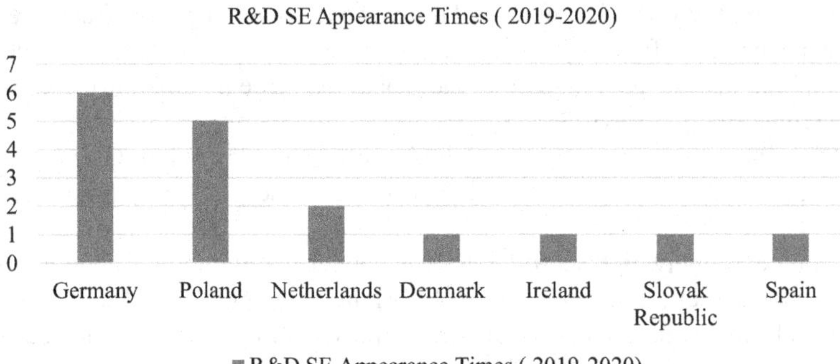

FIGURE 2.9 R&D efficiency appearance times. (Authors' Calculations.)

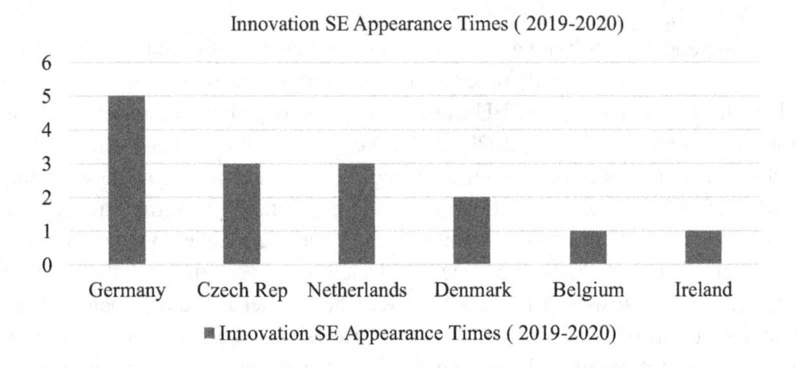

FIGURE 2.10 Innovation SE appearance times. (Authors' Calculations.)

two SE values, whereas Ireland, Slovak Republic and Spain have one appearance time(Figure 2.9). EPO statistics show a 4.3% increase in patent applications from Poland compared to the decrease in total applications for year 2020 (EPO, 2021b).

For the innovation economic efficiency, again Germany with five appearance times is the leader, followed by three times of Netherlands and Czech Republic. Belgium has two times appearance whereas Ireland has a value of one time appearance(Figure 2.10).

The overall results can be shown in Figure 2.11. The top five performers for years 2019–2020 are Germany, Netherlands, Poland, Czech Republic and Denmark.
Improvement ratios can be enlightening to understand the comparative status of the sectors and to see how much effort is needed to be efficient. The result of the improvement ratios shown in Tables 2.9 and 2.10 urges that, in terms of budget allocation, hydro sector could perform well with a 14%–22% less amount for the analyzed years. The same is true with a personnel decrease of around 12% for year 2020 for hydro sector. In terms of the filed patents, as some countries in the reference set were a way behind the SE levels, the overall improvement should be 4–5 times of today.

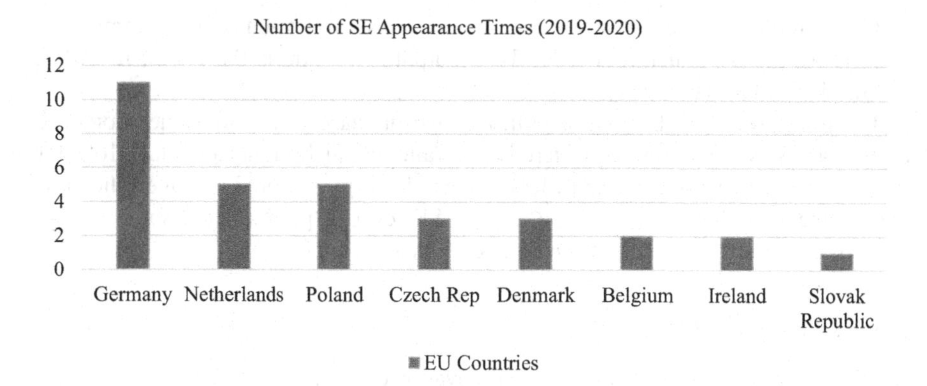

FIGURE 2.11 Number of SE appearance times of EU countries in both categories. (Authors' Calculations.)

TABLE 2.9
EU Countries' Average Improvement Ratios for Inputs and Outputs (Stage 1)

	Hydro		Solar		Wind	
Inputs and Outputs	**2019**	**2020**	**2019**	**2020**	**2019**	**2020**
R&D Budget	−13.56%	−21.46	−1.5%	−19.08%	−26.82%	−6.73%
R&D Personnel	-	−11.68%	-	−7.81%	−18.60%	-
Filed Patents	471.41%	315.35%	276.72%	189.45%	130.79%	205.01%

Source: Authors' calculations.

TABLE 2.10
EU Countries' Average Improvement Ratios for Inputs and Outputs (Stage 2)

	Hydro		Solar		Wind	
Inputs and Outputs	**2019**	**2020**	**2019**	**2020**	**2019**	**2020**
Filed Patents	−9.82%	−9.71%	−4.72%	−5.71%	−5.28%	−5.24%
Export	408.75%	463.05%	440.55%	576.77%	566.56%	648.88%

Source: Authors' calculations.

For the solar sector, with a 20% less budget and 8% less personnel, the same level of efficiency can be achieved. With a different perspective, when the improvements in the number of patent filings are analyzed, they should be 3–4 times of today.

The wind R&D budget can be decreased by 27% for year 2019 and the personnel can be decreased by 19%. The filed patents number should be improved by 2–3 times.

Compared to R&D efficiency improvement ratios, innovation economic improvement ratios require much higher levels of output. The ratio is the lowest in hydro, followed by solar and wind.

For hydro sector in EU, the overall average increase in export value should be around 435% when 2 years considered, this value is much higher in solar 510% for the analyzed 2 years and around 610% for wind energy sector. The huge efficiency gap is understandable because of the high SE level of premier renewable energy equipment exporters (Germany, Denmark, Netherlands, etc.).

2.5 CONCLUSIONS

In this research, we have examined the innovation economic efficiency in two stages. In the first stage, we have looked at the R&D efficiency levels. The first phase outputs were taken as inputs for the second phase and with DEA, we have investigated the economic results of all these efforts. For R&D efficiency levels, EU countries on average have good efficiency scores.

The hydro sector is the most efficient one for the years 2019 and 2020, followed by wind and solar. It can be interpreted as the hydro sector is producing more patent fillings compared to the other renewable sources with the given inputs.

When economic aspects are considered, the overall efficiency scores decrease. It means that patents are not creating expected economic values at least for the years studied (2018–2019–2020). In the second stage again, hydro levels are superior to the others. The second efficient sector is solar, which has a record capacity installed from 2019 and 2020 even in the Covid-19 turmoil. It seems that the European solar sector took the opportunity of the temporal closure of factories in China and the problems in transportation and they achieved a high-efficiency level with the sales from domestic factories.

When this disparity between the two average scores of efficiencies is considered, it is obvious that the EU countries should focus their efforts on innovation economic efficiency improvements. In other terms, the innovations' commercialization process from the patent application to export from the related patents should be analyzed deeply and bottlenecks should be identified. While doing this, Germany can be used as a benchmark which is the leader among the five of six efficiency scores. The country is the forerunner of the efficiency races, except for wind energy innovation economic efficiency of Denmark (which as mentioned above can be attributed to having the world leader in turbine manufacturer in its terrains). The highest values in its scorecard are linkages, innovators and firm investments, which further need scrutiny (European Union, 2022c).

Our findings on Germany are in harmony with other studies comparing the innovation levels of energy sectors (Gökgöz & Güvercin, 2018; Gökgöz & Yalçın, 2022; Aytekin et al., 2022; Mavi & Mavi, 2021).

Poland needs attention as it follows Germany in R&D efficiency in most renewable energy sectors. According to the European Union Innovation Index, Poland has been increasing its score consecutively from 2015 to 2022. The two areas where Poland is at the top are the design applications and job-to-job mobility of HR S&T (European Union, 2022b). As mentioned above, its specific new technology innovation and emission reduction programs can create a stimulating environment for these successes.

The results show that the Netherlands scores well in both solar and wind innovation economic efficiency levels. The country has been the pioneer of solar technology in building fully functional solar systems and patenting some fundamental technology, which is still in use today (Top Sector Energy et al., 2020). Their success can be attributed to their healthy home market, which they continue to develop through integrating PV solutions in buildings, infrastructure and public spaces through sustainable materials and circular design.

Our findings can help to shape the policy focus areas on the way to fight against climate change. At the macro level, to achieve its target of zero-emission continent, the growth of RE is essential. For an incremental growth in RE, innovations are needed.

The empirical results reveal that the innovation economic efficiency levels are low compared to R&D efficiency levels. This is a reflection of a need to increase the export from local manufacturers. The renewable industries' needs for increasing their exports and the bottlenecks they face should be communicated clearly to EU and country-level policymakers.

EC initiative of bringing together the PV manufacturers under one organization can be a good example for the other industries as well. Not only on the EU level but also at the country level, the manufacturers in the supply chain can create platforms for the diffusion of the RE innovations among the countries and facilitate the flow of R&D info and workers between the countries. The governments can create and mobilize existing funds according to the feedback obtained from these alliances.

There can be some specific takeaways from the high performers of our empirical analysis; from Germany, its innovation linkages; from Poland, its researchers' mobility and the intra-industry cooperation and from the Netherlands, its wider integration of renewable energy systems into other industries.

Consequently, our energy efficiency analyses would be fruitful for both institutional and individual decision-makers in analyzing the R&D and innovation economic efficiencies of renewable sources in EU countries.

REFERENCES

Altomonte, C., Gamba, S., Mancusi, M.L., & Vezzulli, A. (2016). R&D investments, financing constraints, exporting and productivity. *Economics of Innovation and New Technology*, 25(3), 283–303.

Andersen, P. & Petersen, N.C. (1993). A procedure for ranking efficient units in data envelopment analysis. *Management Science*, 39(10), 1261–1264.

Andritz Hydro. (2023). https://www.andritz.com/hydro-en/hydronews/hn-europe/andritz-hydro-in-europe. Access date: July 26, 2023.

Ayllon, S. & Radicic, D. (2019). Product innovation, process innovation and export propensity: Persistence, complementarities and feedback effects in Spanish firms. *Applied Economics*, 51(33), 3650–3664.

Aytekin, A., Ecer, F., Korucuk, S., & Karmaşa, Ç. (2022). Global Innovation Efficiency assessment of EU member and candidate countries via DEA-ETAWIOS multi-criteria methodology. *Technology in Society*, 68, 101896.

Balkan Green Energy News. Dönmez: Turkey is world's No. 4 solar panel producer. (2022). https://balkangreenenergynews.com/donmez-turkey-is-worlds-no-4-solar-panel-producer/. Access date: June 19, 2023.

Benitez, R., Coll-Serrano, V., & Bolos, V.J. (2021). deaR-shiny: An interactive web app for data envelopment analysis. *Sustainability*, 13(12), 6774.

Bilbao-Osorio, B. & Rodriguez-Pose, A. (2004). From R&D to innovation and economic growth in the EU. *Growth and Change*, 434–455.

Bodendorf, F., Dimitroc, G., & Franke J. (2022). Analyzing and evaluating supplier carbon footprints in supply networks. *Journal of Cleaner Production*, 372, 133601.

Bresciani, S., Puertas, R., Ferraris, A. & Santoro, G. (2021). Innovation, environmental sustainability and economic development: DEA-Bootstrap and multilevel analysis to compare two regions. *Technological Forecasting & Social Change*, 172, 121040.

Campos-Guzmán, V., García-Cáscales, M. S., Espinosa, N., & Urbina, A. (2019). Life cycle analysis with multi-criteria decision making: A review of approaches for the sustainability evaluation of renewable energy technologies. *Renewable and Sustainable Energy Reviews*, 104, 343–366.

Cassiman, B. & Golovko, E. (2011). Innovation and internationalization through exports. *Journal of International Business Studies*, 42, 56–75.

Chen, C.P., Hu, J.L., & Yang, C.H. (2014). An international comparison of R&D efficiency of multiple innovative outputs: The role of the national innovation system. *Innovation: Management, Policy & Practice*, 13(3), 341–360.

Chen, C.J, Wu, H.L., & Lin, B.W. (2006). Evaluating the development of high-tech industries: Taiwan's science park. *Technological Forecasting and Social Change*, 73(4), 452–465.

Cook, W.D., Zhu, J., & Tone, K. (2014). Data envelopment analysis: Prior to choosing a model. *Omega*, 44, 1–4.

Cooper, W.W., Seiford, L.M., & Tone, K. (2006). *Data Envelopment Analysis a Comprehensive Text with Models, Applications, References and DEA-Solver Software*. Dordrecht: Kluwer Academic Publishers.

Cooper, W., Seiford, L., & Zhu, J. (2011). Data Envelopment Analysis: History, Models, and Interpretations.

CPIA. Czech Power Industry Alliance. (2023). https://www.cpia.cz/en/members/. Access date: 19 June 2023.

D'Angelo, A. (2012). Innovation and export performance: A study of Italian high-tech SMEs. *Journal of Management and Governance*, 16, 393–423.

Dong, G., Kokko, A., & Zhou, H. (2022). Innovation and export performance of emerging market enterprises: The roles of state and foreign ownership in China. *International Business Review*, 31(6), 102025.

EC. (2015). Past Research and Innovation Policy Goals. https://research-and-innovation.ec.europa.eu/strategy/past-research-and-innovation-policy-goals_en. Access date: January, 06, 2023.

EC. (2020a). Research and Innovation Strategy 2020-2024. https://research-and-innovation.ec.europa.eu/strategy/strategy-2020-2024_en. Access date: January, 06, 2023.

EC. (2020b). Turning Europe into a True Innovation Union. 06th October 2010. https://ec.europa.eu/commission/presscorner/detail/en/MEMO_10_473. Access date: January, 06, 2023.

EPO. (2021a). Healthcare Innovation Main Driver of European Patent Applications in 2020. https://www.epo.org/news-events/news/2021/20210316.html. Access date: May, 26, 2023.

EPO. (2021b). Number of European Patent Applications from Poland Increased by over 4% in 2020. https://scienceinpoland.pl/en/news/news%2C86961%2Cnumber-european-patent-applications-poland-increased-over-4-2020.html. Access date: May, 26, 2023.

Euractiv. (2022). How COVID Helped EU Countries Meet Their Renewable Energy Targets. https://www.euractiv.com/section/energy/opinion/how-covid-helped-eu-countries-meet-their-renewable-energy-targets/. Access date: July 27, 2023.

Eurobserver. (2022). The State of Renewable Energies in Europe Edition 2021. https://www.eurobserv-er.org/20th-annual-overview-barometer/. Access date: July, 27, 2023.

Euronews. Spain, Sweden and Belgium: The European Countries Setting New Wind and Solar Records. (2023). https://www.euronews.com/green/2023/05/18/lightning-pace-portugal-produces-over-50-of-electricity-from-wind-and-solar-for-first-time. Access date: June 19, 2023.

European Union. (2022a). European Innovation Scoreboard. https://research-and-innovation.ec.europa.eu/statistics/performance-indicators/european-innovation-scoreboard_en. Access date: January, 06, 2023.

European Union. (2022b). European Innovation Scoreboard. https://ec.europa.eu/assets/rtd/eis/2022/ec_rtd_eis-country-profile-pl.pdf. Access date: June, 13, 2023.

European Union. (2022c). European Innovation Scoreboard. https://ec.europa.eu/assets/rtd/eis/2022/ec_rtd_eis-country-profile-de.pdf. Access date: January, 17, 2024.

Eurostat. (2021). Green Energy Products 2020: EU Imports Exceed Exports. https://ec.europa.eu/eurostat/web/products-eurostat-news/-/ddn-20211202-1. Access date: July, 26, 2023.

Georgescu, M.R., Stoica, E.A., Bogoslov, I.A, & Lungo, A.E. (2022). Managing efficiency in digital transformation - EU member states performance during the COVID-19 pandemic. *Procedia Computer Science*, 204, 432–439.

Gökgöz, F. & Güvercin, M.T. (2018). Energy security and renewable energy efficiency in EU. *Renewable and Sustainable Energy Reviews*, 96, 226–239.

Gökgöz, F. & Yalçın, E. (2022). An environmental, energy, and economic efficiency analysis for the energy market in European Union. *Environmental Progress & Sustainable Energy*, 42, 1–12.

Griliches, Z. (1998). Patent statistics as economic indicators: A survey. *R&D and Productivity: The Econometric Evidence*, by Zvi Griliches, 287–343. Chicago: University of Chicago Press, 1998.

Guan, J. & Chen, K. (2010a). Measuring the innovation production process: A cross-region empirical study of China's high-tech innovations. *Technovation*, 30(5–6), 348–358.

Guan, J. & Chen, K. (2010b). Modeling macro-R&D production frontier performance: An application to Chinese province-level R&D. *Scientometrics*, 82(1), 165–173.

Guan, J. & Chen, K. (2012). Modeling the relative efficiency of national innovation systems. *Research Policy*, 41(1), 102–115.

Hashimotoa, A. & Hanedab, S. (2008). Measuring the change in R&D efficiency of the Japanese. *Research Policy*, 37(10), 1829–1836.

Hollanders, H.J.G.M & Celikel-Esser, F. (2007). *Measuring Innovation Efficiency*. European Commission. 2007 European Innovation Scoreboard. Maastricht: Maastricht University.

Hou, G., Sun, H., Jiang, Z., Pan, Z., Wang, Y., Xiaodan, Z., Zhao, Y., & Yao, Q. (2016). Life cycle assessment of grid-connected photovoltaic power generation from crystalline silicon solar modules in China. *Applied Energy*, 164, 882–890.

IEA. (2021). Renewables 2021. https://www.iea.org/articles/what-is-the-impact-of-increasing-commodity-and-energy-prices-on-solar-pv-wind-and-biofuels. Access data: July, 27, 2023.

IEA. (2022). Clean Air Programme. https://www.iea.org/policies/11538-clean-air-programme. Access date: June 19, 2023.

IEA. (2023). Solar Pv. https://www.iea.org/energy-system/renewables/solar-pv.

IRENA. (2022). IRENA Inspire Patents. https://irena.sharepoint.com/:x:/s/statistics-public/EaUa5nqXqt1Hmm1XEgmOcWQB5MJxYvS_u7eZi8uwJ3EK1A?rtime=jBrQE4Vr20g. Access date: April, 17, 2023.

IRENA. (2023). The Changing Role of Hydro Power, Challenges and Opportunities. https://mc-cd8320d4-36a1-40ac-83cc-3389-cdn-endpoint.azureedge.net/-/media/Files/IRENA/Agency/Publication/2023/Feb/IRENA_Changing_role_of_hydropower_2023.pdf?rev=85b54f8dd8794f8fbc6270b5a1e0b92a. Access date: July, 26, 2023.

Jiang, T., Panjing, J., Yi, S., Zhen, Y., & Jin Q. (2021). Efficiency assessment of green technology innovation of renewable energy enterprises in China: A dynamic data envelopment analysis considering undesirable output. *Clean Technologies and Environmental Policy*, 23, 1509–1519.

Lan, X., Zhe, L., & Wang, Z. (2022). An investigation of the innovation efficacy of Chinese photovoltaic enterprises employing three-stage data envelopment analysis (DEA). *Energy Reports*, 8, 456–465.

Lewis, H.F. & Sexton, T.R. (2004). Network DEA: Efficiency analysis of organizations with complex internal structure. *Computers and Operations Research*, 31(9), 1365–1410.

Li, H., Haiyan, H., Jiefei, S., & Jingjing, C. (2019a). Innovation efficiency of semiconductor industry in China: A new framework based on generalized three-stage DEA analysis. *Socio-Economic Planning Sciences*, 66, 136–148.

Li, Y., Yung-ho, C., & Liang, C.L. (2019b). New Energy Development and Pollution Emissions in China. *International Journal of Environmental Research and Public Health*, 16(10), 1764.

Liu, H.H., Yang, G.L., Liu, X.X., & Song, Y.Y. (2020). R&D performance assessment of industrial enterprises in China: A two-stage DEA approach. *Socio-Economic Planning Sciences*, 71, 100753.

Mavi, R.K. & Mavi, N.K. (2021). National eco-innovation analysis with big data: A common-weights model for dynamic DEA. *Technological Forecasting and Social Change*, 162, 120369.

Moon, H. & Lee, J.-D. (2005). A fuzzy set theory approach to national composite S&T indices. *Scientometrics*, 64, 67–83.

OECD. (2023). OECD Stat Main Science and Technology Indicators. https://stats.oecd.org/viewhtml.aspx?datasetcode=MSTI_PUB&lang=en#. Access: June, 06, 2023.

OECD ilibrary. (2023). IEA Energy Technology RD&D Statistics. https://stats.oecd.org/BrandedView.aspx?oecd_bv_id=enetech-data-en&doi=data-00488-en. Access: May 23, 2023.

OECD & SOEC. (2005). *Oslo Manual*. Manual, Paris: OECD Publishing.

Pv Magazine. (2019). Turkish PV Manufacturer Report Reveals Country's Annual Production Capacity is 5,610 MW/year. https://www.pv-magazine.com/2021/01/19/turkish-pv-manufacturer-report-reveals-countrys-annual-production-capacity-is-5610-mw-year/#:~:text=The%20major%20export%20regions%20for,Europe%20and%20the%20Middle%20East. Access date: June 19, 2023.

Pv-Magazine. (2021). Belgium Deployed 900 MW of PV in 2020. https://www.pv-magazine.com/2021/01/08/belgium-deployed-900-mw-of-pv-in-2020/. Access date: June 19, 2023.

Rabaia, M.K.H., Abdelkareem, M.A., Sayed, E.T., Wilberforce, T., & Olabi, A.G. (2021). Environmental impacts of solar energy systems: A review. *Science of the Total Environment*, 754, 141989.

Reglobal. (2021). Global Wind Turbine Supplier Ranking for 2020. https://reglobal.co/global-wind-turbine-supplier-ranking-for-2020/. Access date: June, 06, 2023.

Sandu, S. & Ciocanel, B. (2014). Impact of R&D and innovation on high-tech export. *Procedia Economics and Finance*, 15, 80–90.

Segarra-Blasco, A., Teruel, M., & Cattaruzzo, S. (2020). Innovation, productivity and learning induced by export across European manufacturing firms. *Economics of Innovation and New Technology*, 31(5), 387–415.

Seiford, L.M. & Zhu, J. (1999). Profitability and Marketability of the Top 55 U.S. Commercial Banks. *Management Science*, 45(9), 1270–1288.

Sharma, S. & Thomas, V. (2008). Inter-country R&D efficiency analysis: An application of data envelopment analysis. *Scientometrics*, 76(3), 483–501.

Silva, G.M., Styles, C., & Lages, L.F. (2017). Breakthrough innovation in international business: The impact of tech-innovation and market-innovation on performance. *International Business Review*, 391–404.

Smit, J., Nacer, E., Sikorski, A., Godard, C., & Magdziarz, W. (2023). Social and Economic Consequences of COVID-19, Publication for the special committee on COVID-19 pandemic: Lessons learned and recommendations for the future (COVI). Policy Department for Economic, Scientific and Quality of Life Policies, European Parliament. Luxembourg.

Solar Generation. (2020). Solar Generation. https://solargeneration.ie/seai-announce-new-solar-pv-scheme-for-2020/. Access date: June 18, 2023.

Solar Alliance EU. (2023). Europe in strong position to exceed goal of 30 GW annual PV manufacturing by 2025, according to the Alliance. https://solaralliance.eu/news/europe-in-strong-position-to-exceed-goal-of-30-gw-annual-pv-manufacturing-by-2025-according-to-the-alliance/. Access date: July, 26, 2023.

S&P Global. (2022). *UDI World Electric Power Plants Database*. London: S&P Global Platts.

Stip Compass. (2021). Poland Country Dashboard. https://stip.oecd.org/stip/interactive-dashboards/countries/Poland. Access date: June 19, 2023.

Tone, K. (2001). Slacks-based measure of efficiency in data envelopment analysis. *European Journal of Operational Research*, 458–506.

Top Sector Energy; Netherlands Enterprise Agency; FME; TKI Urban Energy. (2020). Next Generation Solar Power. Fortelle.

Wang, E.C. & Huang, W. (2007). Relative efficiency of R&D activities: A cross-country study accounting for environmental factors in the DEA approach. *Research Policy*, 36(2), 260–273.

Wang, C.H., Lu, Y.H. Huang, C.W., & Lee, J.Y. (2013). R&D, productivity, and market value: An empirical study from high-technology firms. *Omega*, 41(1), 143–155.

Wang, Q., Ye Hang, L. S., & Zengyao, Z. (2016). Two-stage innovation efficiency of new energy enterprises in China: A non-radial DEA approach. *Technological Forecasting and Social Change*, 112, 254–261.

World Integrated Trade Solution. (2023). Exports by Countries with Headings Numbers. https://wits.worldbank.org/trade/comtrade/en/country/ALL/year/2021/tradeflow/Exports/partner/WLD/product/841280. Access date: May 23, 2023.

Zhong, W., Yuan, W., Li, S.X., & Huang, Z. (2011). The performance evaluation of regional R&D investments in China: An application of DEA based on the first official China economic census data. *Omega*, 39(4), 447–455.

Zoltan, J.A., Anselin, L., & Varga, A. (2002). Patents and innovation counts as measures of regional production of new knowledge. *Research Policy*, 31(7), 1069–1085.

Zuo, Z., Guo, H., Li, Y., & Cheng, J. (2022). A two-stage DEA evaluation of Chinese mining industry technological innovation efficiency and eco-efficiency. *Environmental Impact Assessment Review*, 94, 106762.

3 BIM-Powered Energy Efficiency and Life-Cycle Cost Analyses for Greener Design

Figen Balo
Firat University

Lütfü S. Sua
Southern University

Hazal Boydak
Dicle University

3.1 INTRODUCTION

Energy is the primary driver of social and economic progress and is seen as a crucial criterion for increasing living standards throughout the world. The world's energy consumption is increasing as a result of a growing population and dramatic improvements in living conditions. The availability of energy resources is constrained and using fossil fuels pollutes the environment. Four major sectors that consume energy are industry, construction, agriculture, and transportation. For space heating and cooling, the energy used in construction accounts for the biggest share worldwide (Balo and Polat, 2021; Balo et al., 2022; Balo, 2011).

The energy usage in the construction industry explains the primary part of 40% of the global annual final energy utilization. A big part of the energy is demanded for space cooling and heating; for this purpose, the activity to decrease energy usage in this industry is one of the preferences. Thus, the instruction 2002/91/EC of the Council of 16/Dec./2002 and the European Parliament laid the base for the constructions' energy efficiency regulation. It focalizes on installing a general method, forming minimal norms for regular inspections, energy performance, and energy certification (Hroudová and Zach, 2017).

Green buildings are environmentally friendly structures. The cost of building a green construction is subtly more than those of a traditional structure, but the operating and maintenance costs are lower and offer good ecological benefits. The primary difficulty is earning these advantages at a fair expense and within budget limitations.

DOI: 10.1201/9781003403456-3

Green buildings are comparable to conventional buildings in that they serve the same purpose, but the distinctions are in the plan principles employed and the structure processes utilized to protect the environment and nature. Green New Buildings provide both concrete and intangible benefits. The quantifiable benefits include a 30%–50% reduction in water usage through rainwater collection channels (i.e., collecting rainwater from the roof and filtering it and using it), gray water recovery system (i.e., using the collected gray water in toilet cisterns and cleaning after simple purification), and replacing the faucets with photocell ones in order to save water (if heat and flow adjustable armatures are used in bathrooms, sinks, kitchens, and reservoirs, water savings of up to 67% can be achieved per year). Another benefit is a 20%–30% reduction in energy consumption through supplying hot water to the building with a solar collector, using an electric heating system by giving up fossil fuels used in heating, designing an insulation system to reduce the effect of hot air in summer and cold air in winter, keeping indoor air quality at a high level with a natural ventilation system, using materials with low toxicity rate saving energy with the double glazing system, and illuminating the dark parts of the house with daylight without wasting energy. Intangible benefits include better air quality, tenant health, outstanding daylighting, well-being, comfort, conservation of restricted country sources, and other protective advantages (IGBC, 2015).

Building information modeling (BIM) is one of the most favorable current improvements in the engineering, architecture, construction, and operation sector. The structures utilize a tremendous amount of electricity, water, and raw materials each day, leaving a significant environmental footprint. The conclusion that buildings must become greener and more environmentally friendly was reached by this fact. The review method must be practical, simple to repeat countless times, and produce findings in less time to apply alterations in buildings and assess building efficiency. BIM can be utilized to perform this research. BIM's tool is specifically designed to facilitate energy analysis applications to determine the likely loss or gain of energy for the construction, as well as to estimate and detect its maintainability early in the notional planning process. As a result, the design group and owner will be able to begin sustainability evaluation from the very beginning of project delivery (Ebrahim and Wayal, 2019; Karahan et al., 2022; Boydak et al., 2022). BIM is a contemporary technique that is used to produce and organize numerical data during the life cycle of a structured plan, encompassing the stages of construction, planning, maintenance, operation, and destruction (Becker et al., 2018). Plan development, decision analysis, document management, cost calculations, operational management, and efficiency management are just a few of the many applications of information handled utilizing BIM (Mohammed, 2020). For these reasons, BIM is considered a full tool for a building's constructional data. Azhar et al. explored the BIM's possibility to superimpose multi-disciplinary data within the modeling and adapt with energy efficiency to define the LEED potential of a nonresidential construction (Azhar et al., 2012). BIM is extensively utilized in the construction sector all over the world because of its capacity to correct faults in early design phases with precise construction dates, construction sequencing, conflict detection, and facilitating complicated projects. Developed actual-time collaboration among different partners via a digital arena is a huge benefit of BIM, since it may lead to fewer mistakes owing to greater teamwork.

Furthermore, BIM provides an information-driven strategy for building design, maintenance, and construction. The study of project partners is simplified and made more efficient since BIM allows them to concurrently communicate, update, and acquire the construction's functional and physical data on one construction planning. To guarantee people study jointly during a structure's life cycle, it is critical in sustainable design to involve end consumers and their great concern in choosing greener properties (El-Diraby et al., 2017). BIM is a great tool for +695 Sustainable design because of its capacity to test, evaluate, and improve again and over. The "building performance analysis" is the phrase for this (Otuh, 2016). There are several software solutions for energy analysis in the market today, such as IES-VE, Ecotect, EnergyPlus, Equest, and Green Building Studio, which may be utilized in conjunction with BIM simulation programs like ArchiCAD, Bentley, Graphisoft, and Revit.

Many researchers have carried out an in-depth investigation in the area of sustainability and BIM. In Solla et al., the BIM's potential application in green building authentication and the prospect of establishing sustainability credit were investigated with the aid of adapting BIM equipment and analysis simulation (Solla et al., 2016). BIM can provide feedback for alterations in material, orientation, and design reasonably quickly (Rajendran et al., 2012). The combination of BIM equipment with Microsoft Visual Studio, according to Raffee, Hassan, and Karim, will promote sustainability and make it easier for the planning group and other partners to meet their challenging maintainability goals (Raffee et al., 2015). Because of the speed and accuracy of evaluating even the most complicated modelings, results demonstrate that BIM energy efficiency evaluation can be a substitute for doing manual building energy consumption calculations (Stundon et al., 2015). Buildings can be designed with high performance by integrating sustainable design techniques with BIM technology, which can revolutionize the way traditional design processes are done (Jalaei and Jrade, 2013). Because a schematic model is created before a full construction model, the planner can obtain a more proper appraisal of the design, which enhances the project's performance and overall quality (Kubba, 2012; Becerik-Gerber and Rice, 2010; Sauba et al., 2015). Early project decisions have an important effect on a building's life-cycle costs (Dahl et al., 2005; Lam et al., 2004). Using BIM, several material combinations were evaluated to find alternative, environmentally friendly ways to cut down on operational energy use (Shoubi et al., 2015; Longo et al., 2013; Balo and Polat, 2022). Effective utilization of renewable energy can reduce the utilization of man-made sources, increasing a construction's energy efficiency (Kaplan et al., 2012; Guerrisi et al., 2012; Khan and Halder, 2016). Another attempt has been done to increase the maintainability of construction and so turn it into a green structure, depending on the investigation evaluated from diverse publications on providing maintainability in BIM. BIM-based literature research is given in Table 3.1.

This study's major goal is to transform a designed structure into an energy-efficiency structure by introducing energy ideas that promote sustainability. The aim is accomplished by developing a three-dimensional BIM of the structure and evaluating the construction's energy efficiency using Autodesk Revit. In the cafe designed in this study, based on the usage directions of the green wall on the wall

TABLE 3.1
BIM-Based Studies

Research Topic	Research Method	Source
Building energy optimization	Modeling	Kim and Kim (2017)
Energy renovation of historic buildings	Case study	Piselli et al. (2020b)
Energy-efficient building operations	Model and case study	Gokce and Gokce (2013)
Building processes and energy efficiency	Modeling	Costa et al. (2013)
Building processes and energy efficiency	Case study	Petri et al. (2017)
Trade-offs between operational and embodied energy	Model and case study	Shadram and Mukkavaara (2018)
Workflow design	Case study	Chen and Tang (2019)
The facility energy management's application	Case study	Piselli et al. (2020a)
Integration of BIM and LCA	Case study	Cheng et al. (2020)
Tradeoffs between embodied and operational energy	Mixed	Venkatraj et al. (2020)
GHG emissions from concrete	Review	Wu et al. (2014)
Integration of LCA and BIM	Case study	Wang et al. (2018)
Greenhouse gas emissions	Modeling	Xu et al. (2019)
Structure and demolition waste disposal sector	Model and case study	Shi and Xu (2021)

facades of the building (no green wall was used on any facade, green wall was used on all facades, only the green wall on the north facade, the green wall only on the south facade, the green wall only on the east facade, and the green wall only on the west facade) of the building's energy performance, annual energy cost, life-cycle cost, and life-cycle energy values were evaluated. In all alternative scenarios of the analyzed building in this study, if the roof is a tile roof or a green roof, the change values in all investigated parameters are interpreted together. In the analysis, the building material (gas concrete, pumice block) and insulation material (rock wool, expanded polystyrene (EPS)) were selected and applied to the exterior walls of the cafe building, by choosing the two most commonly used materials in the climatic conditions of the building. In this way, the change in the energy performance parameters of the cafe building was analyzed with the help of Green Building Studio simulation, in the case of adding green walls to different aspects of the building facades rather than the fixed building, insulation, and roof materials. The aim of this study is to guide researchers, contractors, designers, and engineers on the improvement of performance in terms of both energy efficiency and all parameters related to energy cost, in case of applying the most performance green wall direction or aspects in green building designs to be newly built in the researched region. This study has been presented to the literature as an exemplary research in terms of constructing more energy-performing buildings with designs made with the help of Green Building Studio simulation before construction, taking into account the climatic data of different regions.

The rest of the paper is organized as follows: Section 3.2 provides background information on the design stages of the building, including the climatic data and technical properties of the materials and the building. Section 3.3 provides the results of the analysis which is followed by the conclusions section.

3.2 DESIGN STAGES OF A CAFE IN TEKIRDAG CITY

3.2.1 Climatic Data in Tekirdag City

For the facility forecast humidity and temperature are the most critical climate characteristics. The climatic information of Tekirdağ city of Turkey is displayed in Figure 3.1. The monthly design information (a threshold of 2%) is provided in Figure 3.1a. The wind impact assessment is displayed in Figure 3.1b. This figure displays that the main wind directions are northeast-north. In a yearly frequency distribution, the wind speed (knots) is found to be between 3.1% (winter Jan–Mar) and 2.3% (summer July–Sept). Green Buildings studio analysis displays that the wind blows from north to northeast direction in winter and indicates the most predominant wind from north-northeast to northeast in summer. The dry-bulb cumulative and the dry frequency distributions in the analysis are shown in Figure 3.1c and d, respectively. The wind is a passive methodology for cooling in the summer months and promotes natural ventilation.

The relative humidity frequency distribution (yearly) and the impact of dew point frequency distributions in the assessment are indicated in Figure 3.1e and f, respectively. The choice of bio-climate developments is dependent on the climatic assessment of Tekirdag and the sensibility study.

The area coated through the clouds in the sky is known as sky cover. The percentage of the sky covered with clouds requires to be defined to compute it. During the period of the investigation, the cloud fields that do not cover the sky entirely are added to this computation. The whole field in the sky is supposed to be ten/ten in climatology investigations. Therefore, a scale between 10 and 0 is utilized. According to this scale, a value of 10 indicates a sky that is entirely covered with clouds, while a zero value indicates a sky with no clouds. While analyzing the building energy savings under investigation here, yearly overall sky cover frequency distributions that are thought within the perspective of Green building studio are significant for a total evaluation (DMI, 2005).

The impact of total direct normal irradiation and sky cover frequency distributions (yearly) on assessments is represented in Figure 3.1g and h, respectively. The directly normal irradiation is the amount of sun radiation found per unit area through a superficies that is forever held perpendicular (or normal) to the lights directly from the way of the sun at its present circumstantial in the skies (Lee et al., 2017). Reliable information on solar radiation must be supplied as the most significant input to the energy performance software. The most advantageous solar irradiation information is worldwide horizontal irradiation but diffuse or directly normal horizontal irradiation is also critical.

For instance, irradiation on the superficies of a solar cell or solar collector is defined when either direct normal or diffuse horizontal irradiation is supplied apart

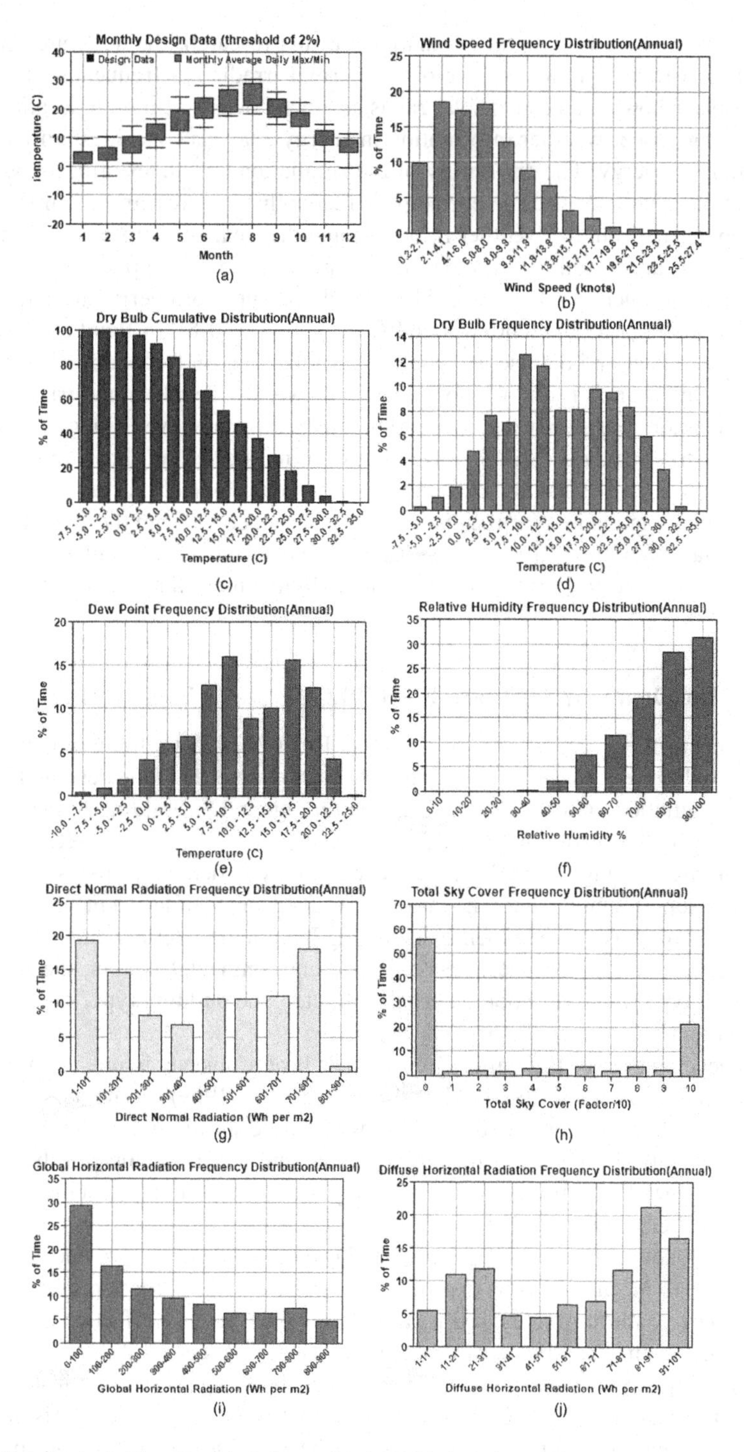

FIGURE 3.1 (a–j) The climatic information in Tekirdağ city of Turkey.

from the worldwide horizontal irradiation. In this research, the software employs these three irradiation values. The direct normal irradiation frequency dispersion chart displays how much direct solar ray is advanced over a 12-month duration.

Selecting proper weather information in energy analysis concludes with a better estimation of energy. The credibility of the simulation concludes utilizing weather information (characteristic-meteorological-12 months) was researched in the planning step. Apart from the weather information, other realistic improvement matters require to be addressed. The angle and temperature of the appliance also affect the electric production's output. BIM-sourced measures of energy are suitable to utilize in this regard (Orzechowski and Orzechowski, 2018). In Revit analysis, the worldwide impact and diffuse horizontal yearly irradiation frequency distributions are shown in Figure 3.1i and j, respectively. The worldwide horizontal irradiation is the whole amount of shortwave irradiation released from above through horizontal superficies to the world. This involves both normal direct and horizontal diffuse irradiation and is of specific relevance to photovoltaic panel placements. Diffuse horizontal irradiation is solar irradiation that does not come from the sun on a direct path but has been emitted in the sky through particles and molecules and comes uniformly from overall directions. It is the enlightenment that reaches the clouds and the blue sky.

3.2.2 The Design by Autodesk Revit 2021

The project of the designed cafe by Autodesk Revit 2021 in Tekirdağ city of Turkey is shown in Figure 3.2. The interior and exterior appearances of the designed cafe by Lumion simulation in Tekirdağ are displayed in Figure 3.3. Building information of the designed cafe is given in Table 3.2.

Factors influencing thermal efficiency are widely understood process expenses, the energy supply's mode and kind, air-conditioning and ventilation mechanism efficiency and type, insulation expenses including labor and materials, the used insulation type, climatic region characteristics, and environmental factors, the insulated component type (building floor, external wall, etc.), the building type, size, use, and construction (Karlsson et al., 2013).

The development of the building envelope's heat efficiency is one of the energy performance measures of the building (Roucoult et al., 1999). A characteristic construction envelope is a combined one with two or more sheets, and where there is primarily a weighty sheet (such as the construction material) that composes the thermal values' big part of building external wall components and a slight sheet that serves as heat isolator (Gounni et al., 2017).

Thermal isolation in construction also plays a significant role in decreasing the reliance on HVAC mechanisms to process energy-saving, buildings comfortably, and the associated natural sources. It is regarded as one of the most effective methodologies for decreasing heat transfer by constructions (Kameni et al., 2017).

To accomplish an energy-efficient and green building model, environmentally friendly materials can be used in the construction process. The comparison of the energy statistics for several factors between the building and insulation materials shows the effectiveness of BIM from an energy performance perspective.

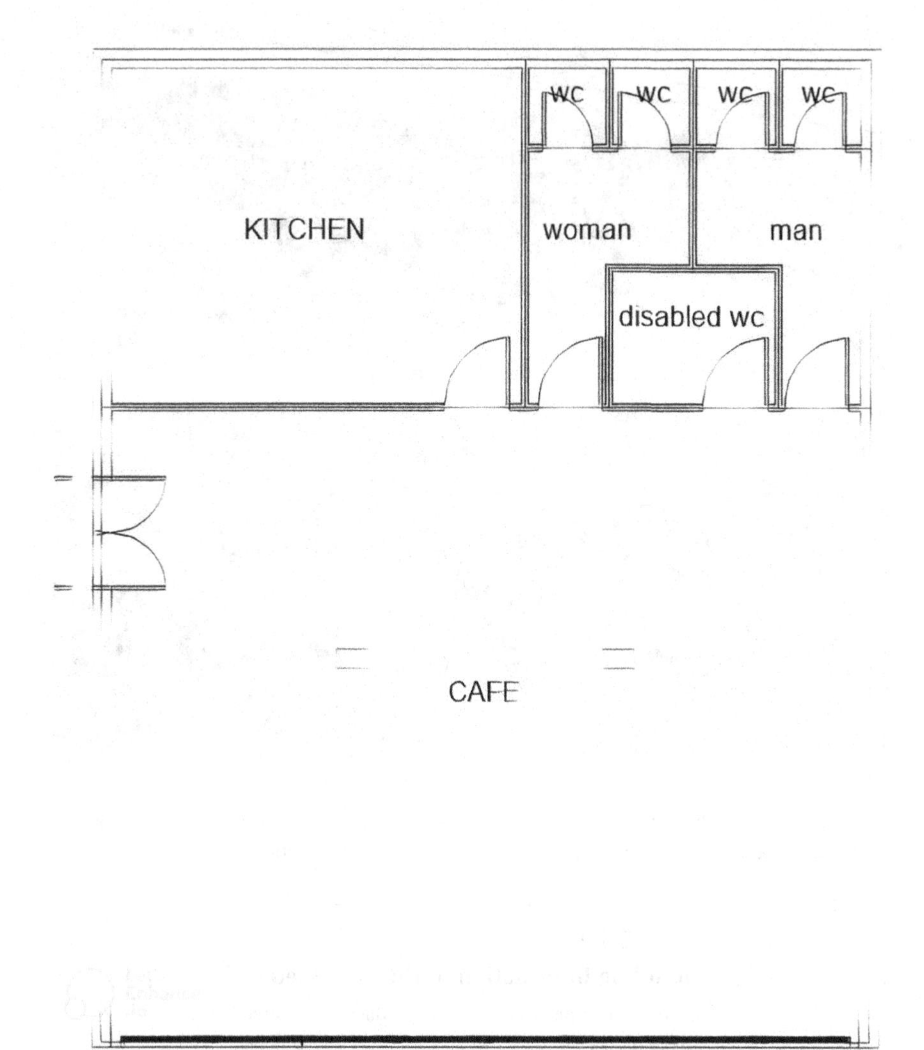

FIGURE 3.2 The project of the designed cafe by Autodesk Revit 2021 in Tekirdağ.

The building envelope components according to the scenarios of the designed cafe at the sandwich wall structure in Tekirdağ are displayed in Figure 3.4.

Building envelope layers and thicknesses are given in Table 3.3.

In this study, the effect of the green wall at different directions added to the outermost layer of the building envelope on the building energy performance was investigated. In BIM-supported Autodesk Revit simulation, a café was designed, for this purpose. The climatic conditions of Tekirdag province were taken into account in the analysis. While natural gas was used as the heating system source in the reference building, electricity was used as the cooling system source. The structure of the building envelope is designed as a sandwich wall. Two different building materials (Gas concrete, Bims block), two different insulation materials (Rock wool, EPS),

FIGURE 3.3 (a) The interior appearance of the designed cafe by Lumion simulation. (b) The exterior appearance of the designed cafe by Lumion simulation.

TABLE 3.2
Building Information of the Designed Café

Carrier mechanism	Masonry construction
Floor number	Ground floor
Height of story	3.5 m
Ways	12 m * 15 m
Gross field	180 m^2
Net field	157 m^2
Wall thickness:	30 cm

and two different roofing materials (Green Roof, Tile Roof) are used in the building envelope. Eight different types were created by using different construction and insulation materials for each wall type used.

It is planned as a green roof between Type 1 and Type 4 and as a tile roof between Type 5 and Type 8. A green wall is not used in the outer layer of the building envelope of the first eight types.

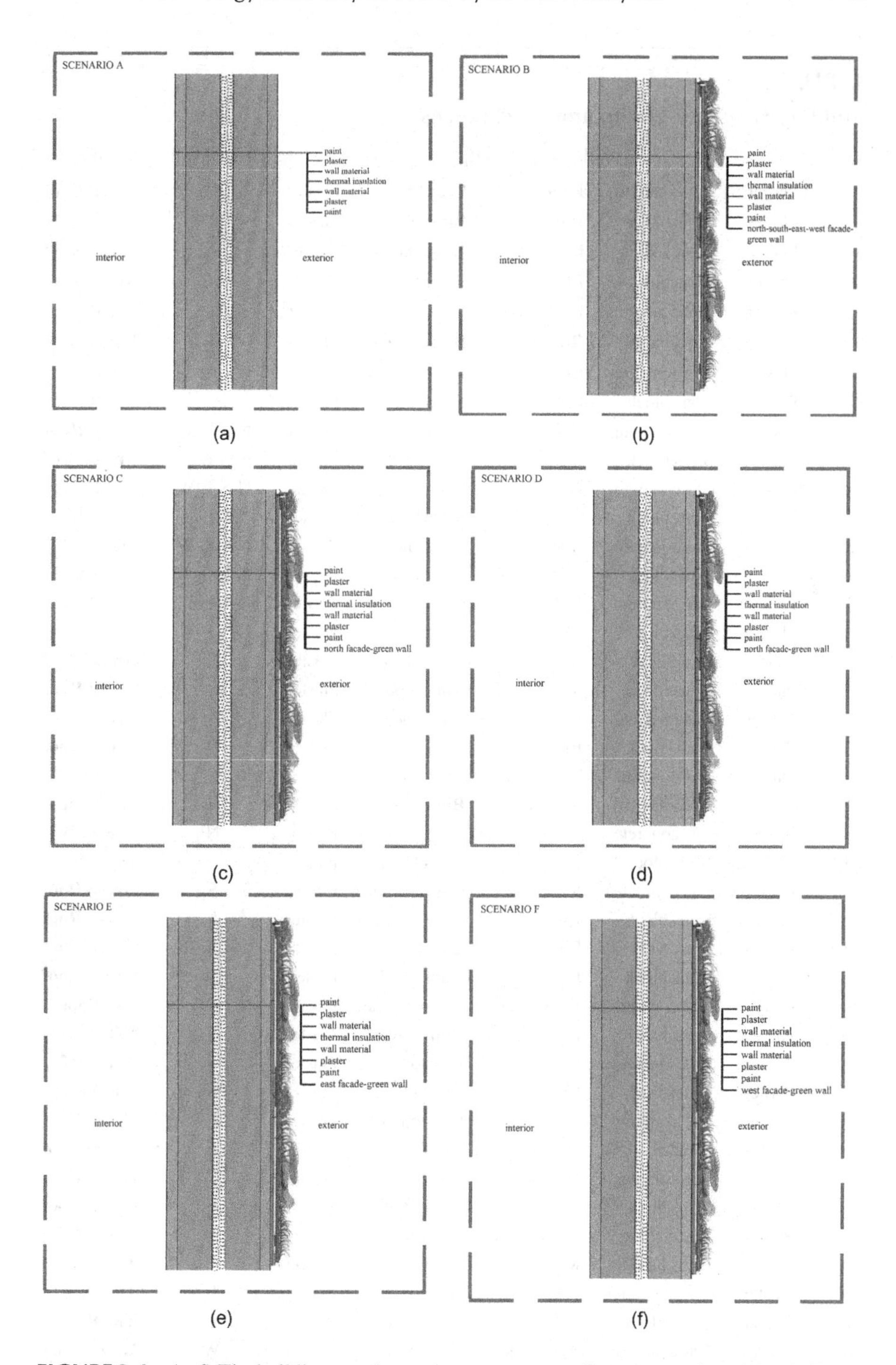

FIGURE 3.4 (a–f) The building envelope components according to scenarios of the designed cafe at the sandwich wall structure.

TABLE 3.3
Building Envelope Layers and Thicknesses

Type	Wall (cm), (in-out)					Green Wall	Roof
1	Plaster[a]	Gas concrete[b]	Rock wool[c]	Gas concrete	Plaster	×	**Green Roof**
2	Plaster	Bims block[b]	Rock wool	Bims block	Plaster	×	**Green Roof**
3	Plaster	Gas concrete	EPS[c]	Gas concrete	Plaster	×	**Green Roof**
4	Plaster	Bims block	EPS	Bims block	Plaster	×	**Green Roof**
5	Plaster	Gas concrete	Rock wool	Gas concrete	Plaster	×	**Tile Roof**
6	Plaster	Bims block	Rock wool	Bims block	Plaster	×	**Tile Roof**
7	Plaster	Gas concrete	EPS	Gas concrete	Plaster	×	**Tile Roof**
8	Plaster	Bims block	EPS	Bims block	Plaster	×	**Tile Roof**
9	Plaster	Gas concrete	Rock wool	Gas concrete	Plaster	N-S-E-W	**Green Roof**
10	Plaster	Bims block	Rock wool	Bims block	Plaster	N-S-E-W	**Green Roof**
11	Plaster	Gas concrete	EPS	Gas concrete	Plaster	N-S-E-W	**Green Roof**
12	Plaster	Bims block	EPS	Bims block	Plaster	N-S-E-W	**Green Roof**
13	Plaster	Gas concrete	Rock wool	Gas concrete	Plaster	N-S-E-W	**Tile Roof**
14	Plaster	Bims block	Rock wool	Bims block	Plaster	N-S-E-W	**Tile Roof**
15	Plaster	Gas concrete	EPS	Gas concrete	Plaster	N-S-E-W	**Tile Roof**
16	Plaster	Bims block	EPS	Bims block	Plaster	N-S-E-W	**Tile Roof**
17	Plaster	Gas concrete	Rock wool	Gas concrete	Plaster	N	**Green Roof**
18	Plaster	Bims block	Rock wool	Bims block	Plaster	N	**Green Roof**
19	Plaster	Gas concrete	EPS	Gas concrete	Plaster	N	**Green Roof**
20	Plaster	Bims block	EPS	Bims block	Plaster	N	**Green Roof**
21	Plaster	Gas concrete	Rock wool	Gas concrete	Plaster	N	**Tile Roof**
22	Plaster	Bims block	Rock wool	Bims block	Plaster	N	**Tile Roof**
23	Plaster	Gas concrete	EPS	Gas concrete	Plaster	N	**Tile Roof**
24	Plaster	Bims block	EPS	Bims block	Plaster	N	**Tile Roof**
25	Plaster	Gas concrete	Rock wool	Gas concrete	Plaster	S	**Green Roof**
26	Plaster	Bims block	Rock wool	Bims block	Plaster	S	**Green Roof**
27	Plaster	Gas concrete	EPS	Gas concrete	Plaster	S	**Green Roof**
28	Plaster	Bims block	EPS	Bims block	Plaster	S	**Green Roof**
29	Plaster	Gas concrete	Rock wool	Gas concrete	Plaster	S	**Tile Roof**
30	Plaster	Bims block	Rock wool	Bims block	Plaster	S	**Tile Roof**
31	Plaster	Gas concrete	EPS	Gas concrete	Plaster	S	**Tile Roof**
32	Plaster	Bims block	EPS	Bims block	Plaster	S	**Tile Roof**
33	Plaster	Gas concrete	Rock wool	Gas concrete	Plaster	E	**Green Roof**
34	Plaster	Bims block	Rock wool	Bims block	Plaster	E	**Green Roof**
35	Plaster	Gas concrete	EPS	Gas concrete	Plaster	E	**Green Roof**
36	Plaster	Bims block	EPS	Bims block	Plaster	E	**Green Roof**
37	Plaster	Gas concrete	Rock wool	Gas concrete	Plaster	E	**Tile Roof**
38	Plaster	Bims block	Rock wool	Bims block	Plaster	E	**Tile Roof**
39	Plaster	Gas concrete	EPS	Gas concrete	Plaster	E	**Tile Roof**
40	Plaster	Bims block	EPS	Bims block	Plaster	E	**Tile Roof**
41	Plaster	Gas concrete	Rock wool	Gas concrete	Plaster	W	**Green Roof**
42	Plaster	Bims block	Rock wool	Bims block	Plaster	W	**Green Roof**
43	Plaster	Gas concrete	EPS	Gas concrete	Plaster	W	**Green Roof**

(Continued)

TABLE 3.3 (*Continued*)
Building Envelope Layers and Thicknesses

Type	Wall (cm), (in-out)					Green Wall	Roof
44	Plaster	Bims block	EPS	Bims block	Plaster	W	**Green Roof**
45	Plaster	Gas concrete	Rock wool	Gas concrete	Plaster	W	**Tile Roof**
46	Plaster	Bims block	Rock wool	Bims block	Plaster	W	**Tile Roof**
47	Plaster	Gas concrete	EPS	Gas concrete	Plaster	W	**Tile Roof**
48	Plaster	Bims block	EPS	Bims block	Plaster	W	**Tile Roof**

[a] (2.00 cm), [b] (10.00 cm), [c] (6.00 cm).

TABLE 3.4
Technical Properties for Energy Analysis by the Green Building Studio Simulation of Building Envelope Components

Wall Materials	Thermal Conductivity	Density
	W/(m K)	kg/m³
Gas concrete	0.11	350
Bims block	0.23	770
Plaster	1.0	1,800
Insulation Materials		
Rock wool	0.045	150
EPS	0.035	15
Slab Materials		
Slab on grade – Reinforced concrete slab	2.5	2,400
Bedding mortar – Reinforced screed	1.4	2,000
Protective concrete	1.65	2,200
XPS	0.04	35.00

A green wall was added to all facades of the building envelope of the first eight types between Type 9 and Type 16. A green wall was added only to the northern facades of the building envelope of the first eight types between Type 17 and Type 24. A green wall was added only to the southern facades of the building envelope of the first eight types between Type 25 and Type 32. A green wall was added only to the eastern facades of the building envelope of the first eight types between Type 33 and Type 40. A green wall was added only to the western facades of the building envelope of the first eight types between Type 41 and Type 48.

The amount of energy consumed by the building for one year was analyzed using the Green Building Studio simulation for all types. The needed technical properties for energy analysis by the Green Building Studio Simulation of building envelope components are given in Table 3.4. The *U* values of building envelope components are given in Table 3.5. The layers and thickness of slabs and roofs are given in Table 3.6.

TABLE 3.5
The *U* Values of Building Envelope Components

	U Value W/(m² K)
Floor Slab	0.7729
Suspended Slab	8.6420
Roofing Type	
Green Roof	0.1205
Tile Roof	0.3300

The project is created in Autodesk Revit 2021. The BIM-based 3D modeling of the designed cafe is displayed in Figure 3.5. Figure 3.5 shows the appearance of a green wall added to all facades of the building envelope. Green wall material was selected automatically by the Green Building studio tool. This material was determined according to climatic conditions in Tekirdağ city of Turkey. The green wall material used on the building facades was chosen as the most suitable alternative among the plant alternatives available in the Green building studio simulation for the climatic conditions of the region where the cafe is designed and used in the analyses. Table 3.7 provides the technical properties of the green wall materials used in building facades.

3.3 RESULTS AND DISCUSSION

The assessment mainly provides the annual energy usage, which is a gauge of how much power and fuel the project may use over the course of an average year according to annual cooling and annual heating demands.

According to the results of the analysis, the highest and lowest annual total energy consumption were obtained with Type 30 (the green wall on the south façade of the building envelope) and Type 35 (the green wall on the east façade of the building envelope) as 347,907 and 326,782 kWh, respectively. Type 30 is designed using "Bims block + Rock wool + Bims block" as a building material and Tile roof as a roofing material. Type 35 was designed using "Gas concrete + EPS + Gas concrete" as a building material and Green roof as a roofing material. A difference of 6.072% was found between the highest and lowest annual total energy consumption.

The highest and lowest annual cooling energy consumption were obtained with Type 8 (no green walls on any facade of the building envelope) and Type 10 (with green walls on all facades of the building envelope) as 8,151 and 7,127 kWh, respectively. Type 8 is designed using "Bims block + EPS + Bims block" as a building material and Tile roof as a roofing material. Type 10 is designed using "Bims block + Rock wool + Bims block" as a building material and Green roof as a roofing material. A difference of 12.562% was obtained between the highest and lowest annual cooling energy consumption.

The highest and lowest annual heating energy consumption were obtained with Type 30 (the green wall on the south facade of the building envelope) and Type 16 (the green wall on all facades of the building envelope) as 88,228 and 78,570 kWh,

TABLE 3.6
The Layers and Thickness of Slabs and Roofs

			Floor Slab (cm)							
Slab on grade (50.00 cm)	Waterproofing (0.10 cm)	Bedding mortar (5.00 cm)	XPS (4.00 cm)	Protective concrete (3.00 cm)	Reinforced screed (5.00 cm)	Granite covering (1.20 cm)				
			Suspended Slab (cm)							
Wall paint (0.10 cm)	Plaster (2.00 cm)	Reinforced concrete slab (15.00 cm)	Reinforced screed (5.00 cm)							
			Green Roof							
Vapor Control Layer	Mineral Wool	Protan SE Titanium	Extensive Green Roof Layers of sedum/drainage							
			Tile Roof							
Vapor Barrier	Isolation – Wierer – Rock Wool	VaporRetarder–Wierer–Divoroll Universal 25	Wood – Tile Batten	Fabric – Wierer	Wood – Countern Batten	Vapor Retarder – Wierer	Fabric – Wierer	Wood – Tile Batten	Tile Coat.	

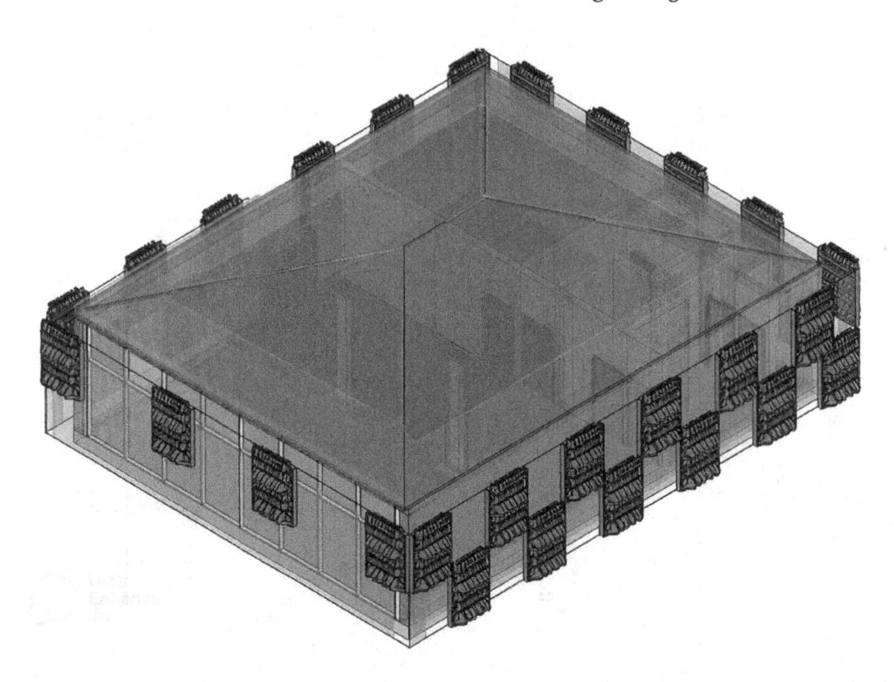

FIGURE 3.5 The BIM-based 3D modeling of the designed café.

TABLE 3.7
The Technical Properties of the Green Wall Materials

	Green Facade Technical Values
Porosity	0.01
Permeability [ng/(Pa s m2)]	3.42
Emissivity	0.95
Density [kg/m3]	920
Specific heat [J/g °C]	2.09
Thermal conductivity [W/(m K)]	0.33

respectively. Type 30 is designed using "Bims block + Rock wool + Bims block" as a building material and Tile roof as a roofing material. Type 16 is designed using "Bims block + EPS + Bims block" as a building material and Tile roof as a roofing material. A difference of 10.946% was defined between the highest and lowest annual heating energy consumption.

According to the results obtained, the heating loads of the building are much higher than the cooling loads.

The study was evaluated in terms of "the best alternatives" for the following six main groups according to the application of the green wall to the building facades.

Type 1–Type 8 (Green wall is not used in the outer layer of the building envelope.)

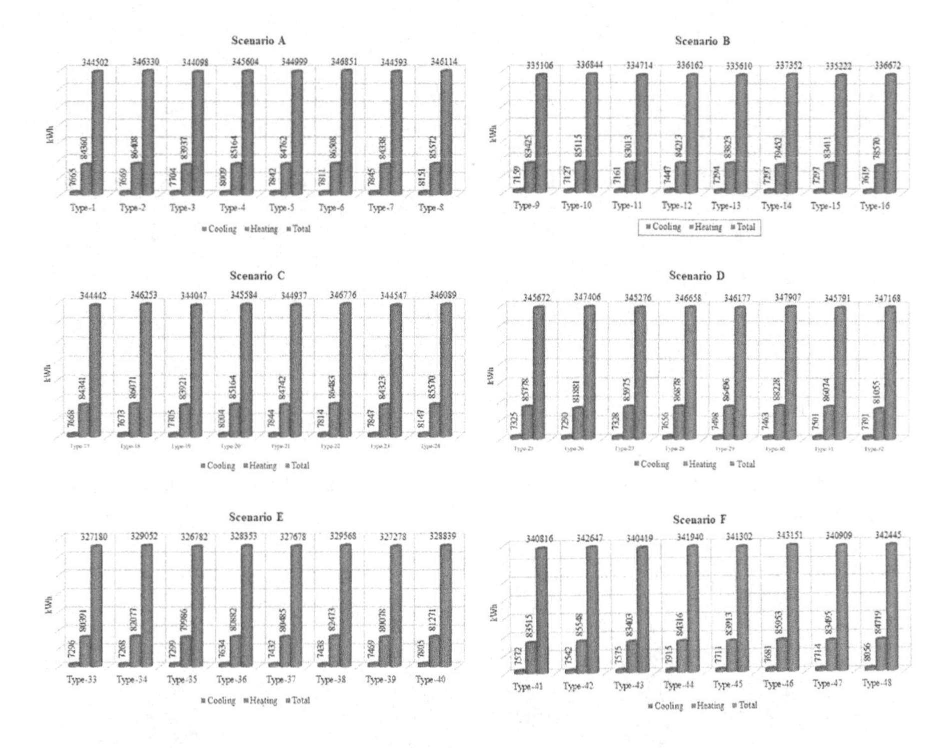

FIGURE 3.6 The energy consumption values obtained according to the Green Building Studio simulation. (a) Green wall added to nothing facades of the building envelope. (b) Green wall added to all facades of the building envelope. (c) Green wall added only to the northern facades of the building envelope. (d) Green wall added only to the southern facades of the building envelope. (e) Green wall added only to the eastern facades of the building envelope. (f) Green wall added only to the western facades of the building envelope.

Type 9–Type 16 (Green wall has been added to all facades of the building envelope.)

Type 17–Type 24 (Green wall has been added only to the northern facades of the building envelope.)

Type 25–Type 32 (Green wall has been added only to the south facades of the building envelope.)

Type 33–Type 40 (Green wall has been added only to the eastern facades of the building envelope.)

Type 41–Type 48 (Green wall has been added only to the western facades of the building envelope.)

In Figure 3.7, in each of the six main groups mentioned above, the type with the lowest “annual total cooling energy consumption amount” was determined as “the most positive alternative”.

Among the best alternatives of all scenarios, Type 17 (the green wall on the north facade of the building envelope) is obtained as the most negative scenario according to annual total heating energy consumption amounts.

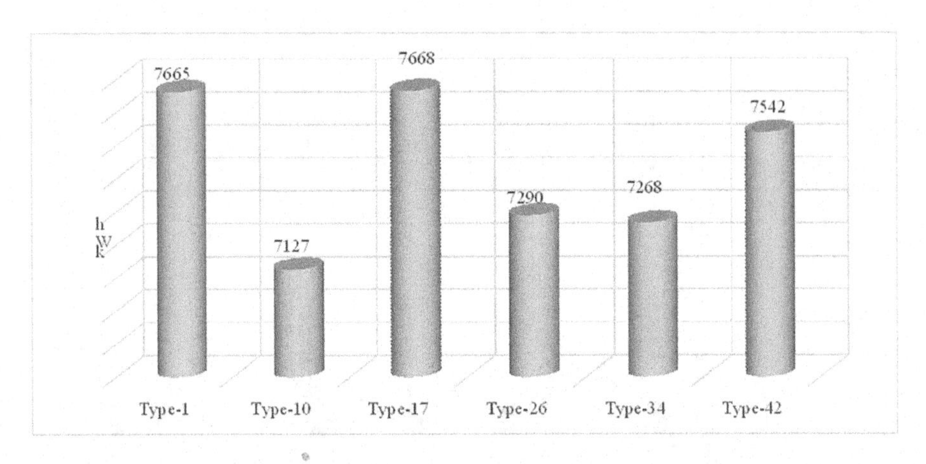

FIGURE 3.7 Annual total cooling energy consumption amounts for the best alternatives of all scenarios.

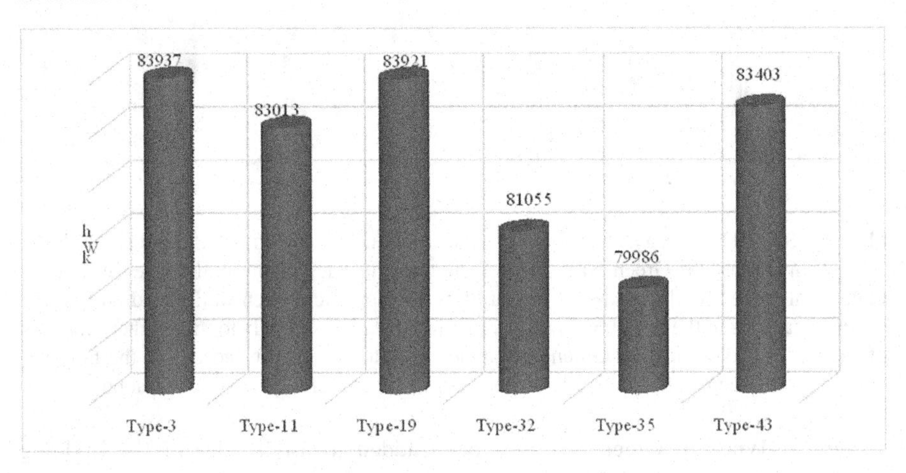

FIGURE 3.8 Annual total heating energy consumption amounts for the best alternatives of all scenarios.

In Figure 3.8, in each of the six main groups mentioned above, the type with the lowest "annual total heating energy consumption amount" was determined as "the most positive alternative".

Among the best alternatives of all scenarios, Type 3 (the green wall is not used in the outer layer of the building envelope) is obtained as the most negative scenario according to annual total heating energy consumption amounts.

In Figure 3.9, in each of the six main groups mentioned above, the type with the lowest "annual total energy consumption amount" was determined as "the most positive alternative".

Among the best alternatives of all scenarios, Type 27 (the green wall on the south facade of the building envelope) is obtained as the most negative scenario according to annual total energy consumption amounts.

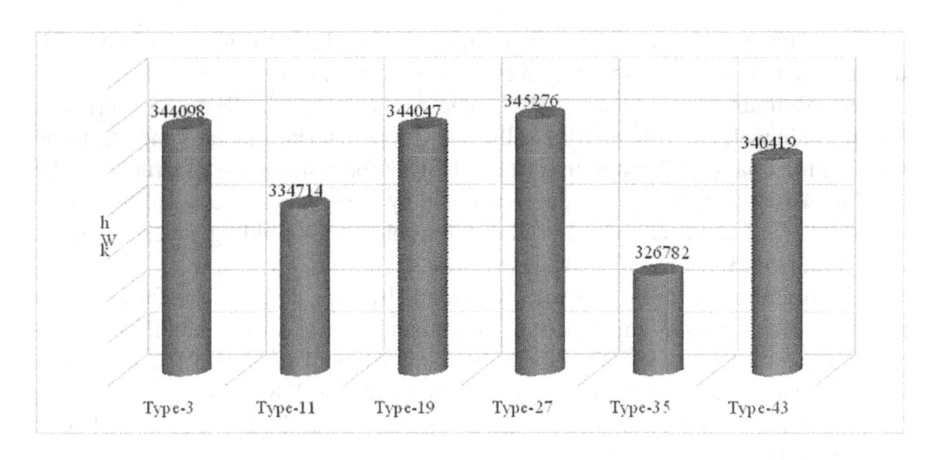

FIGURE 3.9 Annual total energy consumption amounts for the best alternatives of all scenarios.

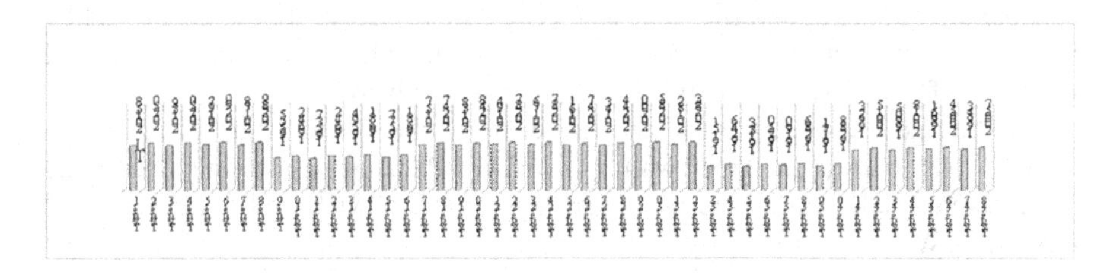

FIGURE 3.10 Annual energy cost values of a cafe building in Tekirdağ.

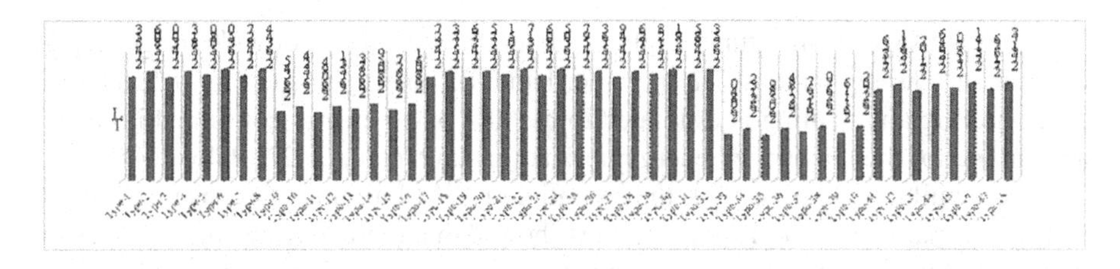

FIGURE 3.11 Life-cycle cost values of a cafe building in Tekirdağ (Turkey).

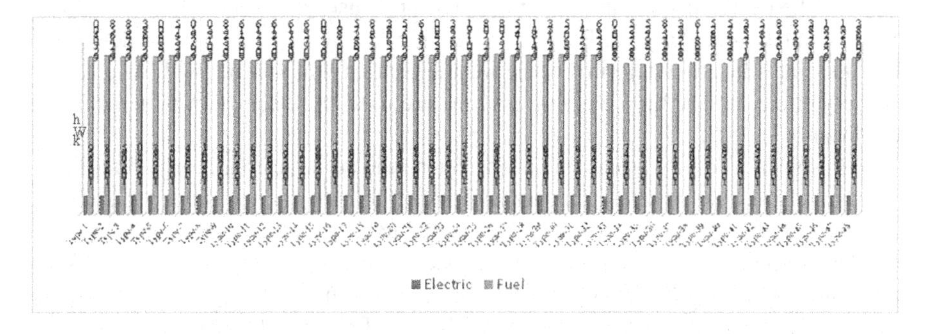

FIGURE 3.12 Life-cycle energy (electric and fuel) of a cafe building in Tekirdağ (Turkey).

Annual energy cost, annual energy cost, and life-cycle cost values of a cafe building designed in Tekirdağ are given in Figures 3.10–3.12, respectively.

The minimum annual energy cost value was obtained by Type 35 (only east façade green wall by utilizing EPS as the insulation material and gas concrete as the building material and green roof) as 1,013 USD at the sandwich wall structure. The maximum annual energy cost value was found by Type 6 (non-green wall by utilizing rockwool as an insulation material and bims block as a building material and tile roof) as 1,075 USD at sandwich wall structure.

The minimum life-cycle cost value was found by Type 35 (only east façade green wall by utilizing EPS as an insulation material and gas concrete as a building material and green roof) as 13,807 USD at the sandwich wall structure. The maximum life-cycle cost value was obtained by Type 6 (non-green wall by utilizing rockwool as an insulation material and bims block as a building material and tile roof) as 14,642 USD at sandwich wall structure.

The lowest life-cycle electric and fuel energy values were defined through Type 16 (all façades green wall by utilizing EPS as an insulation material and bims block as a building material and tile roof) and Type 30 (only south façade green wall by utilizing rockwool as an insulation material and bims block as a building material and tile roof) as 1,087,907 and 9,386,025 kWh, respectively.

The highest life-cycle electric and fuel energy values were determined through Type 35 (only east façade green wall by utilizing EPS as an insulation material and gas concrete as a building material and green roof) at sandwich wall as 1,013,372 and 8,789,725 kWh, respectively.

3.4 CONCLUSIONS

The energy usage in structures explains about 40% of the worldwide usage, especially in developing nations. On the other hand, the energy needed for a building's space heating and cooling is estimated to be 60% of all energy consumed in them, accounting for the majority of energy utilization (Balo et al., 2022; Balo, 2015).

Before building construction, the potential of BIM buildings to consume energy concerning parameters and their locations showed that there may be opportunities to investigate alternative energy-related options. When a building is assessed manually or using more conventional methods, BIM offers excellent opportunities to prevent errors. When a facility is already operational or in use, it is very difficult to accurately simulate mistakes. Yet BIM has a strong capacity for making wise decisions. The building's location determines the weather station that will be used. As a result, the default city and online mapping systems are used to select the weather stations and locations.

In this study, a cafe was designed considering the climatic conditions of Tekirdag city, one of the second climate zones of Turkey. The main purpose of the study is to determine the effect of the green wall added to different aspects of the building facades on the building's energy performance. As a result, Type 35 (326,782 kWh) was found to be the most energy-efficient design. The most energy-efficient designs were obtained by adding a green wall on the east façade of the building envelope. The worst energy efficiency values were found when the green wall was used on the south façade of the building envelope. When a green wall is added to the south façade, it can be thought that the low energy efficiency is due to the inability to make efficient use of the solar radiation coming to the south façade. In Tekirdağ city climatic conditions, the most efficient building components were EPS as insulation materials, gas concrete as a building material, and green roof as roof material among used building elements in the analysis. The main contribution of this study is creating scenario groups with different alternative types according to the sandwich wall structure through different construction, insulation, and roof materials of the modeled café. All types were created by detailing all facades of each scenario group for four different facades, all without green walls, all with green walls, and only one facade with green walls.

REFERENCES

S. Azhar, M. Khalfan, and T. Maqsood. Building information modeling (BIM): Now and beyond. *Australasian Journal of Construction Economics and Building*, (2012), 12(4), 15–28. https://doi.org/10.5130/ajceb.v12i4.3032.

F. Balo. Energy and economic analyses of insulated exterior walls for four different cities in Turkey. *Energy Education Science and Technology Part A: Energy Science and Research*, (2011), 26(2), 175–188.

F. Balo. Feasibility study of "green" insulation materials including tall oil: Environmental, economical and thermal properties. *Energy and Buildings*, (2015), 86, 161–175.

F. Balo, H. Boydak, and H. Polat. (2022). Chapter 8: An energy-efficient green design and modelling of a health clinic located in a cold climatic zone. Book Series on *Mitigating Climate Change*. Sprınger. pp. 187–200. ISBN: 978-3-030-92148-4.

F. Balo, and H. Polat. (2021). Chapter 6: Energy productivity with an efficient design: Case of medical waste storage. Book Series on *Engineering for Sustainable Development and Living (Preserving a Future for the Next Generation to Cherish)*. Brown Walker Press (Unıversal Publıshers Inc.). pp. 181–214. ISBN: 978-1-59942-614-3.

F. Balo and H. Polat. (2022). Chapter 5: The impact of traditional natural stones on energy efficiency for sustainable architecture: The case of an authentic restaurant in Harput region. Book series on *Renewable Energy for Mitigating Climate Change*. Taylor & Francis Group (CRC Press). pp. 85–116. ISBN: 9781003240129.

F. Balo, S. Sua, and H. Boydak. (2022). Energy efficient ski lodge design: Case of Erzurum. Book Series on *Building Information Modelling (BIM): Industry Trends, Benefits, and Challenges*. New York: NOVA SCIENCE Publishing.

B. Becerik-Gerber, and S. Rice. The perceived value of building information modeling in the U.S. building industry. *Journal of Information Technology in Construction*, (2010), 15, 185–201.

R. Becker, V. Falk, S. Hoenen, S. Loges, S. Stumm, J. Blankenbach, S. Brell-Cokcan, L. Hildebrandt, and D. Vallée. Bim - towards the entire lifecycle. *International Journal of Sustainable Development and Planning*, (2018), 13(1), 84–95. https://doi.org/10.2495/SDP-V13-N1-84-95.

H. Boydak, H. Polat, and F. Balo. (2022). Book Series on Chapter 3: Designing a Green Hospital Building Using Insulation Material Alternatives with the Green Building Studio. Danbury: Begell House Publishing Corporation.

C. Chen and Tang, L. BIM-based integrated management workflow design for schedule and cost planning of building fabric maintenance. *Automation in Construction*, (2019), 107, 102944.

B. Cheng, J. Li, V. Tam, M. Yang, and D. Chen. A BIM-LCA approach for estimating the greenhouse gas emissions of large-scale public buildings: A case study. *Sustainability*, (2020), 12, 685.

A. Costa, M. Keane, J. Torrens, and E. Corry. Building operation and energy performance: Monitoring, analysis, and optimisation toolkit. *Applied Energy*, (2013), 101, 310–316.

P. Dahl, M. Horman, T. Pohlman, and M. Pulaski. Evaluating design-build operate- maintenance delivery as a tool for sustainability. *Proceedings of the Construction Research Congress*, San Diego, 2005.

DMI. Klimatoloji - I, Çevre Ve Orman Bakanliği Devlet Meteoroloji İşleri Genel Müdürlüğü, Ankara Mart - 2005 DMI Yayinlari Yayin No: 2005/1. https://www.vaisala.com.

A. Ebrahim, and A.S. Wayal. BIM based building performance analysis of a green office building. *International Journal of Scientific & Technology Research*, (2019), 8(08), 566–573.

T. El-Diraby, T. Krijnen, and M. Papagelis. BIM-based collaborative design and socio-technical analytics of green buildings. *Automation in Construction*, (2017), 82, 59–74. https://doi.org/10.1016/j.autcon.2017.06.004.

H. Gokce and K. Gokce. Holistic system architecture for energy efficient building operation. *Sustainable Cities and Society (SCS)*, (2013), 6, 77–84.

A. Gounni, M. El Alami, M.T. Mabrouk, and A. Kheiri. The optimal allocation of the PCM within a composite wall for surface temperature and heat flux reduction: An experimental approach. *Applied Thermal Engineering*, (2017), 127, 1488–1494.

A. Guerrisi, M. Martino, and M. Tartaglia. Energy saving in social housing: An innovative ICT service to improve the occupant behavior. *International Conference on Renewable Energy Research and Applications*, Nagasaki, Japan, Nov 2012.

J. Hroudová, and J. Zach. The possibilities of modification of crop-based insulation materials applicable in civil engineering in low-energy and passive houses. *Procedia Engineering*, (2017), 180, 1186–1194.

IGBC Green New Buildings. (2015). Retrieved from: https://ibgc.in/igbc/redirectHtml.htm?redVal=showGreenNewBuildingsnosign.

F. Jalaei, and A. Jrade. An automated BIM model to conceptually design, analyze, simulate, and assess sustainable building projects. *Journal of Construction Engineering*, (2013), 2014, 1–21.

N. M. Kameni, P. Ricciardi, S. Reiter, and A. Yvon. A comparative study on optimum insulation thickness of walls and energy savings in equatorial and tropical climate. *International Journal of Sustainable Built Environment*, (2017), 6(1), 170–182.

O. Kaplan, U. Yavanoglu, and F. Issi. Country study on renewable energy sources in Turkey. *International Conference on Renewable Energy Research and Applications*, Nagasaki, Japan, 11–14 Nov 2012.

A. Karahan, F. Balo, and Ü. Yılmaz. (2022). Book Series on *In the Light of Manisa Academic Research, Manisa slate stone as wall covering material*, Vol 4. Berikan Publishing, Ankara. pp. 509.

J. Karlsson, L. Wadsö, and M. Öberg. A conceptual model that simulates the influence of thermal inertia in building structures. *Energy and Buildings*, (2013), 60, 146–151.

I. Khan and P. K. Halder. Electrical energy conservation through human behavior change: Perspective in Bangladesh. *International Journal of Renewable Energy Research*, (2016), 6(1), 43–52.

S. Kim and S. Kim. Lessons learned from the existing building energy optimization workshop: An initiative for the analysis-driven retrofit decision making. *KSCE Journal of Civil Engineering*, (2017), 21, 1059–1068.

S. Kubba. (2012). *Handbook of Green Building Design and Construction: LEED, BREEAM, and Green Globes*. Butterworth-Heinemann, Oxford, UK.

K.P. Lam, N.H. Wong, A. Mahdavi, K.K. Chan, Z. Kang, and S. Gupta. SEMPER-II: An internet-based multi-domain building performance simulation environment for early design support. *Automation in Construction*, (2004), 13(5), 651–663.

H.J. Lee, S.Y. Kim, and C.Y. Yun. Comparison of solar radiation models to estimate direct normal irradiance for Korea. *Energies*, (2017), 10, 594.

M. Longo, D. Zaninelli, M. Roscia, and G.C. Lazaroiu. Smart planning for ecoefficient cities. *International Conference on Renewable Energy Research and Applications*, Madrid, Spain, 20–23 October 2013.

A.B. Mohammed. Collaboration and BIM model maturity to produce green buildings as an organizational strategy. *HBRC Journal*, (2020), 16(1), 243–268. https://doi.org/10.1080/16874048.2020.1807815.

T. Orzechowski and M. Orzechowski. Optimal thickness of various insulation materials for different temperature conditions and heat sources in terms of the economic aspect. *Journal of Building Physics*, (2018), 41(4), 377–393.

N.F. Otuh. (2016). *BIM-based Energy/Sustainability Analysis for Educational Buildings – A Case Study*. HAMK University of Applied Sciences, Finland.

I. Petri, S. Kubicki, Y. Rezgui, A. Guerriero, and H. Li. Optimizing energy efficiency in operating built environment assets through building information modeling: A case study. *Energies*, (2017), 10, 1167.

C. Piselli, A. Guastaveglia, J. Romanelli, F. Cotana, and A.L. Pisello. Facility energy management application of HBIM for historical low-carbon communities: Design, modelling and operation control of geothermal energy retrofit in a real Italian case study. *Energies*, (2020a), 13, 6338.

C. Piselli, J. Romanelli, M. Di Grazia, A. Gavagni, E. Moretti, A. Nicolini, F. Cotana, F. Strangis, H. Witte, and A. Pisello. An integrated HBIM simulation approach for energy retrofit of historical buildings implemented in a case study of a medieval fortress in Italy. *Energies*, (2020b), 13, 2601.

S.M. Raffee, Z. Hassan, and M.S.A. Karim. Enhancement of sustainability assessment of building projects using Building Information Modelling (BIM). *The 3rd National Graduate Conference (Natgrad2015)*, Universiti Tenaga Nasional, Putrajaya Campus, pp. 252–258, 8–9 April 2015.

P. Rajendran, S.T. Wee, and G.K. Chen. Application of BIM for managing sustainable construction. *Proceedings International Conference of Technology Management, Business, and Entrepreneurship 2012*, Renaissance Hotel, Melaka, Malaysia, pp. 305–310, 2012.

J.M. Roucoult, O. Douzane, and T. Langlet. Incorporation of thermal inertia in the aim of installing a natural nighttime ventilation system in buildings. *Energy and Buildings*, (1999), 29, 129–133.

G. Sauba, Jos van der Burgt, A. Schoofs, C. Spataro, M. Caruso, F. Viola, and R. Miceli. Novel energy modelling and forecasting tools for smart energy networks. *4th International Conference on Renewable Energy Research and Applications*, Palermo, Italy, pp. 1669–1673, 22–25 Nov 2015.

F. Shadram and J. Mukkavaara. An integrated BIM-based framework for the optimization of the trade-off between embodied and operational energy. *Energy and Building*, (2018), 158, 1189–1205.

Y. Shi and J. Xu. BIM-based information system for econo-enviro-friendly end-of-life disposal of construction and demolition waste. *Automation in Construction*, (2021), 125, 103611.

M.V. Shoubi, M.V. Shoubi, A. Bagchi, and A.S. Barough. Reducing the operational energy demand in buildings using building information modeling tools and sustainability approaches. *Ain Shams Engineering Journal*, (2015), 6, 41–55.

M. Solla, L.H. Ismail, and R. Yunus. Investigation on the potential of integrating bim into green building assessment tools. *ARPN Journal of Engineering and Applied Sciences*, (2016), 11(4), 2412–2418.

D. Stundon, J. Spillane, J.P.B Lim, P. Tansey, and M. Tracey. Building Information Modelling energy performance assessment on domestic dwellings: A comparative study. *Proceedings for 31st Annual ARCOM Conference, ARCOM*, pp. 671–679, 2015.

V. Venkatraj, M.K. Dixit, W. Yan, and S. Lavy. Evaluating the impact of operating energy reduction measures on embodied energy. *Energy and Buildings*, (2020), 226, 110340.

J. Wang, H. Wu, H. Duan, G. Zillante, J. Zuo, and H. Yuan. Combining life cycle assessment and building information modelling to account for carbon emission of building demolition waste: A case study. *Journal of Cleaner Production*, (2018), 172, 3154–3166.

P. Wu, B. Xia, and X. Zhao. The importance of use and end-of-life phases to the life Cycle Greenhouse Gas (GHG) emissions of concrete-A review. *Renewable and Sustainable Energy Reviews*, (2014), 37, 360–369.

J. Xu, Y. Shi, Y. Xie, and S. Zhao. A BIM-based construction and demolition waste information management system for greenhouse gas quantification and reduction. *Journal of Cleaner Production*, (2019), 229, 308–324.

4 Improving Energy Efficiency of Tall Buildings Using Innovative Environmental Systems

Paul Armstrong
University of Illinois at Urbana-Champaign

Kheir Al-Kodmany
University of Illinois at Chicago

4.1 INTRODUCTION

Buildings are responsible for a large share of the energy demand in urban areas. According to a study, buildings consume around 40% of the total energy used in cities [1]. This energy is used primarily for heating and cooling, and electricity. Heating and cooling systems include boilers and chillers that provide thermal comfort to the occupants of the buildings. Electricity systems include lighting, communication devices, and building management systems (BMSs) that control and monitor various functions of the buildings. Most of the energy used by these systems comes from non-renewable sources, such as fossil fuels, which have negative impacts on the environment and the climate. Although conventional MEP (mechanical, electrical, and plumbing) systems have become more effective over the years, sustainable tall buildings frequently supplement these systems with alternative energy-efficient technologies that are integrated with other building components and systems to attain high performance. Reducing the energy footprint of tall buildings is a challenge that many architects are passionate about. They aim to design "zero-energy" structures that generate all the power they need from within without relying on external sources [2,3]. However, this ideal is still hard to realize in practice. A more feasible approach is to combine old and new techniques to optimize the energy efficiency of skyscrapers. For example, some architects use natural ventilation, passive solar heating, and green roofs to reduce their buildings' heating and cooling costs [4,5]. Others incorporate renewable energy sources such as solar panels, wind turbines, and geothermal systems to produce electricity and heat [6,7].

The design of skyscrapers is increasingly moving towards bioclimatic and zero-carbon solutions, thanks to the development of green building materials and ecological systems, as well as the advancement of alternative renewable energy

DOI: 10.1201/9781003403456-4

sources. Many architects who are dedicated to eco-friendly design argue that traditional technologies are not enough to achieve high-performance tall buildings in the future. For example, 4 Times Square, completed in 1999 and designed by Fox and Fowle, was the first high-rise building in North America to receive an environmental and sustainable certification. The tower uses various strategies to reduce its carbon footprint [4,5], such as photovoltaic (PV) panels, wind turbines, fuel cells, and gray water recycling. Another example is the Pearl River Tower in Guangzhou, China, which was completed in 2011 and designed by Skidmore, Owings, & Merrill. The tower is intended to be one of the most energy-efficient skyscrapers in the world, with features such as a double-skin facade, a solar chimney, a geothermal heat pump, and a wind farm [6,7]. A third example is the Bosco Verticale in Milan, Italy, completed in 2014 and designed by Stefano Boeri Architetti. The tower is a vertical forest that hosts more than 900 trees and 20,000 plants on its balconies, creating a natural habitat for birds and insects and reducing air pollution and noise [8,9].

4.1.1 Goals and Objectives

This chapter aims to accomplish the following:

1. Explain how integrating environmental systems with other building systems and spaces enhances the degree to which they are mutually dependent on one another and the performance of the building as a whole.
2. Provide some examples of innovative technologies being developed to lessen dependency on non-renewable energy sources.
3. Explore how Automatic Building Systems and parametric design tools are used to improve the performance of the building and lower the expenses of its operation and maintenance over its lifetime.
4. Illustrate how environmental systems and infrastructure that are constructed bio-climatically might result in more sustainable buildings and cities that are carbon neutral.

4.1.2 Methodology

It may appear to be a straightforward effort to collect data on the environmental performance of a large structure; nevertheless, this is not the case. It can be challenging to locate essential data and evaluate and verify the actual performance of tall buildings [5,6]. The energy use of each system, like the heating, ventilation, air conditioning, lighting, pumps, computers, and elevators, as well as the contribution of passive systems, is not available. It is impossible to obtain information regarding the breakdown of the energy demand that a large building has for its systems. Consequently, the performance of tall buildings is not evaluated in this chapter because there is a lack of necessary data on the actual energy performance of buildings. Instead, it investigates fresh concepts that aim to enhance their performance [7,8].

This chapter presents case studies of sustainable tall buildings that combine advanced environmental systems with innovative curtain wall systems, sky gardens, and atria. These systems enhance the buildings' performance in terms of thermal comfort, energy conservation, and livability. The main theme of this chapter is the importance of integrating building systems for achieving high performance throughout the design process. The chapter explores various alternative eco-friendly technologies that are used in modern tall buildings, such as wind turbines, PV panels, fuel cells, and others.

Moreover, the chapter advocates for organic and passive design approaches, such as bioclimatic design, which takes into account the interaction between climate, hydrology, geology, and biology. These natural systems provide the basis for designing buildings that can mimic and incorporate these processes into their designs to reduce their energy dependence and create more sustainable eco-environments. The chapter analyzes nine case studies of sustainable tall buildings to assess the effectiveness of their environmental systems and to illustrate how these systems are integrated to ensure a high level of functionality and efficiency.

4.2 ENVIRONMENTAL SYSTEMS

The term "environmental system" refers to the combination of abiotic (non-living) and biotic (living) components that interact within a specific environment. It encompasses the interactions between the atmosphere, hydrosphere, lithosphere, and ecosystems. Environmental systems involve various processes such as energy capture, movement, storage, and utilization. In the context of tall buildings, the term "environmental system" often refers to the MEP systems that regulate the interior microclimate and contribute to the thermal comfort of occupants. These systems are crucial in maintaining a comfortable and healthy indoor environment. Some key points related to environmental systems in tall buildings are as follows [10,11]:

Mechanical Systems: Mechanical systems include heating, ventilation, and air conditioning (HVAC) systems that control temperature, humidity, and air quality. These systems ensure that occupants have a comfortable environment regardless of external weather conditions.

Electrical Systems: Electrical systems power various building components, including lighting, appliances, and equipment. They also incorporate energy-efficient measures such as intelligent lighting controls and management systems to optimize energy usage.

Plumbing Systems: Plumbing systems manage water distribution throughout the building, including supply and drainage systems. They ensure the availability of clean water for consumption and efficient disposal of wastewater.

Thermal Comfort: The environmental systems in tall buildings aim to maintain optimal thermal comfort for occupants by controlling temperature, humidity, and ventilation. This involves balancing heat gains and losses, proper insulation, and adequate air circulation.

Energy Efficiency: Environmental systems in tall buildings should prioritize energy efficiency by incorporating technologies like energy-efficient HVAC systems, smart controls, and renewable energy sources. This helps reduce energy consumption and lower the environmental impact of the building.

Overall, environmental systems in tall buildings encompass the MEP systems that regulate the interior microclimate, ensure occupant comfort, and contribute to the building's energy efficiency. These systems are crucial for creating a healthy and sustainable built environment within the building.

Maintaining a comfortable environment in a building involves more than just temperature control. Factors such as humidity, airflow, lighting, acoustics, dust, scents, and aesthetics also play significant roles in occupant comfort and well-being. These factors need to be carefully considered and integrated into the design of the building's environmental systems. During the design phase, architects and engineers collaborate to ensure that different building systems work together seamlessly. This includes considering the placement and integration of HVAC systems, lighting fixtures, acoustic treatments, ventilation systems, and other components that contribute to the overall environmental quality of the building. Some key points related to the integration of environmental systems in building design are as follows[9,10]:

Comprehensive Approach: Designing a comfortable and functional environment requires a comprehensive approach that considers all relevant factors. Architects and engineers collaborate to address temperature, humidity, airflow, lighting levels, noise control, air quality, and aesthetics.

Coordination of Systems: Different building systems, such as HVAC, lighting, and acoustics, need to be coordinated to achieve optimal performance and occupant comfort. This involves considering the placement of ductwork, light fixtures, and sound-absorbing materials to minimize conflicts and ensure efficient operation.

Early Integration: The integration of environmental systems starts early in the design phase to allow for proper planning and coordination. Architects and engineers work together to determine the optimal placement of systems and develop strategies for their seamless integration.

Building Performance Analysis: Advanced building performance analysis tools and simulations are used to assess the interaction of environmental systems and optimize their performance. These tools help evaluate different design scenarios and make informed decisions to create a comfortable and efficient building environment.

Sustainable Design: Integration of environmental systems also involves incorporating sustainable design principles. This includes energy-efficient systems, renewable energy sources, rainwater harvesting, daylight harvesting, and other strategies to reduce the building's environmental impact.

By carefully integrating and coordinating environmental systems during the design phase, architects and engineers can create buildings that provide occupants with a comfortable, healthy, and aesthetically pleasing environment. This collaboration ensures the building's systems work harmoniously, contributing to overall occupant satisfaction and well-being.

The term "power membrane" describes the interstitial space between the ceiling and the floor of an office, as defined by Abalos et al. [3] in their book Tower and Office. This space hosts various building elements and systems essential for human

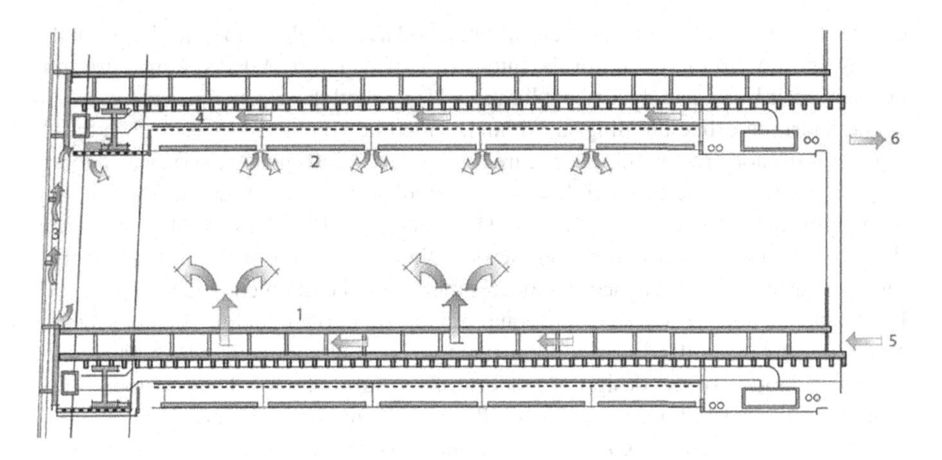

FIGURE 4.1 MEP, lighting, structural, and fire suppression systems are all housed in the Pearl River Tower's "Power Membrane," which runs between the raised floor and fallen ceiling. (Diagram by P. Armstrong.)

comfort and floor/ceiling support (Figure 4.1). They consist of MEP systems, artificial lighting, fire suppression systems, electrical conduits, data transmission cables and wires, and structural components. This power membrane needs to be spatially efficient in tall buildings to reduce material and construction costs and optimize operating costs throughout the building's life cycle. For instance, a small difference in the floor-to-floor height can significantly increase the total construction cost in a supertall skyscraper. Therefore, careful design and a high level of integration are necessary to ensure the adequate performance of the systems in the power membrane while saving space. Therefore, engineers may incorporate an integrated power membrane that combines various systems such as electrical, HVAC, and communication networks. Integrating these systems within the power membrane not only saves valuable space but also improves the overall efficiency and functionality of the building. This careful design and integration allow for optimized construction costs and enhanced operational performance throughout the building's lifespan.

An early example of systems engineering, the General Motors Technical Center in Warren, Michigan, was designed by Eero Saarinen and finished in 1955. It is a 710-acre campus with 21,000 employees where all the building components are integrated to work together to achieve both spatial and performance efficiencies [4]. The campus consists of 38 buildings, including a research center, a design center, an engineering center, a testing center, and a styling center. Underground tunnels and above-ground walkways connect the buildings. The campus also features a lake, a water tower, a reflecting pool, and several sculptures [2]. Saarinen was aware that a modern office building needed an open floor plan that maximized flexibility, expansive walls of glass to let in natural light, and a relatively deep space between the ceiling and floor/roof system for the integration of structure, MEP systems, fire suppression sprinklers, and fluorescent lighting. He applied these principles to his design of the Saarinen office building in Finland, which he completed in 1959. The building is a four-story rectangular structure with a central atrium that serves as a lobby and circulation space. The building has a modular steel frame that allows for easy reconfiguration of the interior spaces. The exterior walls are made of glass panels that provide natural ventilation and views of the surrounding landscape. The building also has a flat roof that accommodates mechanical equipment and skylights [11]. His team was able to construct a rational, highly functional, and physically efficient workspace by using the "form follows function" design approach, which went on to become a model for the design of modern office buildings [4]. The Saarinen office building is considered one of the most influential examples of modern architecture in Finland and has been recognized as a national heritage site [11].

4.2.1 Mixed-Use Tall Buildings

A tall multi-use building is a type of skyscraper that can host a variety of functions within its vertical space. For example, a tall multi-use building may have offices, hotels, retail, residential, and cultural facilities on different floors. The environmental design of each function may differ according to the tenant's demands and expectations, the space and functional requirements, and the location within the building. For instance, offices may require more natural lighting and ventilation, while hotels may need more privacy and sound insulation. A common strategy to achieve optimal environmental performance for each function is to divide the building into zones that have similar environmental needs and provide them with a separate HVAC system. The HVAC system is discretely installed on mechanical floors that are located between the zones. This way, each zone can have its own temperature, humidity, and air quality control [12].

One way to design tall buildings is to use a tripartite structure, which consists of three parts: a base or podium, a shaft, and a top. The base or podium usually contains spaces for commercial and retail activities, such as shops, restaurants, offices, or banks. It also has lobbies that welcome visitors and provide access to the elevator cores and service cores, which are the vertical circulation and utility systems of the building. The base or podium also has parking levels that are above the ground level, which require a vehicle ramp to connect them to the street. The ramp needs to have enough space and a slope to accommodate different types of vehicles. The shaft is the central part of the building, where residential, commercial, or hotel units are stacked vertically. The units can vary in size, shape, and layout, depending on the function and preference of the occupants. The shaft can also have setbacks or projections to create terraces or balconies for the units. Office floors need more elevators in a building with an extensive service core and large open floor areas. When the building rises and floors are added, the number of elevators and the area of the service core both increase. The distance from the building's core to the exterior walls of an office skyscraper in the United States is generally between 12 and 15 meters (40 and 60 feet). On the other hand, residential and hotel buildings can be serviced with fewer elevators, which leads to a reduction in the size of the service core as well as the depth of the floor plates. This results in increased access to natural light and ventilation for each individual unit in the structure [13,14].

The top is the final part of the building, where mechanical equipment, amenities, or architectural features are located. The mechanical equipment includes ventilation systems, water tanks, generators, or antennas. The amenities include pools, gyms, gardens, or lounges. The architectural features include spires, domes, crowns, or sculptures. The top can also have different shapes and styles to create a distinctive skyline for the building. The top or crown of the structure may contain a spire. It could be purely decorative, as is the case with the Chrysler Building in New York City, or it may be structural, as in the case of the New York Times Headquarters also in New York City.

A skyscraper may have several floors below the ground level, which will be used for different functions. Some of these functions are loading docks, where trucks can deliver and pick up goods; mechanical floors, where the building's systems such as heating, ventilation, and electricity are located; storage areas, where the building's occupants can keep their belongings or equipment; and extra parking, where the building's visitors or residents can park their vehicles. The number of floors below the ground level may vary depending on the conditions of the soil and water under the surface. These conditions affect how deep the building's foundation can be. For instance, skyscrapers in Manhattan can be built on a layer of hard rock close to the surface, while Chicago buildings need deeper and stronger foundations because of the city's soft and wet soil [15].

Therefore, the design of a skyscraper that accommodates different functions is often influenced by the vertical distribution of those functions. A typical pattern is to have a large base that serves as a retail or commercial space, medium-sized office floors that provide workspaces for various businesses, and smaller residential units on top that offer living spaces for people who want to live in the city center. This has not always been the case. In 1916, New York became the first city to adopt zoning laws that limited the height of buildings after the Equitable Building was built with 40 stories and 164 meters (538 feet) of uninterrupted elevation from the street. New York had these regulations because it faced a unique challenge. The city authorities realized that if more buildings were erected without any setbacks, they would eventually block the sunlight and air from reaching the narrow streets of Manhattan, which are laid out in a grid pattern. This would have a negative impact on the people living there, as well as on the aesthetic and environmental quality of the city. The zoning rule was based on a calculation that related lot size to building height. This calculation is better known as the floor area ratio (FAR). As the building gets taller, the allowed FAR decreases, creating a setback triangle along the street [16]. The setback triangle is a geometric shape that defines how much a building must recede from the street as it rises, creating a stepped or tapered appearance. This rule ensured that tall buildings would not cast long shadows or create a canyon effect on the streets below.

4.2.2 Bioclimatic Design

Bioclimatic design is a practice that aims to create buildings that are in harmony with the natural systems of their surroundings. It takes into account the complex interactions among climate, hydrology, geology, and biology that shape the weather, landforms, and life of a place. By understanding and mimicking these processes, bioclimatic design can reduce the energy consumption of buildings and create more sustainable eco-environments [17]. To illustrate how bioclimatic design works in practice, some of the aspects of this approach include using local materials, optimizing natural ventilation and lighting, integrating renewable energy sources, and enhancing biodiversity and water conservation.

Bioclimatic tall structures are green skyscrapers that adapt to local environments, such as climate, topography, and culture, and use energy efficiently, such as by minimizing heat loss, maximizing daylight, and generating renewable power. They also satisfy the needs of communities and cities, such as by providing social amenities, enhancing public health, and reducing carbon emissions. They create more sustainable and livable neighborhoods by using environmental solutions for their design and appearance, such as by integrating natural elements, materials, and colors, and by utilizing natural features, such as sun exposure, natural airflow, native plants and greenery, and so on [17]. Architect Ken Yeang, a pioneer of bioclimatic skyscrapers, envisions the tall building as a vertical urban city that consolidates urban services infrastructure, such as water supply, waste management, and communication networks, vertical mobility, such as elevators, escalators, and skybridges, and living, working, and shopping spaces on smaller land plots to minimize horizontal sprawl. He considers the tall building as a city within a city. He advocates for the formation of vertical communities, districts, communal areas, and streets inspired by the horizontally distributed public spaces and services of traditional cities. He also emphasizes the importance of creating a sense of place and identity for the occupants and visitors of the tall building [17].

Architects can apply bioclimatic principles to design skyscrapers that are compatible with local weather conditions and use less energy than conventional buildings [8]. Bioclimatic skyscrapers use a range of techniques to optimize the natural cycles of the sun, wind, and seasons to create indoor environments that are comfortable for the occupants and friendly for the environment. Some of these techniques are solar shading devices that can be passive, such as louvers or overhangs, or active, such as movable panels or blinds; windows that can be opened by the users, curtain walls that have ventilation openings, and atria cores that act as vertical shafts for fresh air; sky courts and sky gardens that provide green spaces for recreation and relaxation, as well as purify the air by filtering carbon dioxide and other pollutants; rainwater collection systems that store rainwater in tanks or ponds, and gray water reuse systems that treat wastewater from sinks or showers and use it for irrigation or flushing; and energy production systems that use fuel cells, wind turbines, or solar panels to generate electricity from renewable sources. These techniques often influence the aesthetic of the skyscrapers, giving them a distinctive look, especially in high-tech buildings that display their structural, mechanical, and circulation systems on their external surfaces [18].

Ecomimesis is the practice of mimicking the functions and features of natural ecosystems in eco-architecture. One of the main goals of this approach is to design eco-life skyscrapers that integrate nature into their cycles and systems. This requires a holistic and seamless consideration of all aspects of the natural world during the design phase [9]. Another goal of eco-architecture is to achieve zero-energy buildings that use renewable energy sources such as solar, wind, and geothermal energy to produce as much or more energy than they consume from the grid [10]. However, this goal should not overshadow the other factors that affect the sustainability of eco-life skyscrapers, such as the costs of technologies, user comfort, and human and environmental health impacts. These factors should be evaluated using a life cycle assessment method. Many eco-architecture projects are still in development and have room for improvement [11,19,20].

4.3 SUSTAINABLE ENVIRONMENTAL SERVICES AND STRATEGIES

Sustainable building services require three key aspects that should be carefully considered by the designers and engineers. The first one is to assess the environmental impact of the technical systems that are used in the building, such as heating, cooling, lighting, and ventilation. These systems should use renewable energy sources as much as possible, such as solar, wind, or geothermal power, to reduce greenhouse gas emissions and improve energy efficiency. The second one is to have a clear and coherent technical concept that guides the design and engineering decisions and helps to achieve the project requirements and objectives. The technical concept should be based on a thorough analysis of the site conditions, the building functions, and the user needs (Figure 4.2). The third one is to plan carefully for the initial and the life-cycle costs of the construction and operation of the building.

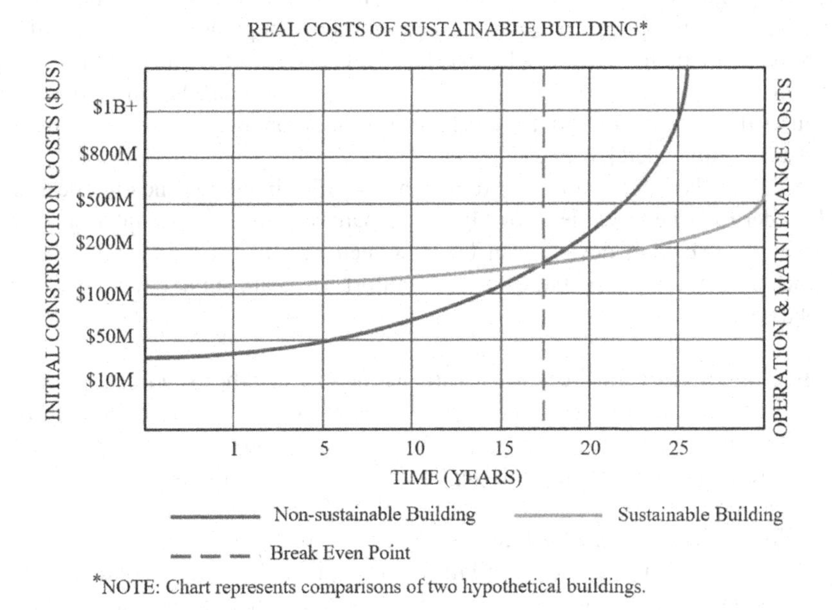

FIGURE 4.2 Along with the initial costs of site purchase and building construction, it is important to consider the operational and maintenance costs of sustainable design goals. Although initially more expensive, sustainable buildings ultimately end up being less expensive than non-sustainable ones [15]. (Diagram by P. Armstrong.)

This includes estimating the capital costs, the maintenance costs, the energy costs, and the environmental costs of the building over its lifespan [15,16,19].

4.3.1 Natural Systems

An evaluation of the site conditions is vital for natural building ventilation, especially in providing adequate air circulation in sheltered areas of the building (such as atria, sky gardens, and sky lobbies) [17,20,21]. Wind roses serve as the basis for comprehensively evaluating wind directions and intensities. They illustrate the predominant winds that occur during each season. Applying vector analysis on virtual models permits a more comprehensive examination. Due to the development of various software packages, architects and engineers can now visualize how wind impacts structures and adjusts their shapes for maximum efficiency. Wind tunnel testing using physical models equipped with sensors that provide precise wind-induced impact measurements on the structure and urban setting can be utilized to conduct more thorough and in-depth experimental research. Various tests can be conducted to reach this objective.

When designing tall buildings, architects should ensure that both positive and negative pressures that act on the surface of the building are considered. This will allow for increased natural ventilation and help architects avoid issues that could arise on the interior of the building, such as the effects of wind pressure on the building's doors and windows. Suppose the wind velocity is above 8 meters per second (18 miles per hour). In that case, natural ventilation of high-rise buildings in temperate regions is a desirable option during the transitional season, which occurs during the fall and spring. To maintain a comfortable temperature within a building when the temperature outside is higher than 22°C (72°F), supplemental mechanical ventilation, especially during the summer months, is essential. When outdoor temperatures drop below 5°C (41°F), mechanical ventilation and heat recovery devices are also recommended [18,22].

Curtain walls are commonly used in high-rise office buildings and provide a sleek and modern aesthetic while allowing ample natural light into the interior spaces. Passive solar shading devices can be integrated into the curtain wall system to enhance energy efficiency and occupant comfort. Here are some key points related to solar shading and BMSs:

Shading Devices: Louvers, solar screens, blinds, or other shading elements can be installed on the outer façade or interior of the building to control solar heat gain and glare. These devices can be adjustable or automated to respond to changing solar angles and optimize daylighting and thermal comfort.

Building Management System: A BMS is a central control system that manages and monitors various building systems, including mechanical ventilation and solar shading devices. It can be programmed to operate shading devices based on factors like time of day, sun position, indoor temperature, and user preferences.

Motorized Operators: Motorized operators can be incorporated into windows, vents, or shading devices to enable automated control. These operators can adjust the opening angles of windows or vents based on outdoor wind conditions and indoor thermal conditions.

Individual Control and Flexibility: A well-designed BMS allows occupants to have individual control over their immediate environment. This means they can override automated systems to adjust temperature, ventilation, or shading according to their preferences while still maintaining energy efficiency.

Energy Monitoring and Alert Systems: The BMS can integrate energy monitoring sensors and software programs to track and optimize the building's energy performance. It can also provide alerts or notifications to building occupants or managers in case of any issues or malfunctions, allowing for timely interventions.

By combining solar shading devices with a comprehensive BMS, building owners and occupants can achieve optimal energy efficiency, indoor comfort, and air quality. These systems ensure that the building's performance is monitored, managed, and adapted to meet the needs of the occupants while minimizing energy consumption and maximizing sustainability.

4.3.2 Daylight Harvesting and Artificial Lighting

Basic considerations for light should include building type and occupancies, core-to-wall distances for access to natural light, the need for supplemental lighting for specific tasks or when natural light is unavailable, and balancing the overall energy consumption and efficiency with natural and artificial lighting.

Natural light is essential for workspaces in both residential and office settings, and it should be maximized whenever possible. To ensure that every employee in an office can benefit from natural light, the distance between the core and the wall should be as small as possible. Additionally, systems that prevent glare and reduce sky brightness or direct sun exposure can be installed. These systems can include passive or active devices such as blinds, louvers, or solar screens that are part of the curtain wall systems or placed outside. Depending on the season, low-E and other glass coatings and films can help regulate the heat from the sun, either by blocking it in the summer or allowing it in the winter.

One way to enhance the natural lighting in a room is to install horizontal light shelves along the window walls below the ceilings. These shelves can reflect daylight deeper into the interior space, reducing the reliance on artificial lighting sources. The shelves should be positioned below the ceilings of rooms that are connected to each other. Moreover, artificial lighting has improved significantly, and LED fixtures are now available with a wide range of color temperatures and brightness levels. LEDs produce very little heat, so they do not affect the room's temperature, and they last much longer than conventional lamps [23]. Furthermore, LEDs are very energy efficient.

4.3.3 Atrium Cores and Passive Mode Design

It is possible for a building to function as its own air shaft, as demonstrated by the Leeza SOHO Tower (see Section 4.4.5), in which a supply of fresh air is piped into the atrium core of the structure from the ground floor and then circulated vertically through the atrium [24]. At the same time, it is vented to the outside through an opening that is located at the very peak of the structure. Using the concepts of passive mode design, architect Ken Yeang incorporates atrium cores into his bio-climatic buildings to reduce air-conditioning loads and increase the supply of fresh air to inhabited levels. According to Yeang, the passive mode refers to the process of designing for increased levels of comfort despite the presence of external conditions without making use of any electromechanical equipment [25].

Passive mode strategies aim to utilize the local climate to optimize building design and minimize the need for mechanical heating, cooling, and ventilation systems. Here are some examples of passive strategies and how they relate to the local climate:

Building Configuration and Orientation: By understanding the local climate, architects can design buildings with appropriate configurations and orientations. For instance, in hot climates, buildings may be designed with narrow floor plans to maximize shading and natural ventilation. In cold climates, buildings may be oriented to capture solar heat gain during winter months.

Facade Design: Balancing solid-to-glazed area ratios is crucial to control heat gain or loss through the building envelope. In hot climates, reducing the amount of glazing and using shading devices can minimize solar heat gain, while in cold climates, more glazing can maximize solar heat gain and natural lighting.

Thermal Insulation: Suitable levels of thermal insulation help reduce heat transfer through the building envelope. Insulation keeps the interior spaces comfortable by preventing heat gain in hot climates and heat loss in cold climates.

Natural Ventilation: Passive cooling and ventilation techniques such as cross ventilation, stack ventilation, and night flushing can be incorporated to take advantage of prevailing winds and temperature differentials. This reduces reliance on mechanical systems and promotes natural airflow.

Vegetation and Greenery: Incorporating vegetation within tall buildings or around them helps create a healthy microclimate. Vegetation provides shade, reduces the heat island effect, improves air quality, and adds aesthetic value. Green roofs and vertical gardens can also enhance building insulation and cooling effects.

By leveraging these passive mode strategies, buildings can minimize energy consumption, enhance occupant comfort, and create sustainable and environmentally-friendly design solutions. The adaptation of these strategies to the local climate ensures that buildings are responsive and well-suited to their specific geographical location.

4.3.4 Heating and Cooling

Thermal comfort is a sensory state of satisfaction with the thermal conditions in a building, which depends on the energy balance of the occupants. The energy balance is the difference between the heat produced by the body and the heat lost to the environment. Several environmental factors influence thermal comfort, such as air temperature, humidity, air movement, and mean radiation temperature [18]. Air temperature is the most obvious factor, as it affects the heat exchange between the body and the air. Humidity is the amount of water vapor in the air, which affects the evaporation of sweat from the skin. Air movement is the speed and direction of the airflow, which can enhance or reduce heat loss by convection and evaporation. Mean radiation temperature is the average temperature of all the surfaces that surround a person, which affects the heat loss by radiation. Moreover, the work performance of workers who need to concentrate or perform complex intellectual tasks is related to a narrow range of temperatures [17]. This range is usually between 21°C and 24°C, depending on the season and clothing. Temperatures outside this range can cause discomfort and reduce productivity. In hot-humid climates, mean radiation temperature is especially important, as it reflects the perceived temperature along with room temperature and humidity. The perceived temperature is the temperature that a person feels, which can differ from the actual temperature due to other factors. Electric chillers are the most widely used cooling method because they have the lowest cost, despite their high energy use. Electric chillers are devices that use electricity to compress a refrigerant and transfer heat from one place to another. Absorption chillers are another option that can use waste heat from different sources (such as district heat or machine installations) to produce cooling energy. Absorption chillers are devices that use a heat-driven chemical process to absorb and release heat from a refrigerant.

4.3.5 Combined Heat and Power

The most effective method for supplying heating energy uses district heat generated by thermal power facilities. Alternatives to massive coal or gas-fired plants and nuclear facilities that are more environmentally friendly include combined heat and power (CHP) stations that burn waste materials. The wood biomass produced by the forest products business is burned at a CHP plant that was created by Ever-Green Energy in St. Paul, Minnesota. It can generate up to 65 watts of heat and nearly 33 megawatts of electrical power. The average amount of renewable electricity supplied to the local electric utility is 25 megawatts (MW) and 153,000 watt-hours (W-h), sufficient to service up to 20,000 dwellings. The production of electricity results in a surplus of thermal energy, which may be used to generate enough hot water to satisfy about 45% of District Energy St. Paul's requirements for the heating and cooling of its 3 million square meters (32 million square feet) of client building space [19,20].

4.4 CASE STUDIES

This section analyzes nine case studies involving tall buildings (Figure 4.3). These towers are selected based on the authors' long-standing teaching, research, and professional experiences. Many of the examined projects have been discussed at academic

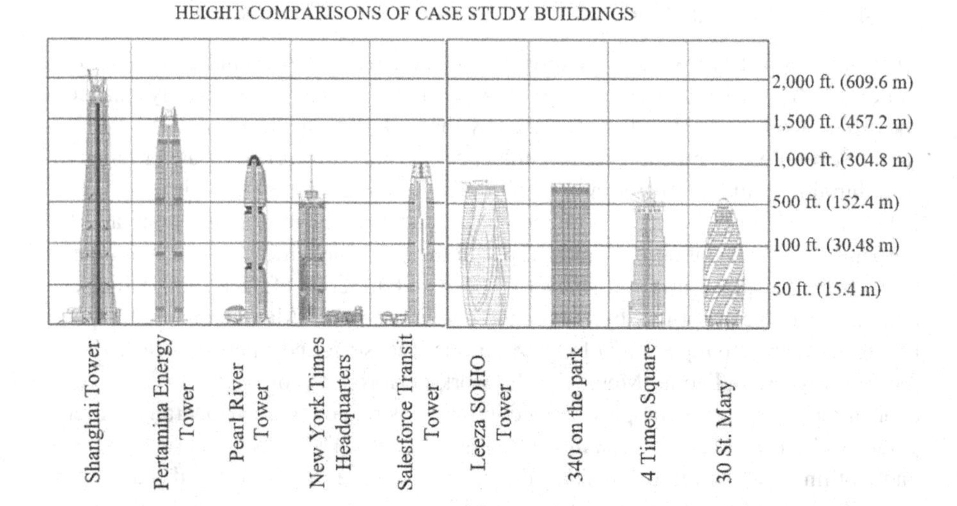

FIGURE 4.3 Height comparisons of case study buildings. (Diagram by P. Armstrong.)

conferences and are well-cited. A quick Google Scholar search reinforces their popularity. Most of the structures are office towers because this type of building consumes the most energy overall. The Shanghai Tower is an example of a multi-functional building that incorporates hotel and business space, making it environmentally responsible. It earned a remarkable LEED Platinum rating for its sustainable features. 340 On the Park is the only residential high-rise type with ABSs (Automated Building Systems) that are more restricted in scope, and it is also the only type with more widespread alternatives for individualized thermal comfort. ABSs are used extensively in modern office towers where sophisticated environmental controls, mechanically ventilated curtain walls, active solar shading systems, and other micro-climate functions are monitored continually and remotely controlled by a centralized BMS.

The Pertamina Energy Tower, which is now in the planning stages, has the potential to be the first zero-carbon building in the world. Its design integrates geothermal energy as well as several other environmentally friendly features. The HVAC systems in large office buildings often account for 30%–40% of the entire energy usage of the facility. It is essential to make tall structures more sustainable, to make these systems more efficient, or do away with them altogether [27]. Therefore, it is important to remember that tall buildings are essential to the urban habitat and that environmental infrastructures must be constructed and modified sustainably to support them. At the urban scale, New Songdo City in South Korea is an excellent example of how a city that is developed with sustainability in mind can produce healthy urban landscapes and ecosystems.

4.4.1 4 Times Square

One of the pioneers in sustainable architecture, 4 Times Square in Manhattan, was designed and built by Fox & Fowle Architects in 1999. It was the first skyscraper in North America to adopt environmental systems and features that improved its performance and efficiency, while reducing its environmental impact. Some of its innovative solutions included recycling chutes for different types of waste, curtain walls with PV panels that generated electricity, natural gas fuel cells that provided backup power, and high-performance glass that reduced heat gain and glare. These features made 4 Times Square a model for "green" buildings around the world [26] (Figure 4.4).

The building is a 52-story skyscraper, of which 47 floors are leased to various tenants who enjoy a spacious and comfortable environment. The total area of the lease space is 148,647 square meters (1.6 million square feet), which is equivalent to about 26 football fields. The building features low-E glazing and task lighting that

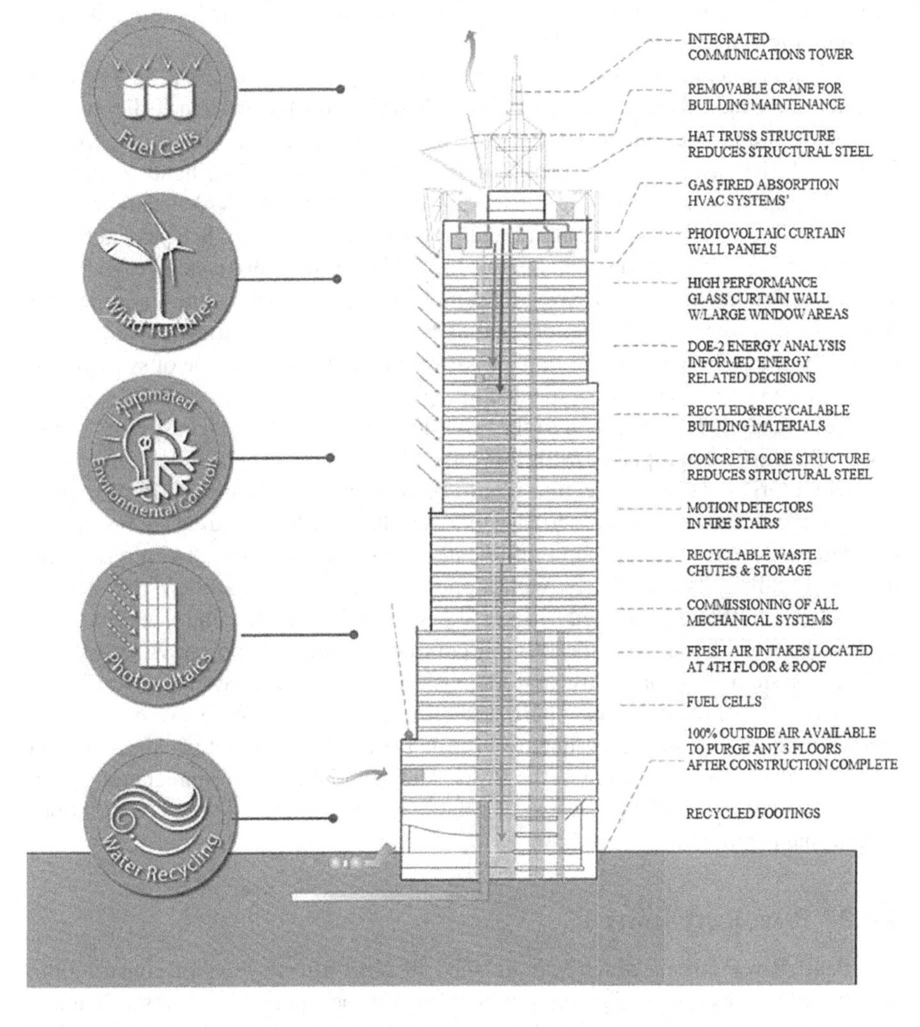

FIGURE 4.4 In 1999, Fox & Fowle Architects designed a building in New York City's Times Square. (Illustration by P. Armstrong.)

allow natural light to enter the rooms, creating a pleasant atmosphere. These features also have several benefits for energy efficiency and environmental sustainability. They minimize heat loss through the windows, which reduces the need for heating in winter. They also reduce cooling expenses in summer, as less artificial lighting is needed. Moreover, they decrease the reliance on central-station electric generation and transmission, which reduces greenhouse gas emissions and improves air quality. The building's occupants can enjoy a comfortable temperature throughout the year, thanks to the main cooling units that are located in a high-bay mechanical room near the top of the building. These units are six gas-fired absorption chillers that use natural gas instead of electricity to produce chilled water for air conditioning. This also helps to lower the amount of electricity consumed by the building and to avoid peak demand charges. The building's design also maximizes the rentable space by placing the exhaust piping in a vertical shaft that runs along the core of the building, instead of occupying valuable floor space.

Moreover, it positions the chillers near the rooftop cooling towers [27]. The total capacity of the chillers is 3,375 metric tons (3,720 U.S. tons), with each chiller having a rating of 562 metric tons (620 U.S. tons). Chilled water at 7°C (45°F) is delivered from the chillers using variable-speed pumps to two specialized air-handler units on each building floor [28]. The return temperature is 14°C (58°F). VAV terminal units are responsible for circulating conditioned air throughout the floor. When switched to the heating mode, the chillers can also be utilized to provide hot water. A free cooling system that integrates a massive and complicated supplementary air-conditioning system for one tenant and monitors actual Btu consumption optimizes the use of secondary water, heat exchanges, and crossover piping [29]. This type of system also allows the tenant to control how much cooling they receive.

Glass spandrels from the 37th to the 43rd floors on the south and east faces of the skyscraper have been replaced with thin-film PV panels to generate electricity [30]. The Durst Organization's objective was to revolutionize environmentally responsible building design and construction while simultaneously providing its occupants with the most energy- and resource-efficient, practically achievable, and ecologically friendly building. Two fuel cells with a combined capacity of 200 kW are used to store the power that is generated by the PV panels and the chiller and then distribute it throughout the building in the form of continuous electrical power. A web-based metering system monitors individual tenants' energy use and the overall structure. Removing 85% of the particulate particles in the outside air requires extensive filtration. Carbon dioxide monitors that are installed permanently in return air ducts and at loading docks are connected to the BMS. The installation of permanent tubing on the floor makes quarterly monitoring of volatile organic compounds and other indoor contaminants more convenient [28].

4.4.2 Pearl River Tower

The Pearl River Tower, which was finished in 2012 after 6 years of construction, is a remarkable example of green architecture that incorporates the most advanced engineering and environmental technologies in a contemporary office building [30]. The tower, which was designed by Adrian Smith of Skidmore, Owings, and Merrill,

LLC, one of the world's leading architectural firms, has a sleek and aerodynamic shape that measures 309 meters (1,014 feet) in height. The shape is designed to direct wind to two large openings at its mechanical floors, where vertical-axis wind turbines capture the wind energy and convert it into electricity for the building. The tower also boasts a number of other innovative green features, such as solar panels on the roof and the south facade, double-skin curtain walls that reduce heat gain and glare, a chilled ceiling system that uses water instead of air for cooling, underfloor ventilation air that delivers fresh air to the occupants, and daylight harvesting that maximizes the use of natural light and reduces artificial lighting. These features all contribute to improving the energy efficiency of the building and reducing its carbon footprint. Moreover, the tower's environmental technologies are not merely added on, but are seamlessly integrated with its core systems to provide optimal comfort and indoor air quality for the occupants. The tower also has a smart control system that monitors and adjusts the performance of the building according to the changing weather conditions and occupancy patterns. The Pearl River Tower is a leading project that demonstrates how architecture can harmonize with nature and create a better environment for people (Figure 4.5).

The multiple layers of low-E coated insulated glazing that make up the outside thermal barrier are designed as a 300-millimeter (12-inch) unitized system on the north and south sides, with a 240-millimeter (9.5-inch) cavity between them [31]. A solitary, monolithic glass panel protects the interior-occupied space. A building automation system (BAS) with motorized sunshades controlled by photocells that track the sun automatically manages glare. On the east and west sides, motorized blinds and permanent external sunshades measuring 750-millimeter wide (29.5 inches) are used to provide solar shading and flare control. Importantly, low-level inlets in the monolithic glass allow air from the occupied room to automatically ventilate the cavity wall. The BAS mixes non-exhausted air to reduce the moisture content of the system's cooled air with the air from exterior ventilation. Return air ducts spaced 2.7 meters (9 feet) apart move warm air through the cavity and deliver it to air handling devices on each mechanical floor [31]. Human comfort is preserved by separating the ventilation system within a raised access floor from the chilled radiant ceiling system. Leung claims that this HVAC system requires less maintenance and replacement and is more adaptable than conventional fan coil or VAV systems [32]. This mechanical method effectively saved five floors of construction by reducing the floor-to-floor height of the building from 4.2 meters (14 feet) to less than 3.9 meters (13 feet) [31].

Two high-efficiency vertical-axis wind turbines capture the current winds at each mechanical floor. Using four wind tunnels at the mechanical levels, the building's sculpted form maximizes the pressure difference between the windward and leeward sides of the tower. This promotes airflow and increases velocity. The average wind speed in Guangzhou is 4.3 meters per second (14 feet per second) at 50 meters (164 feet) of height and 5.3 meters per second (17 feet per second) at 300 meters (984 feet). According to wind studies, the façade inlets will increase wind velocity by a factor of 2.5, producing more electricity than open-field wind turbines by more than eight times [31]. The Pearl River Tower's cutting-edge integrated design method will result in a significant quantity of electrical power being needed to run the building's lighting, ventilation, dehumidification, and cooling systems.

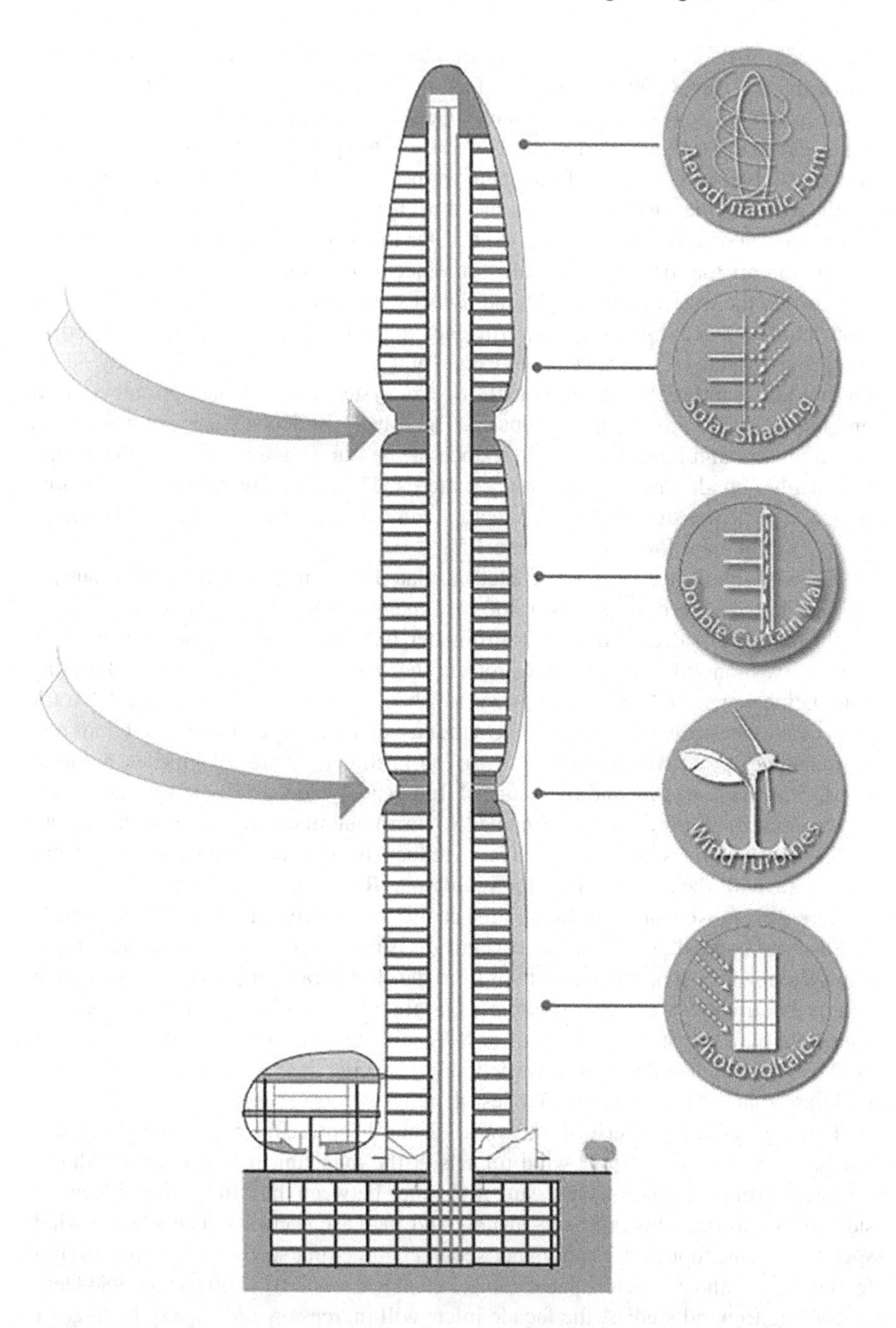

FIGURE 4.5 Pearl River Tower, Guangzhou, China, Skidmore Owings & Merrill, LLC, 2012. (Illustration by P. Armstrong.)

4.4.3 New York Times Headquarters

Italian architect Renzo Piano collaborated with FXFowle Architects and Gensler to create the 52-story New York Times Headquarters between 40th and 41st Streets in New York City. The structure is made of glass and steel and is covered in a thin layer of ceramic rods that act as solar light diffusers (Figure 4.6). Piano has modernized business Modernism for the Internet era. Corporate Modernism peaked in New York in the 1950s and 1960s. The structure has 52 floors and a height of 227 meters

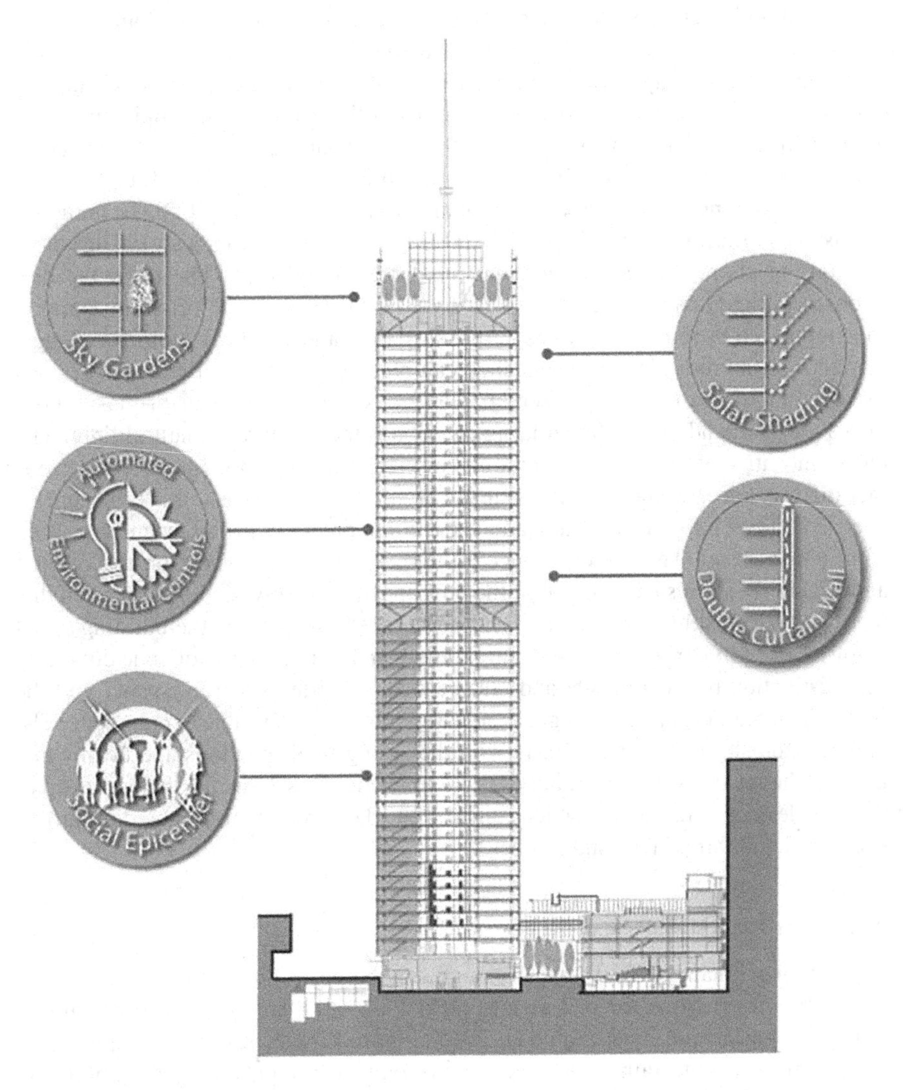

FIGURE 4.6 New York Times Headquarters, New York City, Renzo Piano Building Workshop, 2012. (Illustration by P. Armstrong.)

(744 feet) to the main roof. A 300-foot (91.4 meter) mast rises to a height of 319.4 meters (1,048 feet). Leasable floor space totaling 139,355 square meters (1.5 million square feet) is shared property ownership with the Forest City Ratner Companies. One level is 4.9 meters (16 feet) below grade. A lobby, a storefront, and a glass-enclosed garden are all on the ground floor. The New York Times newsroom occupies the whole five-story pedestal east of the tower. The tower rises 48 more stories above the podium. The building's average story height is about 4.2 meters (14 feet), which gives open office plans an excellent outlook. However, to accommodate equipment and two-story outriggers, the mechanical floors at levels 28 and 51 have floor heights of approximately 8 meters (27 feet) [32].

To support the company's vision of communication, collaboration, and transparency as well as to be ecologically friendly, one of the leading design objectives was to make the interior as light as possible [33]. Thus, flexible cutting-edge reconfiguration of interiors is made possible by quantum lighting controls, which utilize only 0.38 W of lighting power per square foot and save 72% of energy [33]. To maintain a constant overall light level for workers, daylight sensors continuously change electric light levels in accordance with the available natural light. The inner façade's floor-to-ceiling glass is screened by a layer of 175,000 horizontal, off-white ceramic rods supported by thin steel frames [34]. As the light and weather conditions change, the color of the rods also changes. They can block up to 50% of the sun's rays and serve as an energy-efficient sunscreen. The translucent inner layer of glass maximizes views in and out of the building and floods the inside with natural light. The most innovative aspect of the structure is a scrim made of horizontal ceramic rods that diffuse sunlight and give the exterior of the building a neat, uniform appearance [35]. They have the accuracy and texture of a precisely honed machine as seen from a side street. The screens conceal the mechanical equipment on the rooftop and stretch six stories past the top of the building's frame, giving the impression that the tower is vanishing into the sky. The structure's base exhibits the building's best features, including proportion, structural detail, and a sound sense of civic duty [35]. The distinction between inside and out, between the life of the newspaper and the life of the street, is blurred by a glass-enclosed lobby that is visible to onlookers. The lobby's exquisite lightness contrasts with the strong steel spandrels and beams that hold up the tower. A 378-seat auditorium and a café are set against the backdrop of the outside courtyard garden, which can be seen when crossing the lobby. The garden is covered with birch trees and grasses.

4.4.4 Shanghai Tower

The Pudong Financial District of Shanghai is home to the third and last supertall high-rise structure, the 632-meter (2,074-foot) tall Shanghai Tower, which was built by Gensler and inaugurated in 2016 (Figure 4.7). A luxury boutique hotel, class-A offices, entertainment venues, a conference center, an observation deck, and cultural amenity spaces are all located in the 127-story mixed-use tower, which was designed as a "vertical city." Retail is located on the ground floor. The building achieves a near-zero-carbon energy objective through numerous bioclimatic design solutions, resulting in a decreased carbon footprint of 34,000 metric

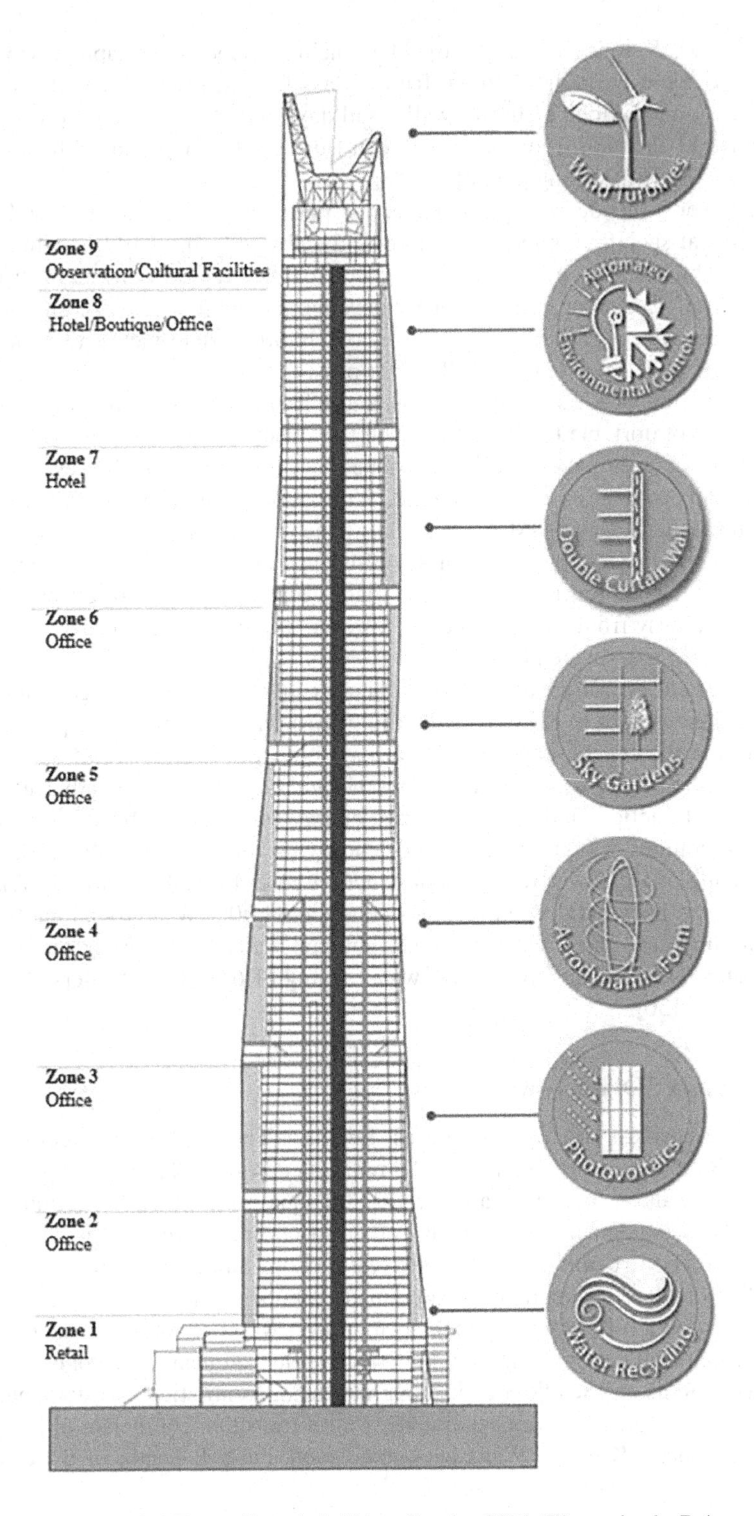

FIGURE 4.7 Shanghai Tower, Shanghai, China, Gensler, 2016. (Illustration by P. Armstrong.)

tons (37,479 U.S. tons) annually [36]. They include indoor landscaping, wind turbines, integrated building controls, fritted glass for sun shading, locally sourced materials, modern double curtain walls, and daylighting. It has received a LEED Platinum Certification from the U.S. Green Building Council and a China Green Building Three Star rating [37,38].

The tower's 120-degree spiral shape was parametrically created to withstand the Shanghai-specific typhoon wind forces, thereby reducing building wind loads by 24%. The result is a 32% material reduction, a simpler, lighter construction with unparalleled transparency. Gensler unveiled a combined exterior and interior curtain wall system with a total glass area of 210,000 square meters (2.26 million square feet) [39]. An extensive, full-height passive atrium system is created in the gap between the façade's two glass skins, and it harnesses natural air convection to maintain a comfortable temperature inside the structure. Only the first 4.6 meters (15 feet) of the atrium are mildly conditioned by the system, which uses perimeter fan coil units to heat or cool each zone. With a combination of the updraft, controlled top exhausts, and spill air on the last zone, the atrium is naturally ventilated for the most part, yielding a 21% energy efficiency and a 12.5% improvement over China's Three-Star Rating. The outside glass wall is staggered to reflect sunlight upward and away from the street below, lowering the light reflectance level to 12% [40]. This prevents severe glare.

The vertical atria's creation of the 4 hectares (10 acres) of "sky garden oasis" is a practical answer to Shanghai's humid environment and poor air quality. The gardens and the cutting-edge curtain wall system work together to filter airborne particles, offer occupants ventilation and thermal comfort, create gathering areas for social interaction, and reduce building energy consumption by 7%. The building's top-mounted wind turbines provide energy for the outside lighting. The rotational shape of the tower reduced construction costs by $58 million (USA). The 45 integrated horizontal wind turbines generate 54,000 kW hours annually [36]. The building consumes 40% less fresh (potable) water by recycling non-potable gray water. This equates to a yearly water saving of 673,803,298 liters (178 million gallons) [36].

4.4.5 Leeza SOHO Tower

The Fentai business district in southwest Beijing is anchored by the Leeza SOHO Tower (Figure 4.8), which Zaha Hadid Architects designed. The skyscraper, which has 45 floors and a total floor area of 172,800 square meters (1.86 million square feet), satisfies the demand for adaptable office space from small- and medium-sized businesses. The structure is divided into two sections and enclosed in a glass shell. The 637-foot tallest atrium in the world is made possible by the space between, which spans the tower's full height – 194 meters (636 feet) [41]. The double-insulated low-E glass curtain wall system creates highly effective environmental control by stepping the glazing units on each floor at an angle with small ventilation corridors that suck outside air through operable cavities [41]. A heat transition coefficient of 2.0 Watts per square meter K (0.186 Watts per square foot) and a B-Factor of 0.4 ensure a

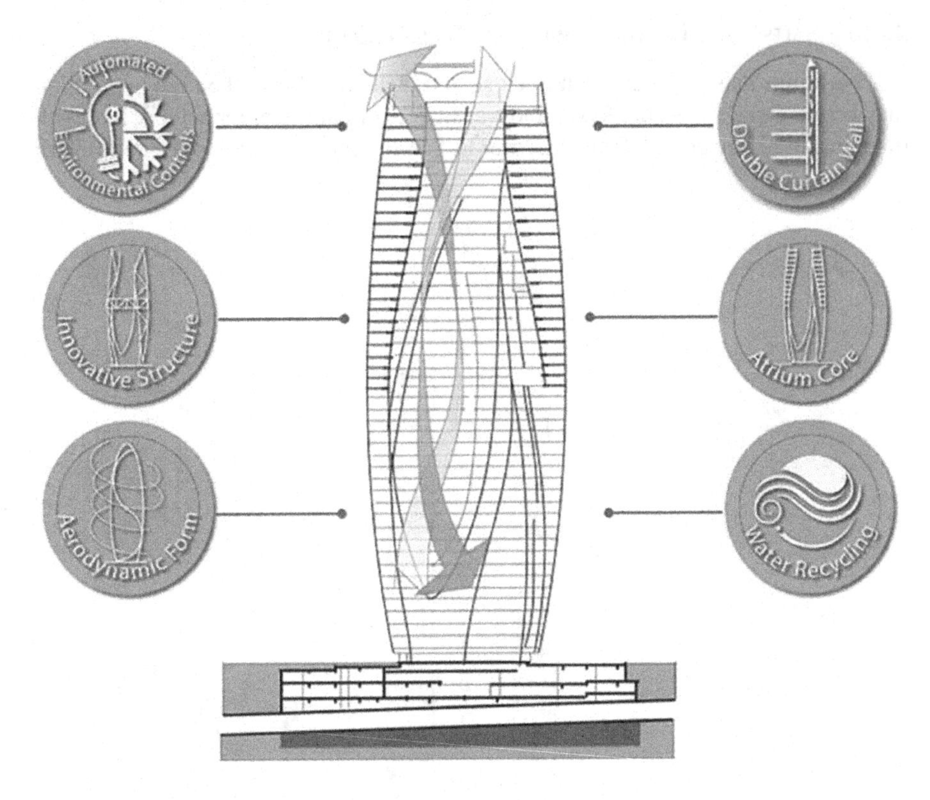

FIGURE 4.8 Leeza SOHO Tower, Beijing, China, Zaha Hadid Architects, 2019. (Illustration by P. Armstrong.)

comfortable indoor climate. The tower shell's u-value is 0.55 Watts per square meter (0.051 Watts per square foot) [41].

As the skyscraper rises to realign the upper floors with Lize Road to the north, the atrium revolves around the structure. The atrium lets in natural light throughout the entire structure. It serves as a thermal chimney with an integrated ventilation system that filters the air inside the tower and maintains positive pressure at a low level to prevent air infiltration. The building's two parts are connected by skybridges on levels 13, 24, 35, and 45, providing sweeping city views. Computations were visualized into 3D energy simulation models using BIM to reduce energy use and emissions. The U.S. Green Building Council's LEED Gold certification is attained with a 3D BIM energy management system that continuously monitors environmental control and energy effectiveness. These systems also have high-efficiency pumps, fans, chillers, boilers, lighting, controls, and heat recovery from exhaust air. The skyscraper features 2,680 bicycle stands, low-rate fixtures with water collection, gray water flushing, an insulating green roof with a PV array to capture solar energy, and underground charging stations for electric vehicles [41].

4.4.6 Salesforce Transit Tower and Transit Center

Tall buildings must be situated at or close to transit hubs due to the high population they house. Sometimes, the tower and transit network are interconnected, like with the Salesforce Tower and Transit Center in San Francisco (Figure 4.9). Pelli Clark

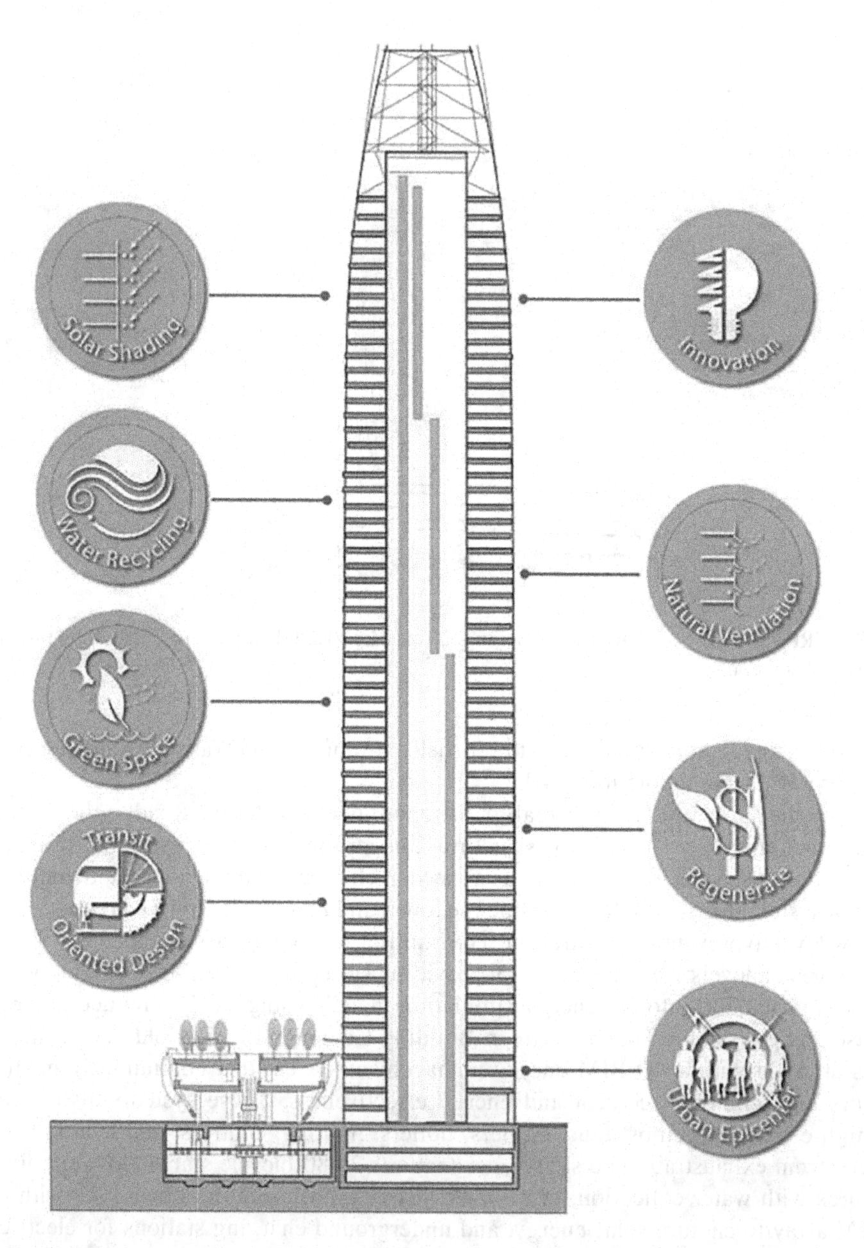

FIGURE 4.9 Salesforce Tower and Transit Center, San Francisco, Pelli Clark Pelli, 2018. (Illustration by P. Armstrong.)

Pelli designed the 61-story Salesforce Tower and Transit Center at Transbay Joint Powers Authority's request to revitalize the Mission District, improve sustainability, concentrate public transportation options, foster neighborhood development, and be financially feasible. The 1.4 million square foot (130,000 square meters) office tower, which was completed in 2018, is 326-meter (1,070-foot) tall and is covered in clear glass with pearlescent metal accents [42]. The tower is the tallest building in San Francisco and the second tallest in the West Coast of the United States. It features a nine-story atrium lobby, a sky deck with panoramic views of the city and the bay, and a public art installation by Jim Campbell that displays images on the top six floors of the building. The transit center has a four-block-long elevated park, and it connects various bus and rail lines, as well as future high-speed rail and Caltrain services. It also includes retail spaces, gardens, playgrounds, and an amphitheater [43]. The Salesforce Tower and Transit Center is an example of how tall buildings can integrate with transit hubs to create a vibrant and sustainable urban environment.

The metal sunshades built into the tower's floors are each tailored to optimize light and vistas while minimizing solar gain. Low-emissivity, high-performance glass also lessens the need for cooling. The building's foundation is covered in heat-exchanging coils that help chill the interior. Residents on each floor receive fresh air from high-efficiency air handlers. Both the transit center and the tower recycle water. The tower has a connection to the 1.5 million square foot (140,00 square meter) Transport Center, which has a 5.4 acre (2.2 hectares) public park on its roof and 11 multimodal Bay Area transit networks [42]. The Great Hall and transportation levels are illuminated naturally by "light columns," which are massive hyperbolic diagrid columns with skylights atop them. The rooftop park has more than a dozen entrances and boasts a cafe, an amphitheater that seats 800 people, a playground, and spots for quiet relaxation. The park has many bays, from oak trees to swamp marshes. The annual energy usage of the LEED Gold-rated building is anticipated to be 25% lower than California's 2008 Title 24 Energy Efficiency Requirements [42].

4.4.7 340 On the Park

On the Park, a Solomon Cordwell Buenz 340 design was Chicago's first residential skyscraper to achieve a LEED Silver rating. The 64-floor building is on Randolph Street at the northern edge of Millennium Park and reaches a height of 204.8 meters (672 feet) [43,44]. It was finished in 2007 (Figure 4.10). The transit-oriented skyscraper is situated in the center of the Chicago Loop Central Business Area, close to Lake Michigan's attractions, shopping, dining, and entertainment. A two-and-a-half-story winter garden on the 25th level leads to a pleasant outdoor balcony. The Blue Cross-Blue Shield Building and Millennium Station to the west are connected to the lower level by a pedway. About 10% of energy expenditures are saved via sustainable environmental systems. The employed environmental systems reduce energy consumption by 10%. These include solar energy, PVs, passive solar shading with overhangs and balconies, insulating glass, and energy-efficient mechanical and lighting systems [45]. The architects called for countertops made

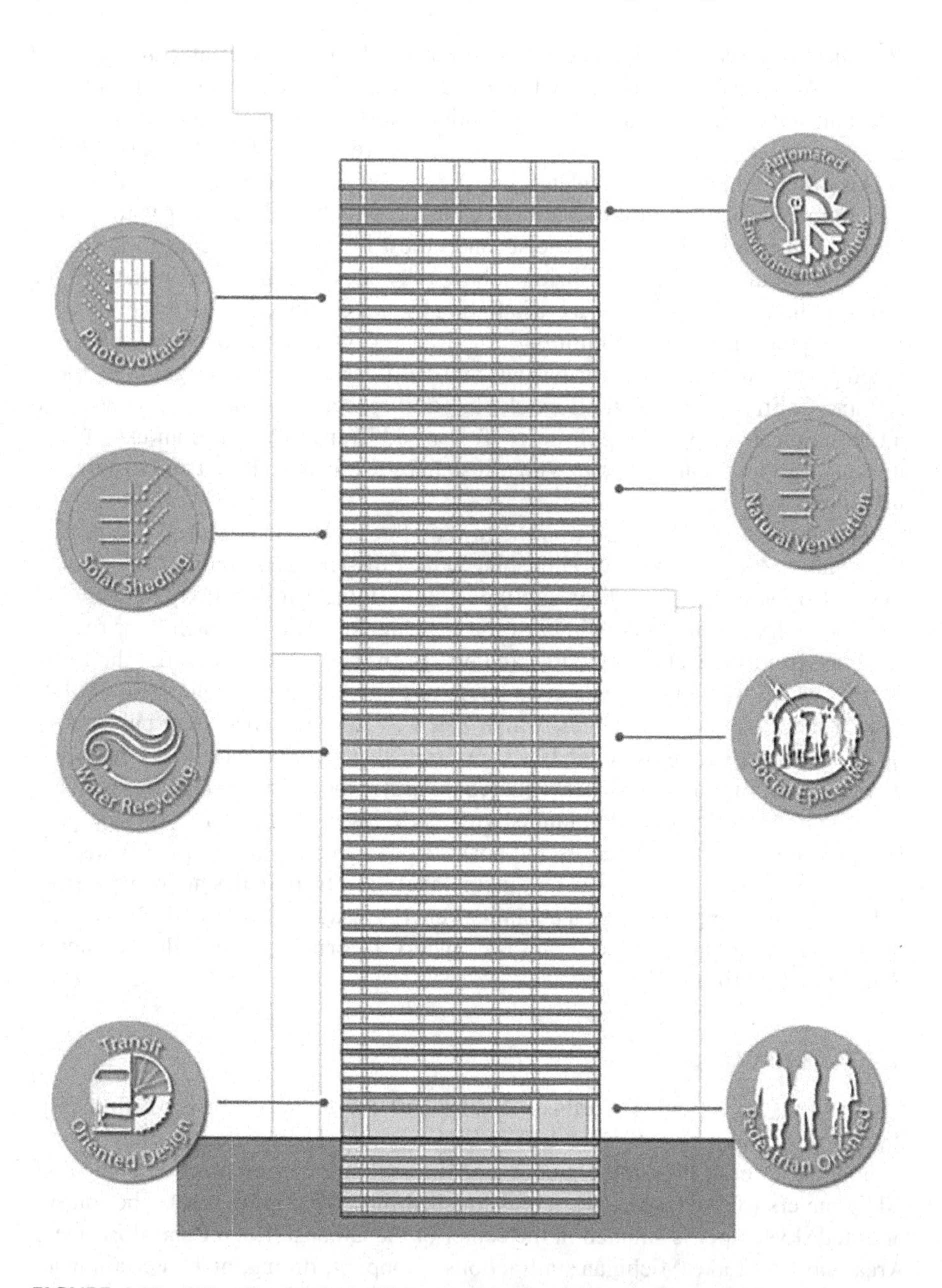

FIGURE 4.10 340 On the Park, Chicago, Solomon Cordwell Buenz, 2007. (Illustration by P. Armstrong.)

of recycled materials and floors made of bamboo, both of which are renewable green materials. 80% of the construction trash was recycled, and a 41,640-liter (11,000-gallon) tank in the basement collects and stores rainwater.

4.4.8 Mary Axe

Buildings are being shaped into sleek aerodynamic shapes using BIM to improve structural efficiency and reduce lateral wind loads. The 180-m (591-foot) tall) 30 St. Mary Axe in London's bullet shape (Figure 4.11) permits winds to flow around the structure without forming unfavorable vortices, minimizing the tower's sway, and improving the comfort of those occupying the public plaza at its base. The Foster Partners-designed tower, which was finished in 2004, is located on a 1.4-acre site in the City of London's Financial District, close to the London Underground. Together with an arcade of shops and cafes, it offers 46,000 square meters (494,445 square feet) of net office space [46].

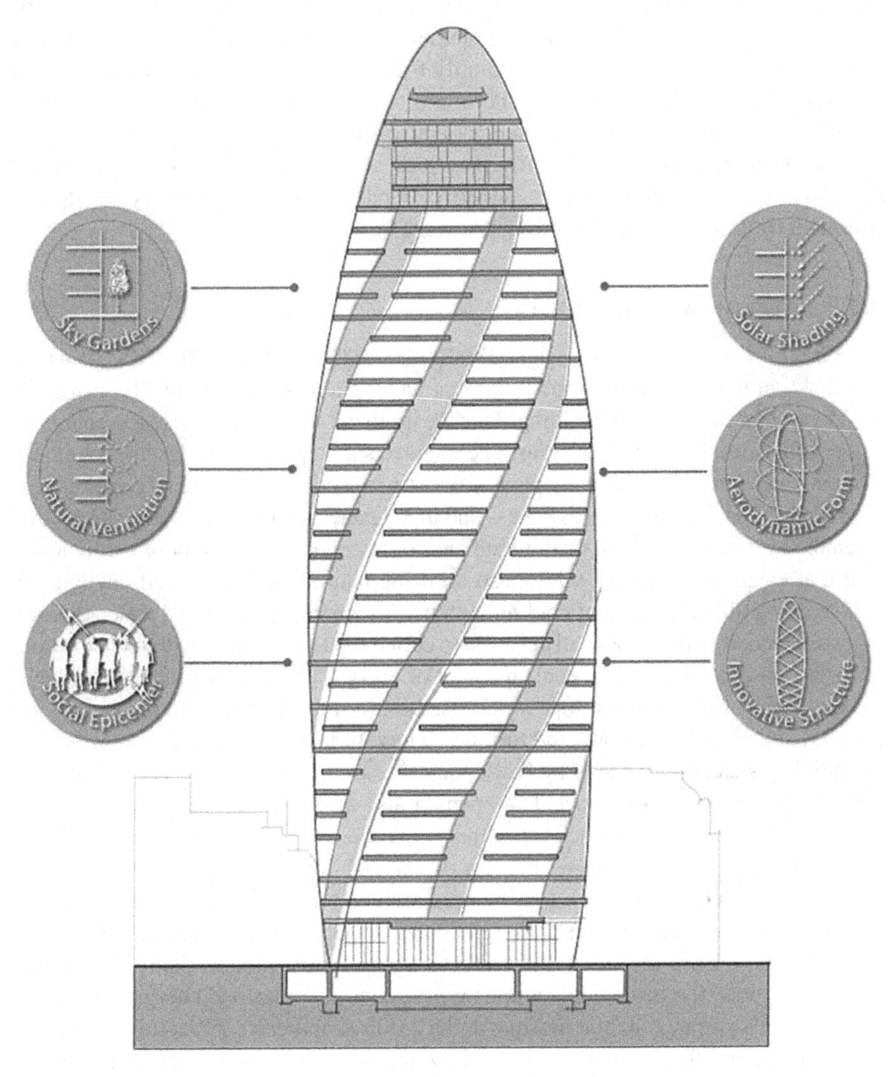

FIGURE 4.11 30 St. Mary Axe, London, Foster Associates, 2007. (Illustration by P. Armstrong.)

Enhancing the public realm at the street level, encouraging building occupants to use public transportation as much as possible, providing flexible services, high-spec, user-friendly, column-free office spaces with access to natural light, ensuring good physical and visual interconnectivity between floors, and reducing energy consumption using low-E glass, natural ventilation, and intelligent building control systems are all goals of the design process. The tower's aerodynamic design, which uses an effective diagrid structure with 41 stories and clear inner floor plates, was engineered by Arup. A glass curtain wall with moveable windows covers the diagrid for 40% of the year to provide ventilation [47]. Six light wells serve as a buffer zone to collect fresh air and regulate temperature inside the building. The spiraling vertical gardens that are part of the building's exterior are its most noticeable environmental elements. Each floor's radiating fingers have atriums in between that connect vertically to create several break-out areas. Each area serves as the "lungs" of the structure, dispersing fresh air brought in through opening panels in the façade. It also offers refreshments and meeting spaces. Along with other environmentally friendly practices, this method lessens the building's dependency on air conditioning and utilizes only 50% of the energy needed to operate a traditionally air-conditioned tower [46,47].

4.4.9 Pertamina Energy Tower

The Pertamina Energy Tower, designed by Skidmore Owings & Merrill, LLC, is a proposed skyscraper in Jakarta, Indonesia, that has the ambitious goal of becoming the world's first net-zero-energy building. This means that it will produce as much energy as it consumes over the course of a year. The architects used parametric design techniques to create a curved facade that adapts to Jakarta's tropical climate and its position near the equator. The facade is made of glass and metal panels that have movable shades that resemble leaves. These shades can rotate and tilt to control the amount of sunlight and heat that enters the building, while also allowing natural ventilation and daylighting. The building also uses radiant cooling systems, which circulate chilled water through pipes embedded in the floors and ceilings, to cool the interior spaces without using conventional air conditioners that consume a lot of energy. The most distinctive feature of the tower is its funnel-shaped top, which rises above the 99th floor and creates a wind tunnel that houses 12 vertical wind turbines. These turbines can generate up to 25% of the building's electricity needs by harnessing the wind currents that flow through the opening at the top of the funnel. The opening also acts as a vent that draws air from below and creates a natural stack effect that helps cool the building [47].

The campus will serve as a "beacon of energy" for 20,000 people by embodying a holistic design approach that merges architectural design, campus planning, and structural and environmental engineering services. It will incorporate living, working, and leisure activities and serve as a sustainable model. Using Indonesia's seismic features, the tower, and campus will invest in geothermal energy for the primary HVAC system. A public mosque and a 2,000-seat theatre for lectures and plays are planned, and they will be connected by an energy-generating "Energy Ribbon" walkway with PVs on its roof (Figure 4.12).

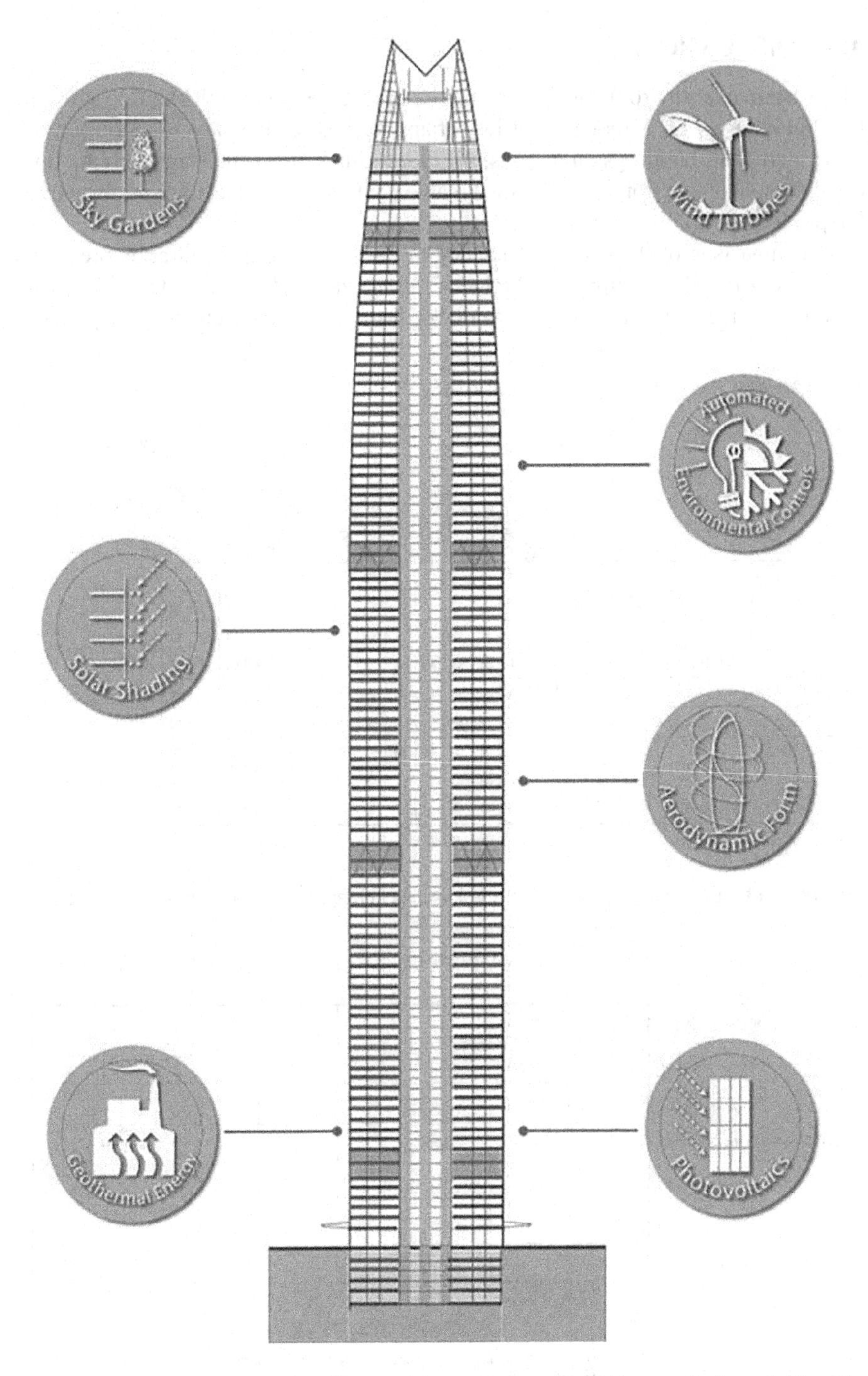

FIGURE 4.12 Pertamina Energy Tower, Jakarta, Indonesia, Skidmore, Owings & Merrill, LLC, on hold. (Illustration by P. Armstrong.)

4.5 DISCUSSION

The sustainable environmental elements of each case study tall building are outlined in Figure 4.13. Although a variety of approaches can be taken to produce more energy-efficient buildings, architects of tall buildings are increasingly relying on a combination of environmentally sustainable systems to create high-performance, "hyper-functional" structures.

A comparison of the energy efficiency of each case study building is presented in Figure 4.14. This comparison excludes the Leeza SOHO and Pertamina Energy Towers; their energy efficiency has not yet been determined but is anticipated to

SUSTAINABLE SKYSCRAPER COMPARISONS

Building	Type	LEED Rating	Aerodynamic Form	Innovative Structure	Vertical Atria	Daylight Harvesting	Natural Ventilation	Solar Shading	Double Skin Curtain wall	SkyGardens / Parks	Water Recycling	Photovoltaic Panels	Wind Turbines	Fuel Cells	Geothermal / Cogeneration	Automated Building Systems	Transit Oriented Design (TOD)	Social Epicenters
4 Times Square	O	ES				●	●	●			●	●	●	●		●	●	
30 St. Mary Axe	O	P	●	●	●	●	●	●		●	●		●			●	●	●
340 On The Park	R	S					●	●			●					●	●	●
New York Times Hdqtrs	O	G				●	●	●	●	●	●		●		●	●	●	●
Shanghai Tower	OH	P	●	●	●	●	●	●	●	●	●	●	●		●	●		●
Sales Force Tower	O	P				●	●	●			●	●	●			●	●	●
Leeza SOHO	O	G	●	●	●	●	●	●	●	●	●	●	●		●	●		●
Pearl River Tower	O	P	●	●		●	●	●	●		●	●	●			●		
Pertamina Energy Tower	O	P	●	●	●	●	●	●		●	●	●	●		●	●		●

Type: O = Office R = Residential O/H = Office/Hotel LEED : P = Platinum G = Gold S = Silver ES = Energy Star

FIGURE 4.13 Environmental sustainability skyscraper comparisons. (Diagram by P. Armstrong.)

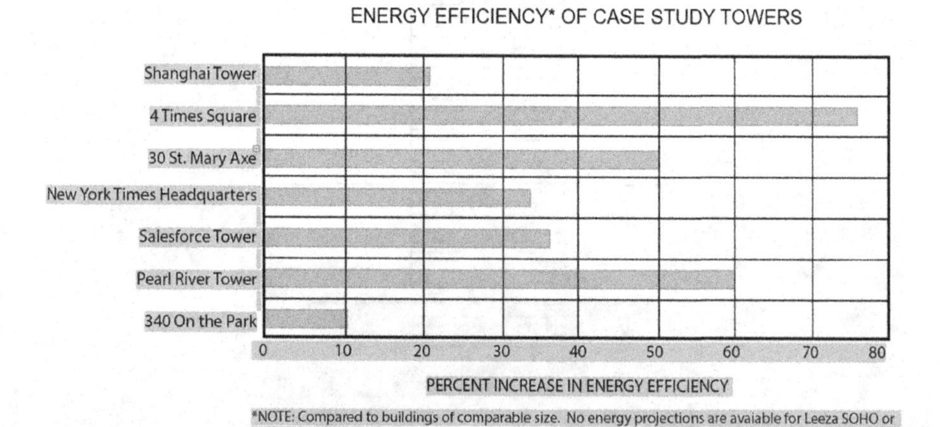

FIGURE 4.14 Energy efficiency comparisons of case study tall buildings [47]. (Diagram by P. Armstrong.)

be relatively low. Most of the time, the percentages indicate the increased energy efficiency compared to other structures of a corresponding height and kind. Mark Frisch, Principal and Technical Design Lead at Solomon Cordwell Buenz in Chicago, explains that given the large scale of tall buildings, increasing a tall building's efficiency by just 10% can be significant when compared with the equivalent volume and energy consumption of multiple single-family houses. 340 On the Park, the lone residential structure that was reviewed, is representative of the achievements that have been made to include more environmentally sustainable features into high-rise buildings to improve their livability and efficiency. Within the context of the sustainability triad, social sustainability is equally essential to economic and environmental sustainability. Cities cannot function without adequate transportation, and the presence of transportation hubs or centers within walking distance makes it possible to construct towering buildings. Alternative forms of transportation focused on transit, such as taxis, buses, and bicycles, help alleviate traffic congestion, reduce air pollution, and make communities healthier and more livable. Access to natural light is an essential component of office buildings. For maximum accessibility to urban services and amenities, high-rise structures like the Salesforce Tower and 30 St. Mary Axe need to be situated at or near transportation hubs like subway stations or train stations. They are constructed on brownfield sites near city centers and promote pedestrian and transit-oriented design opportunities. As a result, they lessen the amount of traffic congestion and air pollution and increase the number of neighborhoods that are walkable and pedestrian friendly [48,49].

4.5.1 Curtain Walls

Advancements in curtain wall technology have significantly improved the performance of skyscrapers, enhancing transparency, vistas, daylight harvesting, and the overall indoor climate for occupants. Sustainable office skyscrapers employ various types of low-E (low-emissivity) glass and solar shading systems to reduce glare, minimize heat gain, and decrease the cooling load required. One of the benefits of modern curtain walls is their ability to provide ample natural light while controlling solar heat gain. Low-E glass varieties, which have a thin coating that reflects heat radiation, can help regulate the amount of heat entering or leaving the building. This reduces the reliance on artificial lighting and minimizes the need for excessive cooling or heating. Solar shading systems, such as louvers, blinds, or automated shading devices, can be integrated into the curtain walls to further control the amount of sunlight and heat entering the building. These systems help optimize the balance between daylighting, views, and thermal comfort for occupants. By reducing excessive glare and heat gain, they contribute to a more comfortable and energy-efficient indoor environment.

Natural ventilation is another feature that can be facilitated by curtain walls. Opening windows or vents allow fresh air to enter the building, providing occupants with a connection to the outdoors and reducing the need for mechanical ventilation systems. During colder months, natural ventilation can also help offset the cooling requirements, leading to energy savings. Improvements in glass and curtain wall technologies have led to increased insulation values, reduced solar glare and reflectance, and other advantages. Fritted glass, which has a patterned or textured surface, can enhance privacy, reduce heat loss during winter months, and increase heat gain during warmer seasons. Low-emissivity glass coatings help to maintain comfortable indoor temperatures by reflecting heat back into the building during winter and blocking heat from entering during summer. Overall, the advancements in curtain wall technologies have allowed for better control over energy performance, thermal comfort, and daylighting in sustainable skyscrapers. These innovations contribute to reducing energy consumption, improving occupant well-being, and creating more environmentally friendly high-rise buildings.

4.5.2 ABSs

ABSs have indeed been integrated into curtain walls of many office buildings to enhance their functionality and energy efficiency. ABSs play a crucial role in controlling solar shading devices, such as louvers or automated blinds, which help regulate the amount of sunlight and heat entering the building. By dynamically adjusting the position of these shading devices, ABSs can optimize daylighting and minimize the cooling load required, contributing to energy savings. In addition to solar shading, ABSs can also control the operation of windows or dampers for natural ventilation. By opening and closing windows or adjusting ventilation dampers based on environmental conditions, ABSs facilitate the exchange of fresh air, reduce reliance on mechanical ventilation systems, and provide occupants with access to natural ventilation, improving indoor air quality and reducing energy consumption.

ABSs can also work in conjunction with HVAC systems to increase heating and cooling efficiency. By integrating ABSs with HVAC controls, the BMS can optimize the operation of these systems based on real-time occupancy, outdoor weather conditions, and indoor temperature requirements. This integration allows for more precise control of the indoor environment, minimizing energy waste and improving occupant comfort. Furthermore, renewable energy technologies like solar PV panels, wind turbines, fuel cells, and geothermal systems are increasingly being integrated into tall buildings. These technologies help supplement or replace traditional fuel sources, contributing to the building's energy self-sufficiency and reducing reliance on non-renewable energy. Solar PV panels, for example, can generate electricity from sunlight, while wind turbines harness wind energy. Fuel cells and geothermal systems provide alternative sources of heating and cooling. The integration of these renewable energy technologies into tall buildings aligns with the goals of sustainable and green design, reducing the environmental impact and promoting a more resilient and efficient energy infrastructure. As ABSs and renewable energy technologies continue to advance, we can expect to see their widespread adoption in the construction of future tall buildings. These technologies offer immense potential for improving energy efficiency, reducing carbon emissions, and creating more sustainable and environmentally friendly high-rise structures.

4.5.3 Cogeneration Plants

Cogeneration plants, also known as CHP plants, can play a significant role in supplying heating and cooling to large buildings. These plants generate electricity while simultaneously utilizing the waste heat produced during electricity generation for heating or cooling purposes. By utilizing recycled waste products as fuel, cogeneration plants can further reduce the carbon footprints of tall buildings and contribute to more sustainable energy systems.

4.5.4 Bioclimatic Design

Bioclimatic design methods, including the incorporation of sky gardens, atria, and vertical farms, are increasingly being embraced in the construction of tall buildings. Sky gardens and vertical farms introduce green spaces into vertical structures, providing numerous benefits. They contribute to improved air quality, biodiversity, and aesthetics while creating a closer connection with nature for building occupants. These green elements can also help regulate indoor temperature, filter airborne particles, and enhance natural ventilation, thereby promoting a healthier and more comfortable indoor environment. Vertical atriums, as seen in buildings like the Leeza SOHO, Shanghai Tower, and 30 St. Mary Axe (also known as "The Gherkin"), serve multiple functions beyond their architectural appeal. They act as central circulation spaces that encourage natural ventilation by promoting the movement of air through the building. Vertical atriums can facilitate stack effect ventilation, allowing warm air to rise and exit through upper-level openings, while drawing in fresh air from lower levels. This natural ventilation strategy helps reduce the reliance on mechanical cooling systems, contributing to energy savings and enhanced occupant comfort. Moreover, vertical atriums can serve as social epicenters, providing communal spaces for interaction and fostering a sense of community within tall buildings. They create opportunities for informal gatherings, socializing, and collaboration among occupants.

Overall, the integration of well-designed curtain walls, cogeneration plants, bioclimatic design elements, and vertical atriums exemplifies the holistic approach taken in designing sustainable and healthier micro-environments for tall building occupants. By combining innovative energy systems, green spaces, and social spaces, these design strategies contribute to reducing environmental impact, enhancing occupant well-being, and promoting sustainable living in dense urban environments.

4.6 CONCLUSIONS

This chapter dwells on the concept of "green" skyscrapers and their key aspects. Sustainable tall buildings aim to minimize their environmental impact by incorporating bioclimatic, zero-energy, zero-carbon, and zero-waste principles. The integration of environmental systems with architectural elements such as curtain walls, atriums, and sky gardens is crucial for achieving sustainability goals and improving the overall quality of the building. Curtain walls, for instance, play a significant role in reducing heat loss and gain in tall buildings. They are designed to balance transparency and insulation, allowing natural light to penetrate while providing thermal insulation. This helps in optimizing daylighting, reducing the need for artificial lighting, and improving energy efficiency. Atriums, on the other hand, contribute to the overall sustainability and comfort of tall buildings. They serve as central spaces that facilitate natural light penetration, natural ventilation, and air circulation. Atriums can act as thermal chimneys, allowing hot air to rise and escape, promoting natural ventilation, and reducing the reliance on mechanical cooling systems. Also, sky gardens are an integral part of sustainable tall buildings as they offer multiple benefits. They provide green spaces within the urban environment, improving the overall aesthetics of the building and enhancing biodiversity. Sky gardens also contribute to reducing the urban heat island effect by providing a cooling effect through evaporative processes and shade. Additionally, they can improve air quality by acting as natural filters, absorbing pollutants, and releasing oxygen.

Bioclimatic design principles offer a promising approach to creating environmentally sustainable systems in tall buildings that reduce reliance on fossil fuels. This design philosophy leverages natural materials and processes to harmonize buildings with their surroundings. By incorporating elements such as wind turbines and solar cells for energy generation, along with passive design elements like daylight harvesting, solar shading, and natural ventilation, tall buildings can significantly reduce their energy consumption. Additionally, features like vegetated sky gardens and atria contribute to regulating indoor temperature, lighting, and air quality while creating a connection with nature and enhancing the aesthetic value of the buildings. The integration of these bioclimatic design elements, renewable energy systems, and energy-efficient features creates a holistic approach to sustainability in tall buildings. By optimizing energy use, promoting healthy microenvironments, and incorporating natural elements, these buildings can provide comfortable and sustainable spaces for occupants while minimizing their impact on the environment. Continued research and innovation in this field will be crucial to further advance the development of net-zero-energy buildings and refine the application of bioclimatic design principles. Through collaboration among architects, engineers, researchers, and other stakeholders, the vision of creating highly sustainable tall buildings that operate without relying on fossil fuels can become a reality, contributing to a greener and more resilient urban future.

Advancements in technology and parametric analysis tools have greatly aided architects and engineers in optimizing environmental systems in tall buildings. Through detailed analysis and simulations, these professionals can identify the most efficient design solutions, coordinate environmental systems with other building components, and align them with site-specific conditions and microclimates. This allows for the creation of customized and site-responsive designs that maximize energy efficiency, occupant comfort, and overall sustainability. Overall, the integration of environmental systems with architectural elements in tall buildings is essential for creating sustainable, energy-efficient, and livable spaces. By incorporating bioclimatic design principles and leveraging advancements in technology, the concept of "green" skyscrapers can continue to evolve and contribute positively to the built environment, addressing the challenges of climate change and promoting a more sustainable future.

This chapter also highlighted some key points regarding the importance of creating net-zero-energy buildings with renewable energy sources and bioclimatic architectural principles. Indeed, this is a vital area of research that holds significant potential for minimizing the environmental impact of tall buildings and achieving sustainable design solutions. Architectural firms like Adrian Smith + Gordon Gil Architects, SOM, Gensler, Foster Partners, and others are at the forefront of designing skyscrapers with zero-energy consumption as their goal. Zero-energy skyscrapers strive to balance the amount of energy consumed with the amount of energy produced, ultimately achieving a state of equilibrium and minimal environmental impact. This approach involves not only the construction of new buildings but also retrofitting existing high-rise structures to incorporate energy-efficient features and renewable energy systems. By improving the performance and reducing the carbon footprint of existing buildings, the overall sustainability of the built environment can be enhanced.

Further, this chapter highlighted the importance of ABSs with intelligent or smart features in contemporary high-rise buildings. These systems play a crucial role in regulating and optimizing the building's microclimate and the functioning of various environmental systems, such as the MEP systems, elevators, natural ventilation, solar shading devices, and more. By working together, these systems aim to achieve the highest possible performance standards while adhering to stringent green energy and environmental requirements. The integration of ABSs with other building systems is essential to ensure maximum efficiency in high-rise buildings. This integration is often referred to as an "integrated web," where the environmental systems are fully interconnected with other vital systems. By establishing this interconnectedness, the various components can communicate and coordinate with each other, resulting in a more cohesive and efficient operation. For example, integrating the MEP systems with the environmental systems allows for better control and management of energy usage, temperature regulation, and air quality. This integration enables the optimization of energy consumption, reduces waste, and ensures that the building operates in an environmentally friendly manner. Elevators can also be integrated into the overall system, allowing for more efficient transportation within the building. By coordinating elevator usage with occupancy levels and scheduling, energy consumption can be minimized while still providing convenient and timely transportation for occupants.

Natural ventilation and solar shading devices can be incorporated into the integrated web to enhance the building's energy efficiency. By leveraging smart features and real-time data, these systems can respond dynamically to changing environmental conditions, optimizing the use of natural light and airflow while minimizing the need for mechanical cooling and lighting. By fully integrating these vital environmental systems, high-rise buildings can achieve the highest performance standards and comply with strict green energy and environmental requirements. This integration enables a synergistic approach where each component works in harmony with the others, maximizing efficiency, sustainability, and occupant comfort. Overall, the implementation of ABSs and the establishment of an integrated web in high-rise buildings are critical steps in achieving optimal environmental performance. Through careful coordination and intelligent control, these integrated systems contribute to energy efficiency, sustainability, and the creation of healthier and more comfortable living and working environments.

REFERENCES

1. Nejat, P., Jomehzadeh, F., Taheri, M. M., Gohari, M., & Majid, M. Z. A. (2015). A global review of energy consumption, CO_2 emissions and policy in the residential sector (with an overview of the top ten CO_2 emitting countries). *Renewable and Sustainable Energy Reviews*, *43*, 843–862.
2. Pidwirny, M. (2006). Environmental systems as energy systems. In Petersen, J.F., Dorothy Sack, D., Robert, E., Gabler, R.E. (Eds.) *Fundamentals of Physical Geography*, 2nd ed. Boston, MA: Cengage Learning. Available from https://www.physicalgeography.net/fundamentals/4d.html (Accessed 28 April 2020).
3. Ábalos, I., Herreros, J., & Ockman, J. (2003). *Tower and Office: From Modernist Theory to Contemporary Practice*, p. 852. Cambridge, MA: MIT Press.
4. Hill, E. J., & Gallagher, J. (2003). *AIA Detroit: The American Institute of Architects Guide to Detroit Architecture*. Wayne State University Press, Detroit.
5. Parakh, J., Gabel, J., & Safarik, D. (2017). *The Space Between: Urban Places, Public Spaces & Tall Buildings*. Council on Tall Buildings and Urban Habitat, Chicago.
6. Elbakheit, A. R. (2020, February). Bioclimatic tall buildings. In *Conference Proceeding, Jubail 2nd International City Planning Forum*, Jubail, KSA.
7. Yeang, K. (2002). *Reinventing the Skyscraper: A Vertical Theory of Urban Design*. Academy Press, Lagos, Nigeria.
8. Rynska, E. D., & Solarek, K. (2018). Adaptive urban transformation: Cities in changing health and wellbeing conditions. *WIT Transactions on Ecology and the Environment*, *217*, 247–256.
9. Yeang, K., & Powell, R. (2007). Designing the ecoskyscraper: premises for tall building design. *The Structural Design of Tall and Special Buildings*, *16*(4), 411–427.
10. Al-Kodmany, K. (2014). Green towers and iconic design: Cases from three continents. *ArchNet-IJAR: International Journal of Architectural Research*, *8*(1), 11.
11. Kim, D., Cho, H., Mago, P. J., Yoon, J., & Lee, H. (2021). Impact on renewable design requirements of net-zero carbon buildings under potential future climate scenarios. *Climate*, *9*(1), 17.
12. Fedorov, M., Matys, E., & Kopytova, A. (2018). Strategic and tactical aspect of the relations between the participants of ICP in high-rise construction. In *E3S Web of Conferences* (Vol. 33, p. 03054). EDP Sciences.
13. Malin, N. (2006). A Group Effort. *Green Source, November*, 46–51.
14. Strelitz, Z. (2011). Tall building design and sustainable urbanism: London as a crucible. *Intelligent Buildings International*, *3*(4), 250–268.
15. Ali, M. M., & Armstrong, P. J. (2010). The role of systems integration in the design of sustainable skyscrapers. *International Journal of Sustainable Building Technology and Urban Development*, *1*(2), 95–106.
16. Fischler, R. (1998). The metropolitan dimension of early zoning: Revisiting the 1916 New York City ordinance. *Journal of the American Planning Association*, *64*(2), 170–188.
17. Olgyay, V. (2015). *Design with Climate: Bioclimatic Approach to Architectural Regionalism*. Princeton University Press, Princeton, New Jersey.
18. Olesen, B. W. (2004). International standards for the indoor environment. *Indoor Air*, *14*(7), 18–26.
19. Hollander, J. B. (2013). SYNERGICITY: Reinventing the postindustrial city. *Geographical Review*, *103*(2), 310–312.
20. Breeze, P. (2017). *Combined Heat and Power*. Academic Press, Lagos, Nigeria.
21. Bachman, L. R. (2004). *Integrated Buildings: The Systems Basis of Architecture* (Vol. 9). John Wiley & Sons, Hoboken, New Jersey.

22. Weerasuriya, A. U., Zhang, X., Gan, V. J., & Tan, Y. (2019). A holistic framework to utilize natural ventilation to optimize energy performance of residential high-rise buildings. *Building and Environment, 153*, 218–232.
23. Fedoronko, K., Gharpure, P., Pozo, M., 'The Vortex Tower,' Sustainable Strategies Diagram, High-Rise & Habitat Graduate Design Studio, Illinois School of Architecture, University of Illinois at Urbana-Champaign, Fall 2020.
24. Schumacher, P. (2020). The mega-void: Unleashing the communicative impact of tall buildings. *Architectural Design, 90*(5), 72–81.
25. Yeang, K., & Powell, R. (2007). Designing the ecoskyscraper: Premises for tall building design. *The Structural Design of Tall and Special Buildings, 16*(4), 411–427.
26. Vaghefi, A., Jafari, M. A., Bisse, E., Lu, Y., & Brouwer, J. (2014). Modeling and forecasting of cooling and electricity load demand. *Applied Energy, 136*, 186–196.
27. Gissen, D. (Ed.) (2002). *Big and Green: Toward Sustainable Architecture in the 21st Century*. Princeton Architectural Press, NYC.
28. Wood, A., & Salib, R. (2013). *Guide to Natural Ventilation in High Rise Office Buildings*. Routledge, NYC.
29. Del Percio, S. T. (2004). The skyscraper, green design, & the LEED green building rating system: The creation of uniform sustainable standards for the 21st century or the perpetuation of an architectural fiction. *Environs: Environmental Law and Policy Journal, 28*, 117.
30. Lippe, P. (July-September, 2005). The Conde Nast building. *Urban Land Institute, 5*(15), 5. Available from https://casestudies.uli. org/wp-content/uploads/2015/12/C035015.pdf (Accessed 19 June 2020).
31. Seçluk, S. A., & Ilgin, H. (2017). Performative approaches in tall buildings: Pearl River tower. *Eurasian Journal of Civil Engineering and Architecture, 1*(2), 11–20.
32. Jodidio, P., & Altmeppen, S. (2007). *Architecture Now!*. Taschen, Berlin, Germany.
33. Medio, S., & Murphy, J. (2007). Paper No: 342: Aesthetic vision and sustainability in The New York Times Building ceramic rod facade. *Environment, 3*(9), 11.
34. Barben, B. R., Bonfanti, E. L., Andres, R., Perez, A. R., The New York Times Building: Technical Report #1 (Collegeville, PA: Pennsylvania State University, 2009), p. 4. Available from https://www.engr.psu.edu/ae/thesis/ portfolios/2010/brb5019/Reports/Structural%20Tech%201_Public.pdf (Accessed 20 June 2020).
35. Lutron Electronics. (2009). Case Study: The New York Times Building, pp. 4–5. Available at https://www.lutron.com/technicaldocumentlibrary/New%20York%20Times%20Building_%20New%20York_%20USA_English.pdf (Accessed 20 June 2020).
36. Xia, J., Poon, D., & Mass, D. (2010). Case study: Shanghai Tower. *CTBUH Journal, 2010*, 12–18.
37. Ouroussoff, N. (2007). Pride and Nostalgia Mix in The Times's New Home. *The New York Times*: Art & Design, November 20, 2007. Available from https://www.nytimes.com/2007/11/20/arts/design/20time.html (Accessed 20 June 2020).
38. Williams, B., Alfonso, V., Messenger, W.S., Shanghai Tower: A Symbol of China's Ascension, MSRE 517 Property Report, 2012, pp. 2–12. Available from https://www.josre.org/wp-content/uploads/2012/10/Shanghai- Tower-in-China-Tall-and-near-net-Zero.pdf (Accessed 21 June 2020).
39. Gong, J. (2017). Shanghai Tower. *Frontiers of Engineering Management, 4*(1), 106–109.
40. Hu, G., Zhu, S., Gao, R., & Xiao, L. (2021, November). Modeling and analysis of Shanghai Tower based on Ansys Workbench. In *2021 4th International Symposium on Traffic Transportation and Civil Architecture (ISTTCA)* (pp. 312–319). IEEE, Suzhou, China.

41. Architects, Z. H. (2019). Zaha Hadid Architects Leeza SOHO Tower, Bejing (China). *AV Proyectos*, *96*, 40–41.
42. Çelebi, M., Haddadi, H., Huang, M., Valley, M., Hooper, J., & Klemencic, R. (2019). The behavior of the salesforce tower, the tallest building in San Francisco, California inferred from Earthquake and ambient shaking. *Earthquake Spectra*, *35*(4), 1711–1737.
43. Thomas, D. (2016). *Placemaking: An Urban Design Methodology*. Routledge, NYC.
44. Farouk, A. (2011). High rise buildings and how they affect countries progression. *E-iataorC leader*, 2011, 1–14.
45. Paul, J., Armstrong, R. A., & Ali, M. M. (2008, February). Green design of residential high-rise buildings in livable cities. In *IBS/NAHB Symposium*, Orlando (pp. 13–16).
46. Munro, D. (2004). Swiss Res building, London. *Nyheter Stålbyggnad*, *3*, 36–43.
47. Akın, Ö., & Akın, Ö. (2022). The Swiss Re Tower: Analysis of a seminal case. In Akın, Ö. (Ed.) *Design Added Value: How Design Increases Value for Architects and Engineers*, pp. 111–124, Cham: Springer.
48. Abel, C. (2010). The vertical garden city: Towards a new urban topology. *CTBUH Journal*, 2(1), 20–30.
49. Charney, I. (2007). The politics of design: Architecture, tall buildings and the skyline of central London. *Area*, *39*(2), 195–205.

5 Efficiency of Space Utilization in Supertall Towers with Free Forms

Hüseyin Emre Ilgın
Tampere University

5.1 INTRODUCTION

As the global human population burgeons at an unprecedented pace, poised to soar by a staggering 2.5 billion individuals by the year 2050, the skyscraper emerges as a pivotal solution, embodying the concept of the vertical city paradigm [1]. Considering this pressing concern, political figures, urban planners, and architects have increasingly directed their attention toward this paradigmatic shift [2]. Moreover, the adoption of towering edifices as the predominant architectural typology in many cities across the globe has become an increasingly pervasive trend, firmly rooted in the fabric of the 21st century [3]. Since the 1950s, the architectural landscape of high-rise structures has experienced a profound metamorphosis, catalyzing the birth of extraordinary and emblematic forms [4,5]. Noteworthy illustrations of this trend include the awe-inspiring Merdeka PNB118, towering magnificently at an elevation of 644 m across an impressive 118 stories, characterized by its resplendent crystalline configuration (Figure 5.1). Similarly, the CITIC Tower, commanding attention at a soaring height of 528 m across an equally formidable 108 stories, distinguishes itself through an alluringly vase-like silhouette (Figure 5.2).

In the preliminary design phase, the choice of specific building forms assumes paramount significance, as it necessitates a nuanced response to diverse exigencies encompassing both the symbolic visage of skyscrapers and the compliance with building codes and regulations. The paradigm governing skyscraper design undergoes a transformative shift, fostering a generation of design processes grounded in performance-oriented approaches. By harnessing the amalgamation of analytical tools wielded during the initial stages of design, profound possibilities manifest themselves within the realm of architectural form discovery. Consequently, this empowers architects to emancipate themselves from the constraints imposed by conventional methodologies, propelling them toward uncharted territories of creative exploration.

The advent of cutting-edge design methodologies and innovative digital technologies, particularly within the realm of architecture, has paved the way for the materialization of supertall towers that exhibit exceedingly intricate and audacious forms [6]. The burgeoning fascination with constructing "iconic" skyscrapers in urban landscapes, coupled with the indomitable ardor of architects in their quest to fashion

DOI: 10.1201/9781003403456-5

FIGURE 5.1 Merdeka PNB118 (Wikipedia.)

unrestrained and unconfined architectural expressions, has fundamentally reshaped the contemporary architectural lexicon. This is redefining the very essence of architectural typology in the present era [7].

As the vertical dimension of a building progressively extends, the available options for load-bearing systems diminish [8]. In simpler terms, while a broad spectrum of load-bearing system alternatives exists for low-rise structures, the choices become constrained in supertall towers due to the inherent challenges engendered by heightened elevations [9]. The irregular geometries inherent in unconventional building

FIGURE 5.2 CITIC Tower (Wikipedia.)

forms exacerbate this conundrum, rendering the selection of suitable structural systems even more crucial for the triumphant realization of architectural endeavors [10,11]. Within this context, the precise identification of any freeform tower poses formidable obstacles owing to their complex geometric configurations. The integration of load-bearing systems and architectural forms emerges as a pivotal concern of paramount significance. For instance, triangular geometric units intrinsically derived from diagrid-frame-tube systems, as exemplified by the soaring 98-story, 441-m tall KK100, can effectively embody the essence of any freeform tower without compromising spatial integrity [12].

Within contemporary skyscraper design practices, there exists a propensity among architects to excessively prioritize self-satisfying aesthetic considerations.

This tendency may result in adverse outcomes, particularly due to the dearth of interdisciplinary collaboration in the realm of efficient structural design. Consequently, it becomes increasingly imperative to comprehend the intricate interplay between the prevalent employment of freeform architectural configurations. This represents one of the frequently adopted building typologies, and the various other design parameters that govern the holistic design process. Space efficiency emerges as a fundamental criterion among these parameters, assuming an essential role in the design of freeform skyscrapers. It holds major importance not only for ensuring their long-term sustainability but also for aligning with the investment cost considerations associated with these towering structures [13].

Consequently, the existing literature is deficient in any comprehensive investigations that shed light on the intricate interrelationships between the space efficiency of freeform configurations and the essential structural planning parameters that govern supertall towers. To address this notable knowledge gap, this chapter undertook an in-depth examination of 35 case study towers (see Appendices A–C), meticulously considering their functional attributes, structural systems, and materials employed. It is essential to highlight that the primary criterion employed in the selection of these specific buildings for the study pertains to the availability of crucial data, including core type, structural system, and structural material. Notably, the aftermath of the tragic World Trade Center incident in the United States, which occurred during the deliberate and malevolent attacks on September 11, 2001, has significantly hindered data collection efforts due to the heightened safety concerns of skyscrapers. Hence, it is expected that this study will serve as a significant and informative reference, providing essential guidance and recommendations to stakeholders in the planning, design, and construction of freeform towers. This includes architects, structural engineers, and developers, who will greatly benefit from the valuable insights offered in this chapter.

The segments below were organized in the following manner: Initially, a comprehensive examination of the existing scholarly literature in the field was conducted. Subsequently, methods employed in the study were delineated, and the resultant findings were presented. This was followed by an exploration of case studies, which encompassed pertinent details regarding the overall characteristics and space efficiency aspects of these notable exemplars. Lastly, a discussion as well as conclusion was formulated, incorporating an in-depth analysis and interpretation of the findings, and a discussion of the future studies and the research limitations.

5.2 LITERATURE SURVEY

The scientific literature exhibits a dearth of comprehensive research endeavors dedicated to unraveling the intricacies of space efficiency within the realm of towering structures. Prior investigations in this domain have primarily been constrained in scope, primarily examining a limited selection of tall buildings, with notable exceptions found in the extensive and noteworthy body of work conducted by the author [14–16].

Okbaz and Sev [17] undertook a study encompassing 11 freeform office towers, aiming to elucidate the concept of space efficiency. Their comprehensive analysis encompassed various planning considerations, such as the configuration of the service core and load-bearing elements. The outcomes of their investigation revealed that freeform compositions exhibited a lower level of space efficiency compared to conical forms. This finding underscored the significant influence of building form on the utilization of space, while the height between floors was found to have a negligible impact.

In a separate endeavor, Hamid et al. [18] conducted interviews with architectural firms, aiming to explore the aspect of space efficiency in 60 single-family dwellings in Sudan. The study encompassed an array of factors, including the positioning of courtyards, as well as the arrangement of vertical circulation components. The results illuminated that the corner positioning of the buildings yielded the most effective utilization of land. The optimal placement for vertical circulation elements was identified to be in the middle of building edges.

In another investigation focused on the domain of hotel buildings, Suga [19] delved into the realm of space efficiency. The study underscored the positive impact of space-efficient design on hotel projects, with a particular emphasis placed on the effective arrangement of common areas in relation to the size of guest rooms.

The author embarked on a research initiative exploring the concept of space efficiency within 44 office skyscrapers, considering crucial architectural and structural planning factors [14]. In a parallel effort, he examined the aspect of space efficiency in 27 residential skyscrapers, incorporating similar design parameters [15].

Additionally, the author undertook an exploration of space efficiency of 64 towers characterized by mixed-use functions [16]. The collective findings derived from these studies indicated a prevalent preference for the implementation of a central core typology, with outriggered frame systems emerging as commonly employed load-bearing systems. Moreover, a noteworthy inverse relationship between space efficiency and building height was observed.

Employing regression analysis techniques, Arslan Kılınç [20] investigated the factors influencing the service core design and load-bearing systems in prismatic tall buildings. The study revealed a correlation between building height and the allocation of larger areas to the structural system and service core. However, no significant scientific technical relationship was established between space efficiency and construction material.

Nam and Shim [21] dedicated their research efforts to examining high-rise corner forms and lease spans regarding space efficiency. The study identified square-cut corner forms as having a negative impact on space efficiency, while lease span exerted a significant influence on space efficiency, with corner cuts being found to have a negligible effect.

Sev and Özgen [22] conducted a comprehensive investigation into the realm of spatial efficiency, focusing on 10 tall office towers. Their analysis encompassed factors such as structural material, core typology, floor-to-floor height, and lease spans. The findings underscored the importance of core arrangement and load-bearing systems in achieving optimal space efficiency. Core planning strategies exhibited substantial variations depending on occupant requirements, with the central core typology emerging as the most favored approach for tall office projects.

Saari et al. [23] pursued an analysis examining the interplay between space efficiency and the total cost of tall office buildings. Their findings unveiled a significant influence of increased space efficiency on the attainment of desired levels of indoor climate comfort.

Lastly, Kim and Elnimeiri [13] conducted a comprehensive analysis of space efficiency ratios within a sample of 10 mixed-use tall towers. They underscored the pivotal role played by elevator optimization techniques and the strategic distribution of functional areas in augmenting space efficiency. Furthermore, they emphasized the profound significance of integrating building form and load-bearing systems as crucial factors contributing to the enhancement of space efficiency.

Until now, there has been a notable absence of scholarly research that has delved into the domain of space efficiency within the context of supertall towers, particularly focusing on free forms. Consequently, this investigation endeavors to fill this significant research gap by comprehensively examining and elucidating the intricate nuances of space efficiency in supertall towers, with a specific emphasis on free forms. By undertaking this chapter, it is my intention to shed light on this crucial and timely subject matter, thereby contributing valuable insights into the existing literature.

In this context, the following section introduces research methods based on case studies, which concentrate on architectural and structural design parameters and their correlation with space efficiency. The study draws information from 35 free-form skyscrapers as its dataset (see Appendices A–C).

5.3 METHODS

To explore the concept of space efficiency in freeform supertall towers, a case study methodology was employed, drawing inspiration from established practices in the evaluation of built-environment projects. This approach, widely embraced within the scientific community, facilitates the collection of both quantitative and qualitative data, enabling a comprehensive analysis of the subject matter at hand [24–26]. A meticulous selection process resulted in the inclusion of a total of 35 supertall tower cases with free form, each of which was subjected to a thorough examination. The selected sample for this chapter was distributed across different geographical regions, including 19 towers in Asia (with 15 located in China), 11 towers in the Middle East, 3 towers in Russia, and 1 tower each in the United States and Australia (see Appendix A). Comprehensive details pertaining to each case were meticulously documented and can be found in Appendix B. Notably, most of these architectural marvels, encompassing 80% of the sample, have been completed within the past two decades, featuring renowned landmarks such as Merdeka PNB118 and CITIC Tower. It is important to highlight that supertall towers lacking sufficient and accessible information pertaining to space efficiency or floor plans were intentionally excluded from the case study sample.

In a diligent endeavor, the researcher meticulously scrutinized the floor plans of a diverse range of freeform supertall towers, encompassing typical, low-rise, and ground floors. This rigorous approach ensured the acquisition of dependable and accurate data that would serve as a solid foundation for assessing space efficiency in the sample group. Furthermore, in order to align with the existing body of literature (e.g., [26–30]), the author adopted the comprehensive categorization proposed by Ilgın [31] for key considerations in architectural and structural planning. This decision was based on the complete nature of these categories, as vividly depicted in Table 5.1.

Given its comprehensive nature, the study incorporated a range of building form configurations, namely (Figure 5.3)

a. Prismatic forms (in the context of architectural design denote buildings characterized by symmetrical and parallel figures at both ends, with identical sides and perfectly vertical axes aligned orthogonally to the ground. This configuration ensures that the building maintains consistent geometric proportions throughout its structure.)
b. Setback forms (in architectural terminology describe buildings that feature horizontally recessed sections distributed at various heights along the vertical axis of the structure. These recessed sections create distinct terraces within the building, resulting in a stepped or cascading appearance. The purpose of setback forms is to introduce visual variation, enhance architectural aesthetics, and provide functional benefits such as increased access to natural light and improved views.)
c. Tapered forms (within the realm of architecture, pertain to buildings that exhibit a gradual reduction in their floor plans and surface areas as they ascend vertically. This effect results in the formation of either linear or non-linear profiles, wherein the dimensions and proportions of the building gradually diminish toward the upper levels. The purpose of employing tapered forms is

TABLE 5.1
Core, Structural System, and Structural Material Classifications

Core	**Structural system**
Central core	Shear-frame system
• central	• shear trussed frame
• central split	• shear walled frame
Atrium core	Mega core system
• atrium	Mega column system
• atrium split	Outriggered frame system
External core	Tube system
• attached	• framed-tube
• detached	• trussed-tube
• partial split	• bundled-tube
• full split	Buttressed core system
Peripheral core	**Structural material**
• partial peripheral	Steel
• full peripheral	Reinforced concrete
• partial split	Composite
• full split	

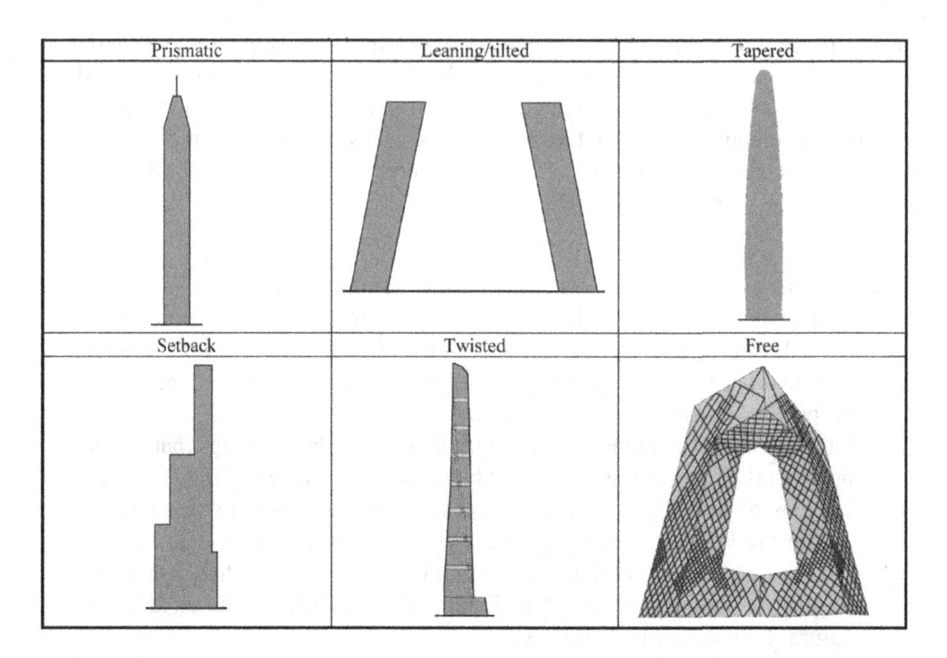

FIGURE 5.3 Building forms.

to create a visually dynamic and aesthetically appealing structure that deviates from a monotonous or uniform appearance. This design approach allows for variations in scale, enhances architectural character, and may offer functional advantages such as improved structural performance.)

d. Twisted forms (in the context of architectural design, pertain to buildings that exhibit a gradual rotation or torsion of their floors or façade as they ascend along a central axis. This rotation occurs in a progressive manner, resulting in a twisted or spiraling effect that imparts a sense of dynamism and visual intrigue to the structure. The twist angle, which represents the degree of rotation between each floor or façade element, is implemented to achieve the desired architectural expression. The utilization of twisted forms allows for the creation of unique building profiles, enhancing the aesthetic appeal and distinguishing the structure from its surroundings. Additionally, the incorporation of twists in the design may offer functional benefits such as optimizing views or providing architectural interest to interior spaces.)
e. Leaning/tilted forms (in architecture denote buildings characterized by an inclined configuration. These structures deviate from the traditional vertical orientation and feature a deliberate tilt in their design. The inclination may be achieved through various architectural techniques such as cantilevering. This intentional leaning imparts a distinct visual appeal and dynamic presence to the building, creating a sense of movement and asymmetry. Leaning forms offer architectural designers the opportunity to create visually striking structures that challenge the norms of verticality. The inclination may serve aesthetic purposes, enhancing the building's uniqueness and making

it a focal point within the urban landscape. Moreover, the inclined form can influence the spatial experience within the building, providing opportunities for innovative interior layouts, and interesting views.), and

f. Free forms (architecture emerges through the application of transformative actions on geometrically basic elements, such as lines or volumes. These actions involve a series of manipulations and modifications carried out by the architect, resulting in a final form that does not conform to the established categories mentioned earlier. The creative process behind the development of freeform architecture may lack clear and predefined sequences, as it involves a more exploratory approach. The resulting form often exhibits a sense of unpredictability, and uniqueness that distinguishes it from conventional architectural typologies. The absence of clear guidelines or predetermined frameworks allows architects to push the boundaries of design and explore uncharted territories, leading to the creation of unconventional architectural expressions.)

These diverse architectural expressions were chosen due to their inherent capacity to offer a more intricate and fully realized structural framework [32,33].

The establishment of a precise criterion for defining the specific number of stories or heights that classify a structure as a supertall tower is still a matter of debate within the scientific community, as there is no universally accepted definition in this regard. Nevertheless, in this research, the classification of a structure as a supertall tower adheres to the criteria provided by the CTBUH database, which designates a supertall structure as one that measures 300 m or more in height [34]. By adopting this criterion, the study ensures consistency and aligns with established industry standards in identifying and categorizing supertall towers for analysis and examination.

Space efficiency encompasses the relationship between the net floor area and the gross floor area (GFA). It holds substantial importance, particularly for investors, as it involves optimizing the usable space within floor plans to attain maximum returns on investment. The degree of space efficiency is primarily influenced by various factors, including the choice of load-bearing system and construction materials, building form, and the arrangement of floor slabs [15]. Additionally, the notion of space efficiency assumes a significant role in the determination of lease span, which pertains to the measurement of the distance between immovable internal components like service core walls and exterior elements such as windows. This aspect directly influences the optimal utilization of space within a given structure [16].

5.4 FINDINGS

5.4.1 Main Architectural Design Considerations: Function and Core Typology

In relation to the purpose of the buildings, the analyzed case study sample prominently featured mixed-use and office developments, which constituted about 90% of the total. Residential occupancy accounted for approximately 11% of the total usage, as depicted in Figure 5.4. The substantial representation of mixed-use towers

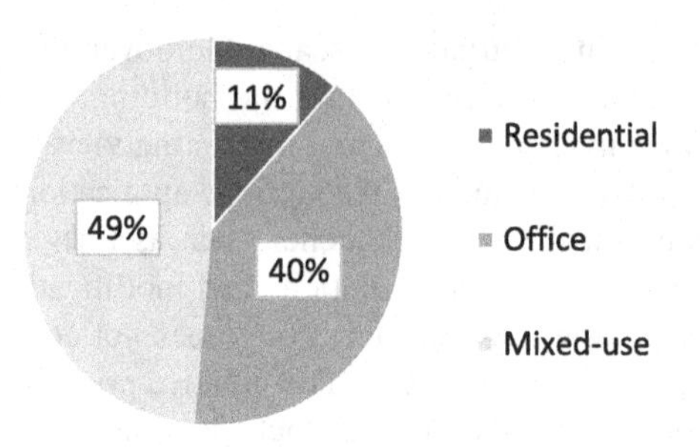

FIGURE 5.4 Freeform supertall towers by function.

can be ascribed to the adoption of vertical communities as a strategic approach to address the challenges posed by population growth and urban sprawl. The preference for mixed-use functionality has further intensified, particularly during market fluctuations, as it allows for enhanced rental profitability and the diversification of the customer base [35]. Conversely, the demand for office towers may be attributed to the clustering of commercial activity zones, propelled by the ongoing process of urbanization in the world.

Among the various design possibilities considered for these structures, the central core strategy was the exclusive choice implemented in freeform supertall towers. The widespread adoption of the central core approach can be attributed to its compact and efficient structural design, which bestows significant benefits in terms of enhancing the overall structural integrity and allowing for streamlined fire escape scenarios, as elucidated by Ilgın et al. [33].

5.4.2 Main Structural Design Considerations: Structural System and Structural Material

Considering the visual representation manifested in Figure 5.5, it becomes apparent that outriggered frame systems have garnered predominant preference, accounting for a majority of more than 70% utilization. In contrast, tube systems constitute a more modest proportion of approximately 11%. The overarching appeal of outriggered frame systems can be attributed to their inherent capacity to confer relative flexibility in the arrangement of the outer columns, thereby endowing architects with enhanced design autonomy when shaping the building envelope, particularly the freedom to articulate the façade design for creating less obstructive view to the outside. This newfound design freedom, in turn, facilitates the exploration of elevated height potentials, making the outriggered frame system an enticing choice in the construction of freeform skyscrapers, as highlighted by Ilgın [31].

As depicted in Figure 5.6, a predominance of composite construction was observed among the analyzed case studies, constituting a dominant share of 69%. Reinforced concrete trailed closely behind, representing approximately 31% of the total proportion. The prevalent adoption of composite construction in freeform supertall towers

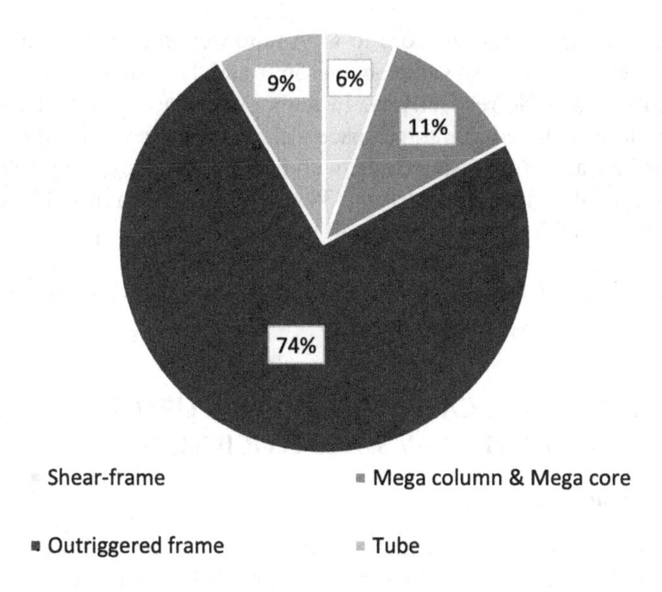

FIGURE 5.5 Freeform supertall towers by the structural system.

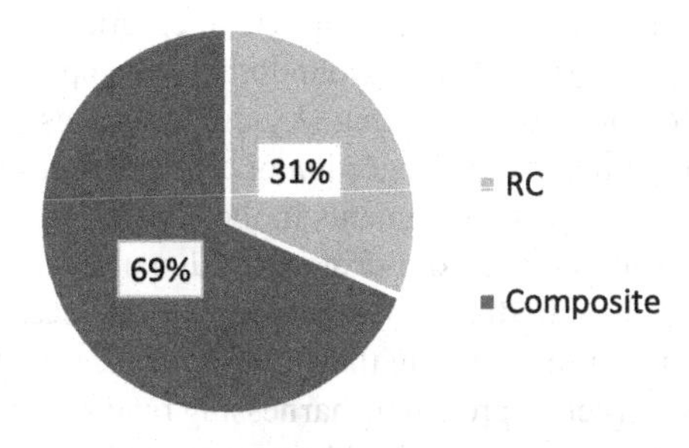

FIGURE 5.6 Freeform supertall towers by the structural material.

can be justified by considering the advantageous combination of steel's high tensile strength and concrete's compressive strength, along with the fire resistance and damping properties inherent to concrete [12]. These factors collectively contribute to the popularity of composite construction as a viable choice for achieving the structural requirements of freeform supertall towers.

5.4.3 Space Efficiency in Freeform Supertall Towers

The threshold ratio for space efficiency in tall towers can be 75%, as proposed by Yeang [36]. In author's study on tall office buildings [14], the average space efficiency and core-to-GFA ratio were determined to be 71% and 26%, respectively, with the lowest values being 63% and 15%, and the highest values reaching 82% and 36%, respectively. Similarly, in author's research on residential towers [15], the average space efficiency and core-to-GFA ratio were found to be 76% and 19%, respectively, with the lowest values being 56% and 11%, and the highest values reaching 84% and 36%, respectively. In author's article focusing on mixed-use supertall buildings [16], the average space efficiency and core-to-GFA ratio were identified as 71% and 26%, respectively, with the lowest values being 55% and 16% and the highest values reaching 84% and 38%, respectively. In this chapter, considering the analysis of 35 freeform supertall towers, the average space efficiency and core-to-GFA ratio were calculated as 72% and 25%, respectively, with the lowest values being 59% and 14% and the highest values reaching 84% and 34%, respectively, as presented in Appendix C.

5.5 REPRESENTATIVE CASE STUDIES EXEMPLIFYING BOTH HIGH AND LOW SPACE EFFICIENCY

5.5.1 Dynamic Tower

Dynamic Tower, also called Rotating Tower, Dynamic Architecture Building, an imposing edifice soaring to a staggering height of 388 m and encompassing an impressive 80 stories, was masterfully designed by David Fisher as a vision project in Dubai. This project is distinguished by its employment of a robust reinforced concrete structure, complemented by the strategic integration of a mega core system. Dynamic Tower is characterized by its groundbreaking approach to design and its strong commitment to sustainability. This skyscraper places a great emphasis on environmental stewardship and the integration of efficient industrial production processes, recognizing them as pivotal elements in the construction of future buildings. The project revolves around three core principles: dynamism, greenery, and industrial production. Firstly, it embodies dynamism by enabling each floor to rotate independently, thereby granting the building the ability to constantly transform its shape. Secondly, it embraces a green approach by harnessing renewable energy sources such as wind and solar power, allowing the building to generate its own energy. Lastly, it employs an industrialized production method, utilizing prefabricated modules that are assembled on-site. Such environmentally friendly buildings, rooted in the principles of sustainability, have the potential to revolutionize city skylines, not only through their dynamic forms but, more significantly, by fostering a harmonious connection with nature [37].

As observed in the low-rise schematic floor plan in Figure 5.7, Dynamic Tower exhibited the highest level of space efficiency, reaching an impressive rate of 84%, and boasted the smallest core-to-GFA ratio among the towers analyzed in the study. This exceptional performance can be attributed to the tower's efficient core design, which focuses on optimizing the organization of service areas and shafts to maintain a compact core space. Furthermore, the integration of a mega core system in the tower played a pivotal role in enhancing its effectiveness. The mega core's ability to bear both vertical and horizontal loads eliminated the need for additional vertical load-bearing elements in the building, further enhancing its efficiency and effectiveness.

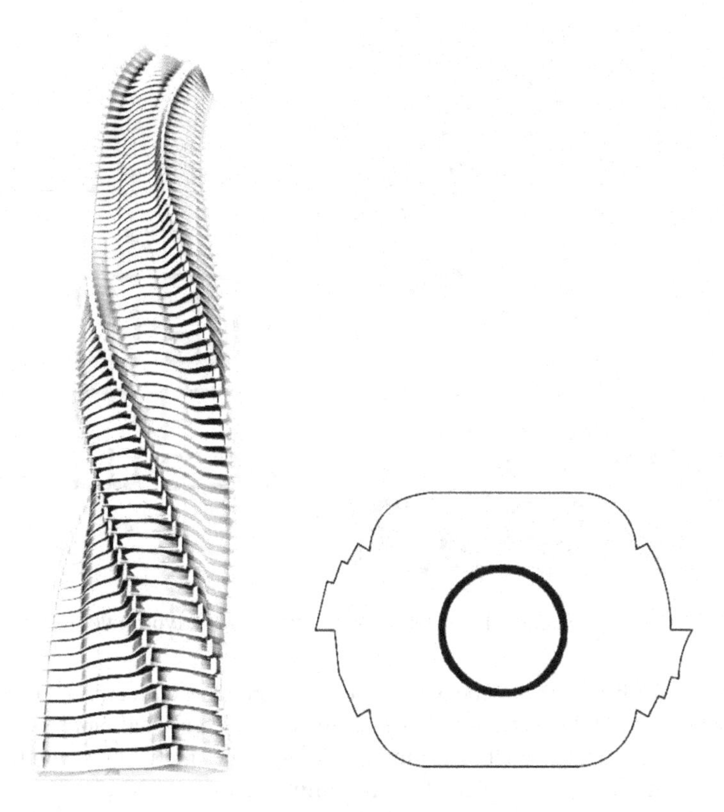

FIGURE 5.7 Dynamic Tower and low-rise schematic floor plan.

5.5.2 MahaNakhon

MahaNakhon, a towering structure rising to a height of 314m and spanning 75 stories, was designed by the Office for Metropolitan Architecture in collaboration with Ole Scheeren, as depicted in Figure 5.8. This impressive architectural feat is characterized by its reinforced concrete construction and features an outriggered frame system. The name "MahaNakhon," derived from the Thai language, conveys the notion of a "great metropolis." In order to overcome the significant challenge posed by the tower's height, the design team employed a pixelation technique, resulting in a globally recognized design approach. This innovative strategy allowed for increased height, facilitated the creation of unique residential layouts with diverse floor plans, enhanced connectivity to the surrounding streetscape, and imbued the structure with an organic form. Moreover, the design incorporated a harmonious blend of indoor and outdoor spaces, fostering a seamless integration with the surrounding environment. Remarkably, as the tower gradually descends toward the ground, its scale undergoes further modulation, giving rise to a multi-level topography comprised of occupiable volumes and cascading indoor and outdoor terraces. This departure from the conventional podium design approach engenders enhanced connections between the building and its immediate surroundings [38].

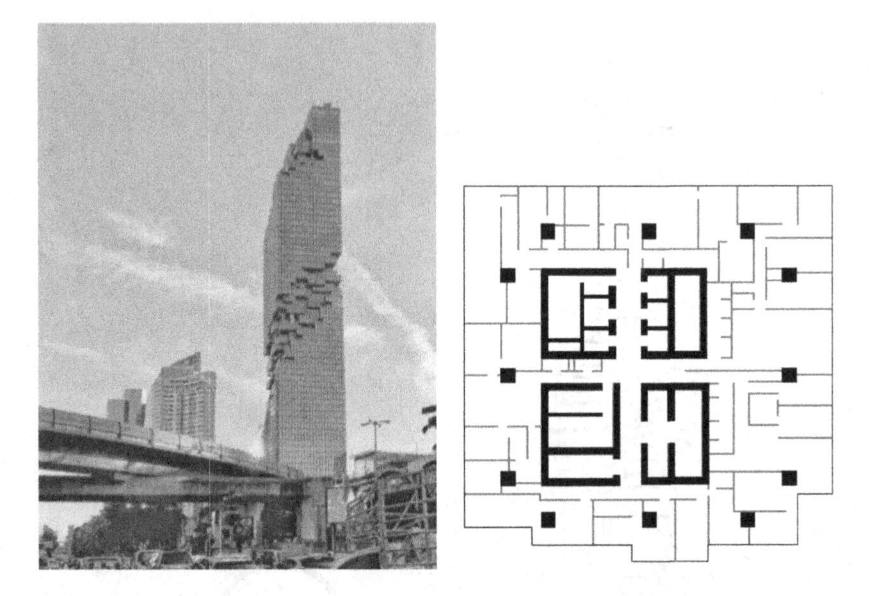

FIGURE 5.8 MahaNakhon, Bangkok, 2016 (Wikipedia) and typical floor plan.

The objective of this endeavor is to establish an architectural prototype that embraces spatial hybrids, deviating from the conventional approach of vertically stacking multi-functional units, through the meticulous implementation of a three-dimensional pixelated form within the urban context. This innovative approach, as an alternative to monotonous repetition, aims to engender a diverse array of indoor and outdoor spaces, setbacks, and balconies throughout the entirety of MahaNakhon. These unique architectural pixels, intricately woven into the mass-breaking form, are thoughtfully designed to cater to the preferences and desires of the occupants, while actively establishing a dynamic connection with the city and its urban fabric. Furthermore, from a structural standpoint, the emphasis lies on maintaining an economical and repetitive form where feasible, rather than focusing solely on architectural embellishment, thus creating opportunities for the deliberate shaping of space [39].

As evident in the low-rise schematic floor plan in Figure 5.9, the MahaNakhon tower demonstrates suboptimal space efficiency, registering a rate of 65%. This can be attributed to the dimensions of its load-bearing elements and the size of the service core area, which deviate from the average norms. These factors impose constraints on the effective utilization of space within the building, resulting in a lower space efficiency compared to the average standards.

5.6 DISCUSSION

The outcomes presented in this chapter exhibited both congruences and distinctions when compared to prior studies, such as the investigations conducted by Oldfield and Doherty [40] and Ilgın [16]. The salient findings derived from this chapter can be succinctly summarized as follows:

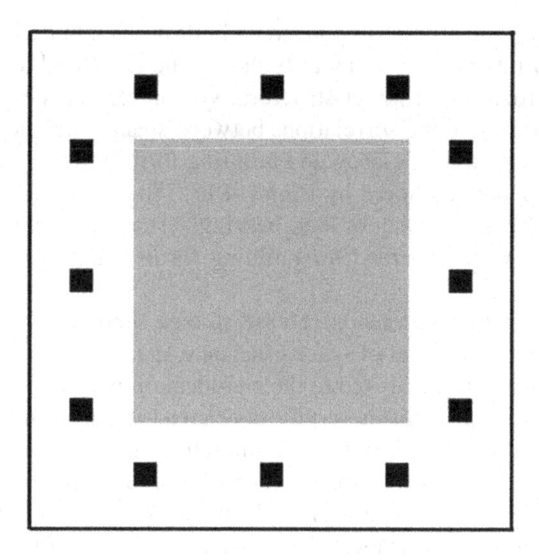

FIGURE 5.9 Low-rise schematic floor plan of MahaNakhon.

The primary findings extracted from this investigation can be outlined as follows:

1. The average space efficiency of the examined towers was 72%, ranging from 59% to 84% across different cases;
2. The average proportion of core area to GFA was 25%, varying from 14% to 34%;
3. In all the examined case studies of freeform structures, a central core typology was consistently utilized, primarily designed to accommodate mixed-use functionalities; and
4. Outriggered frame system emerged as the predominant structural system, while composite materials were the most commonly utilized structural material in the examined cases.

According to Yeang's research [36], which takes into account the 75% space efficiency threshold for tall towers, freeform skyscrapers demonstrate a space efficiency deficit, with an average of 72%. Similarly, studies conducted by the author [14,16] on tall office and mixed-use skyscrapers reported average space efficiencies of 71%, falling below Yeang's threshold. These deficiencies can be primarily attributed to the dimensions of the service core area and structural elements.

Consistent with the research conducted by Oldfield and Doherty [40] and Ilgın [14], the central core design was favored among the sampled buildings. The prevalent purpose of skyscrapers with free form was primarily centered around mixed-use functionalities, corroborated by the findings of Ilgın et al. [33]. Regarding load-bearing systems and structural materials, the outriggered frame systems and composite constructions emerged as extensively employed methods within the case studies, as reported in the investigations conducted by Ilgın [14,16]. As indicated by the studies conducted by Sev and Özgen [22] and

Arslan Kılınç [20], the relationship between building height and space efficiency demonstrated an inverse proportionality due to the increased allocation of core space and the utilization of larger structural system elements at greater heights. The results pertaining to the correlations between space efficiency and structural system, as well as space efficiency and building form, mirrored the conclusions drawn from the papers authored by Ilgın [14,16]. These studies revealed no significant disparity in the impact of load-bearing systems on space efficiency, and similar outcomes were observed for building forms, aligning with the present investigation.

As elucidated in the representative case studies section above, MahaNakhon exemplifies a suboptimal level of space efficiency, manifesting a measure of 65%. This phenomenon can be attributed to the non-ideal dimensions of its load-bearing components and the extent of its service core area, which deviate from the typical mean values. These factors introduce limitations to the efficient utilization of internal space within the tower, thereby contributing to a reduced level of space efficiency relative to established average benchmarks (75%). On the other hand, Dynamic Tower showcased the most elevated degree of space efficiency, achieving a remarkable proportion of over 80%, and featured the most diminutive core-to-GFA ratio among the skyscrapers scrutinized in this chapter. This exceptional achievement can be attributed to the tower's resourceful core design, which strategically optimizes the layout of service zones and vertical conduits to uphold a compact core area. Furthermore, the incorporation of a mega core system within the tower played a pivotal role in augmenting its efficacy. The mega core's capability to bear both vertical and horizontal loads obviated the necessity for supplementary vertical load-bearing elements in the edifice, thereby further amplifying its efficiency and effectiveness. As a result, particularly noteworthy is the discernible impact of service core planning on space efficiency apparent in these representative case studies. Moreover, the choice of an ideal structural system and the fine-tuning of the dimensions of structural components assume prominence in enhancing space efficiency in freeform skyscrapers.

5.7 CONCLUSIONS

The chapter focused on the examination of space efficiency in freeform skyscrapers, encompassing an analysis of 35 selected case study towers. The primary objective was to explore the main design parameters associated with architectural and structural aspects that impact space efficiency. This study provided comprehensive data concerning the general characteristics of the case studies, emphasizing the key design parameters that play a crucial role in determining space efficiency. Furthermore, the interrelationships between these design parameters and their influence on space efficiency were thoroughly investigated and presented in the chapter.

Ensuring high space efficiency in free forms, which are commonly favored by skyscraper architects, particularly for iconic building designs, holds significant importance for sustainability. The architect should take into account structural design and circulation elements to enhance space efficiency, with critical attention to the dimensions of the service core and structural system elements as they play a vital role in achieving this objective.

Future research endeavors could focus on exploring other prevalent tall building forms, such as tapered form designs. By conducting comparative analyses, valuable insights into the correlation between building form and space efficiency can be unveiled.

The author acknowledges the limitations of this chapter. The scope of the data analysis was restricted to a sample of 35 supertall towers with free form, which may not fully encompass the diverse range of skyscrapers present in the region. To enhance the robustness of the findings, future investigations could consider expanding the dataset to include larger case study buildings, thereby providing a more comprehensive and encouraging analysis. Additionally, to broaden the applicability of the research, future studies may also include skyscrapers below the 300-m threshold, enabling the creation of an ample number of subgroups for a more detailed analysis and interpretation.

REFERENCES

1. Karjalainen, M., Ilgın, H.E., Tulonen, L. (2021), "Main design considerations and prospects of contemporary tall timber apartment buildings: Views of key professionals from Finland", *Sustainability*, Vol. 13, No. 12, 6593.
2. Tulonen, L., Karjalainen, M., Ilgın, H.E. (2021), "Tall wooden residential buildings in Finland: What are the key factors for design and implementation?", *Engineered Wood Products for Construction*, 1st ed, IntechOpen. DOI: 10.5772/intechopen.98781.
3. Ilgın, H.E., Karjalainen, M., Pelsmakers, S. (2021), "Finnish architects' attitudes towards multi-storey timber residential buildings", *International Journal of Building Pathology and Adaptation.*
4. Ilgın, H.E., Günel, M.H. (2007), "The role of aerodynamic modifications in the form of tall buildings against wind excitation", *METU Journal of the Faculty of Architecture*, Vol. 24, No. 2, 17–25.
5. Ilgın, H.E. (2023), "A study on interrelations of structural systems and main planning considerations in contemporary supertall buildings", *International Journal of Building Pathology and Adaptation*, Vol. 41, No. 6, 1–25.
6. Ilgın, H.E. (2018), "Potentials and limitations of supertall building structural systems: Guiding for architects", PhD Dissertation, Department of Architecture, Middle East Technical University, Ankara, Turkey.
7. Ilgın, H.E. (2021), "Contemporary trends in supertall building form: Aerodynamic design considerations", *LIVENARCH VII Livable Environments and Architecture 7th International Congress OTHER ARCHITECT/URE(S)*, September 28–30, 2021, Trabzon, Turkey, Volume I, pp. 61–81.
8. Günel, M.H., Ilgın, H.E. (2014), *Yüksek Bina: Taşıyıcı Sistem ve Aerodinamik Form*, METU Faculty of Architecture Press, Ankara, Turkey (in Turkish).
9. Günel, M.H., Ilgin, H.E. (2014), *Tall Buildings: Structural Systems and Aerodynamic Form*, Routledge, London; New York.
10. Moon, K.S. (2011), "Diagrid structures for complex-shaped tall buildings", *Procedia Engineering*, Vol. 14, 1343–1350.

11. Lacidogna, G., Scaramozzino, D., Carpinteri, A. (2020), "Influence of the geometrical shape on the structural behavior of diagrid tall buildings under lateral and torque actions", *Developments in the Built Environment*, Vol. 2, 100009.
12. Ali, M.M., Moon, K.S. (2018), "Advances in structural systems for tall buildings: Emerging developments for contemporary urban giants", *Buildings*, Vol. 8, 104.
13. Kim, H.I., Elnimeiri, M. (2004), "Space Efficiency in Multi-Use Tall Building", Tall Buildings in Historical Cities-Culture and Technology for Sustainable Cities, 10–13.
14. Ilgın, H.E. (2021), "Space efficiency in contemporary supertall office buildings", *Journal of Architectural Engineering*, Vol. 27, No. 3, 4021024.
15. Ilgın, H.E. (2021), "Space efficiency in contemporary supertall residential buildings", *Architecture*, Vol. 1, No. 1, 25–37.
16. Ilgın, H.E. (2023), "A study on space efficiency in contemporary supertall mixed-use buildings", *Journal of Building Engineering*, Vol. 69, 106223.
17. Okbaz, F.T., Sev, A. (2023), "A model for determining the space efficiency in non-orthogonal high rise office buildings", *Journal of the Faculty of Engineering and Architecture of Gazi University*, Vol. 38, No. 1, 113–125
18. Hamid, G.M., Elsawi, M., Yusra, O. (2022), "The impacts of spatial parameters on space efficiency in hybrid villa-apartments in greater Khartoum, Sudan", *Journal Architecture & Planning*, Vol. 34, No. 4, 425–440.
19. Suga, R. (2021), Space efficiency in hotel development, MSc Thesis, "Business administration with a specialization in real estate management and development", MODUL University Vienna, Vienna.
20. Arslan Kılınç, G. (2019), "Improving a model for determining space efficiency of tall office buildings", Ph.D. dissertation, Mimar Sinan Fine Art University, Department of Architecture, Istanbul, Turkey (in Turkish).
21. Nam, H.J., Shim, J.H. (2016), "An analysis of the change in space efficiency based on various tall building corner shapes and lease spans", *Journal of the Architectural Institute of Korea Planning & Design*, Vol. 32, No. 8, 13–20.
22. Sev, A., Özgen, A. (2009), "Space efficiency in high-rise office buildings", *METU Journal of the Faculty of Architecture*, Vol. 26, No. 2, 69–89.
23. Saari, A., Tissari, T., Valkama, E., Seppänen, O. (2006), "The effect of a redesigned floor plan, occupant density and the quality of indoor climate on the cost of space, productivity and sick leave in an office building-A case study", *Building and Environment*, Vol. 41, No. 12, 1961–1972.
24. Saarinen, S., Ilgın, H.E., Karjalainen, M., Hirvilammi, T. (2022), "Individually designed house in Finland: Perspectives of architectural experts and a design case study", *Buildings*, Vol. 12, 2246.
25. Rinne, R., Ilgın, H.E., Karjalainen, M. (2022), "Comparative study on life-cycle assessment and carbon footprint of hybrid, concrete and timber apartment buildings in Finland". *International Journal of Environmental Research and Public Health*, Vol. 19, 774.
26. Ilgın, H.E., Karjalainen, M. (2023), "Freeform supertall buildings", *Civil Engineering and Architecture*, Vol. 11, No. 2, 999–1009.
27. Ilgın, H.E. (2006), "A study on tall buildings and aerodynamic modifications against wind excitation", MSc Thesis, Department of Architecture, Middle East Technical University, Ankara.
28. Gunel, M.H., Ilgın, H.E. (2007), "A proposal for the classification of structural systems of tall buildings", *Building and Environment*, Vol. 42, No. 7, 2667–2675.
29. Taranath, B.S. (2016), *Structural Analysis and Design of Tall Buildings: Steel and Composite Construction*, CRC Press, Taylor & Francis Group, Boca Raton, FL.

30. Ali, M.M., Al-Kodmany, K. (2022), "Structural systems for tall buildings", *Encyclopedia*, Vol. 2, No. 3, 1260–1286.
31. Ilgın, H.E. (2023), "Review on supertall building forms", *Civil Engineering and Architecture*, Vol. 11, No. 3, 1606–1615.
32. Ilgın, H.E., Karjalainen, M., Pelsmakers, S. (2023), "Contemporary tall residential timber buildings: What are the main architectural and structural design considerations?", *International Journal of Building Pathology and Adaptation*, Vol. 41, No. 6, 26–46.
33. Ilgın, H.E., Ay, B.Ö., Gunel, M.H. (2021), "A study on main architectural and structural design considerations of contemporary supertall buildings", *Architectural Science Review*, Vol. 64, No. 3, 212–224.
34. Ilgın, H.E. (2022), "Use of aerodynamically favorable tapered form in contemporary supertall buildings", *Journal of Design for Resilience in Architecture and Planning*, Vol. 3 No. 2, 183–196.
35. Ilgın, H.E. (2023), "Interrelations of slenderness ratio and main design criteria in supertall buildings", *International Journal of Building Pathology and Adaptation*, Vol. 41, No. 6, 139–161.
36. Yeang, K. (2000), *Service Cores: Detail in Building*, Wiley-Academy, London.
37. Fisher, D.H. (2008), "Rotating tower Dubai", *CTBUH 2008 8th World Congress*, Dubai.
38. Beck, K. (2016). "MahaNakhon Thailand's tallest tower", *CTBUH 2016 China Conference*, Trabzon .
39. Ilgın, H.E. (2021), "A search for a new tall building typology: Structural hybrids", *LIVENARCH VII Livable Environments and Architecture 7th International Congress OTHER ARCHITECT/URE(S)*, September 28–30, 2021, Trabzon, Turkey, Volume I, pp. 95–107.
40. Oldfield, P., Doherty, B. (2019), "Offset cores: Trends, drivers and frequency in tall buildings", *CTBUH Journal*, Vol. II, 40–45.

APPENDIX A

Freeform Supertall Buildings

#	Building Name	Country/City	Height (m)	No. of Stories	Completion	
					Date	Function
1	Nakheel Tower	UAE/Dubai	1,000	200	NC	M
2	Merdeka PNB118	Malaysia/Kuala Lumpur	644	118	UC	M
3	CITIC Tower	China/Beijing	528	108	2018	O
4	Evergrande Hefei Center 1	China/Hefei	518	112	OH	M
5	Pentominium Tower	UAE/Dubai	515	122	OH	R
6	Busan Lotte Town Tower	South Korea/Busan	510	107	NC	M
7	TAIPEI 101	Taiwan/Taipei	508	101	2004	O
8	Zifeng Tower	China/Nanjing	450	66	2010	M
9	KK 100	China/Shenzhen	441	98	2011	M
10	Al Hamra Tower	Kuwait/Kuwait City	413	80	2011	O
11	Dynamic Tower	UAE/Dubai	388	80	NC	M
12	PIF Tower	Saudi Arabia/Riyadh	385	72	ATO	O

(Continued)

APPENDIX A (*Continued*)
Freeform Supertall Buildings

#	Building Name	Country/City	Height (m)	No. of Stories	Completion Date	Function
13	Shun Hing Square	China/Shenzhen	384	69	1996	O
14	Burj Mohammed Bin Rashid	UAE/Abu Dhabi	381	88	2014	R
15	Federation Tower	Russia/Moscow	373	93	2016	M
16	St. Regis Chicago	USA/Chicago	362	101	2020	M
17	Almas Tower	UAE/Dubai	360	68	2008	O
18	Greenland Group Suzhou Center	China/Suzhou	358	77	UC	M
19	OKO - Residential Tower	Russia/Moscow	354	90	2015	M
20	Spring City 66	China/Kunming	349	61	2019	O
21	Hengqin International Finance Center	China/Zhuhai	337	69	2020	M
22	Shimao International Plaza	China/Shanghai	333	60	2006	M
23	Sinar Mas Center 1	China/Shanghai	320	65	2017	O
24	Australia 108	Australia/Melbourne	316	100	2020	R
25	MahaNakhon	China/Bangkok	314	79	2016	M
26	Menara TM	Malaysia/Kuala Lumpur	310	55	2001	O
27	Pearl River Tower	China/Guangzhou	309	71	2013	O
28	Fortune Center	China/Guangzhou	309	68	2015	O
29	Jiangxi Nanchang Greenland Central Plaza, Parcel A	China/Nanchang	303	59	2015	O
30	Jiangxi Nanchang Greenland Central Plaza, Parcel B	China/Nanchang	303	59	2015	O
31	Kingdom Centre	Saudi Arabia/Riyadh	302	41	2002	M
32	Capital City Moscow Tower	Russia/Moscow	301	76	2010	R
33	Al Wasl Tower	UAE/Dubai	300	64	UC	M
34	Aspire Tower	Qatar/Doha	300	36	2007	M
35	NBK Tower	Kuwait/Kuwait City	300	61	2019	O

Note on abbreviations: "M" indicates mixed-use; "R" indicates residential use; "O" indicates office use; "UAE" indicates the United Arab Emirates; "UC" indicates Under construction; "NC" indicates Never completed; "OH" indicates On hold.

APPENDIX B
Freeform Supertall Buildings by Core Type, Structural System, and Structural Material

#	Building Name	Core Type	Structural System	Structural Material
1	Nakheel Tower	Central	Mega column	Composite
2	Merdeka PNB118	Central	Outriggered frame	Composite
3	CITIC Tower	Central	Trussed-tube	Composite
4	Evergrande Hefei Center 1	Central	Outriggered frame	Composite
5	Pentominium Tower	Central	Outriggered frame	RC
6	Busan Lotte Town Tower	Central	Outriggered frame	Composite
7	TAIPEI 101	Central	Outriggered frame	Composite
8	Zifeng Tower	Central	Outriggered frame	Composite
9	KK 100	Central	Diagrid-framed-tube	Composite
10	Al Hamra Tower	Central	Shear walled frame	Composite
11	Dynamic Tower	Central	Mega core	RC
12	PIF Tower	Central	Trussed-tube	Composite
13	Shun Hing Square	Central	Outriggered frame	Composite
14	Burj Mohammed Bin Rashid	Central	Outriggered frame	RC
15	Federation Tower	Central	Outriggered frame	Composite
16	St. Regis Chicago	Central	Outriggered frame	RC
17	Almas Tower	Central	Outriggered frame	Composite
18	Greenland Group Suzhou Center	Central	Outriggered frame	Composite
19	OKO - Residential Tower	Central	Outriggered frame	RC
20	Spring City 66	Central	Outriggered frame	Composite
21	Hengqin International Finance Center	Central	Outriggered frame	Composite
22	Shimao International Plaza	Central	Mega column	Composite
23	Sinar Mas Center 1	Central	Outriggered frame	Composite
24	Australia 108	Central	Outriggered frame	RC
25	MahaNakhon	Central	Outriggered frame	RC
26	Menara TM	Central	Outriggered frame	RC
27	Pearl River Tower	Central	Outriggered frame	Composite
28	Fortune Center	Central	Outriggered frame	Composite
29	Jiangxi Nanchang Greenland Central Plaza, Parcel A	Central	Outriggered frame	Composite
30	Jiangxi Nanchang Greenland Central Plaza, Parcel B	Central	Outriggered frame	Composite
31	Kingdom Centre	Central	Shear walled frame	RC
32	Capital City Moscow Tower	Central	Outriggered frame	RC
33	Al Wasl Tower	Central	Outriggered frame	Composite
34	Aspire Tower	Central	Mega core	RC
35	NBK Tower	Central	Outriggered frame	Composite

Note on abbreviation: "RC" indicates reinforced concrete.

APPENDIX C

Freeform Supertall Buildings' Floor Plan, Space Efficiency, and Core/GFA Ratio

Building Name			
Space Efficiency[a]		Core/GFA Ratio[b]	
Nakheel Tower	**Merdeka PNB118**	**CITIC Tower**	**Evergrande Hefei Center 1**
69% 26%	65% 31%	70% 25%	59% 37%
Low-rise floor	Low-rise floor	Ground floor	Typical floor
Pentominium Tower	**Busan Lotte Town Tower**	**TAIPEI 101**	**Zifeng Tower**
73% 21%	70% 27%	72% 25%	71% 28%
Low-rise floor	Low-rise floor	Typical floor	Ground floor
KK 100	**Al Hamra Tower**	**Dynamic Tower**	**PIF Tower**
61% 34%	70% 26%	84% 16%	65% 33%
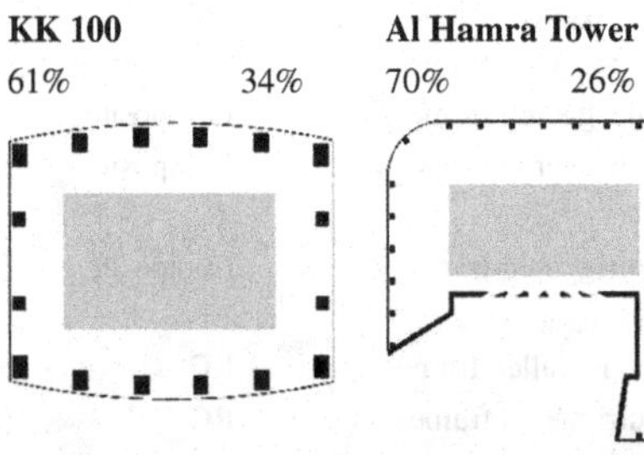		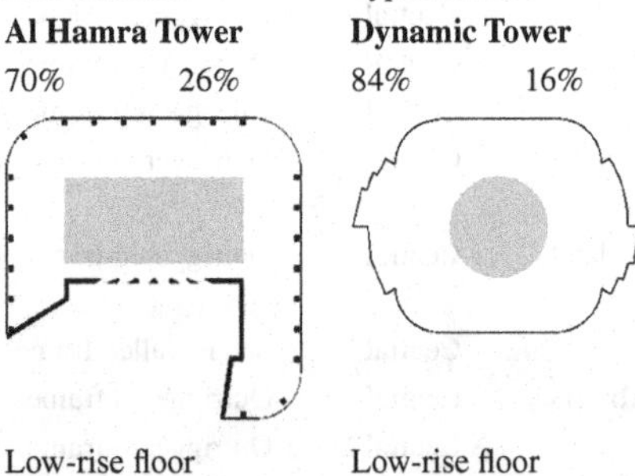	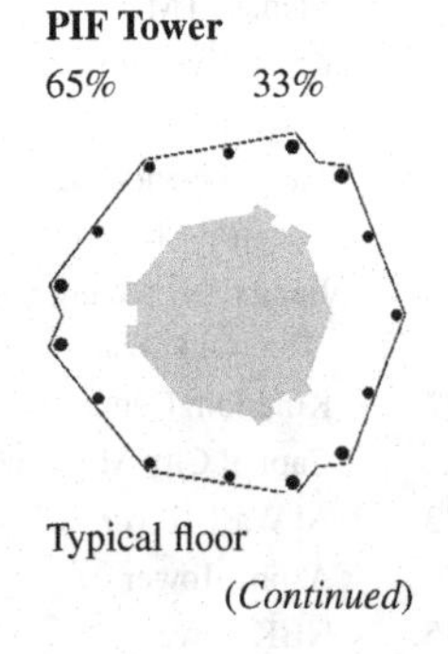
Low-rise floor	Low-rise floor	Low-rise floor	Typical floor

(*Continued*)

APPENDIX C (*Continued*)

Freeform Supertall Buildings' Floor Plan, Space Efficiency, and Core/GFA Ratio

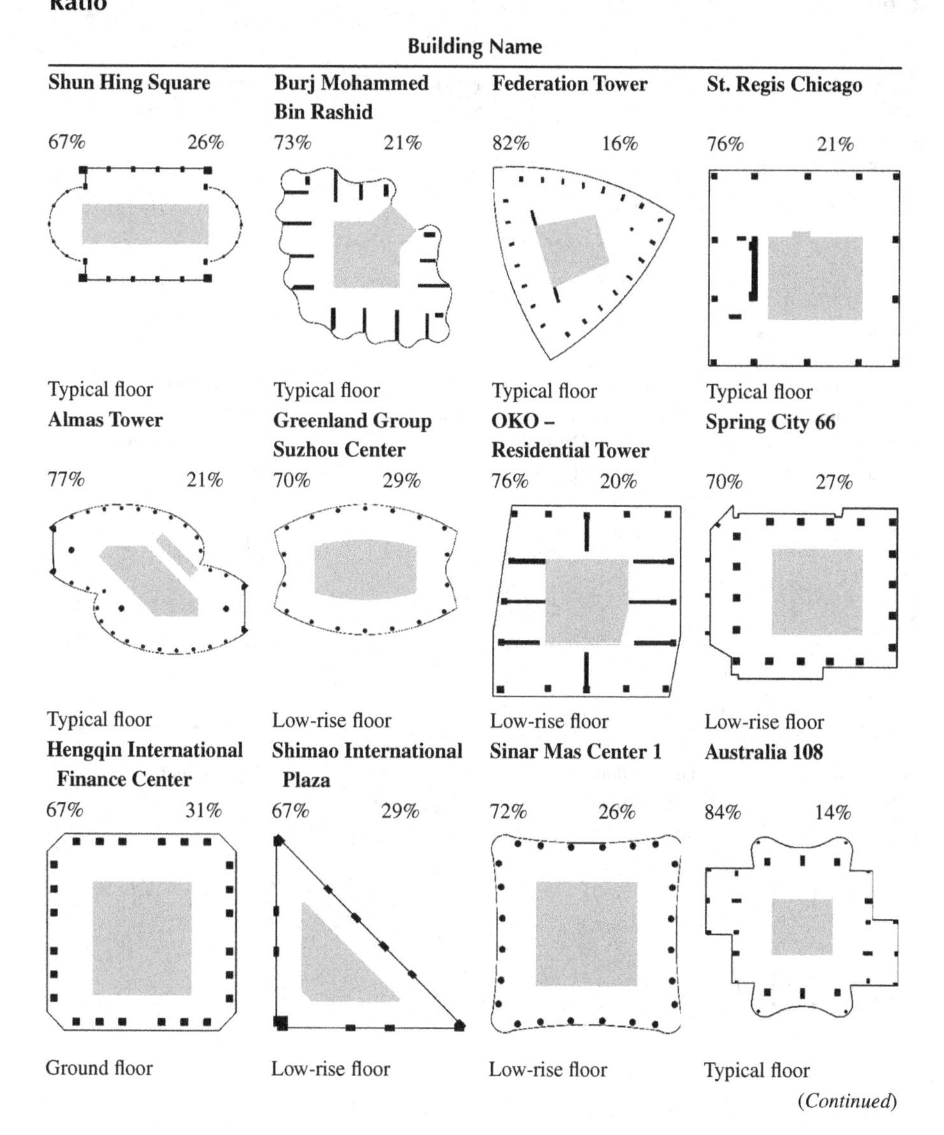

Building Name			
Shun Hing Square	**Burj Mohammed Bin Rashid**	**Federation Tower**	**St. Regis Chicago**
67% 26%	73% 21%	82% 16%	76% 21%
Typical floor	Typical floor	Typical floor	Typical floor
Almas Tower	**Greenland Group Suzhou Center**	**OKO – Residential Tower**	**Spring City 66**
77% 21%	70% 29%	76% 20%	70% 27%
Typical floor	Low-rise floor	Low-rise floor	Low-rise floor
Hengqin International Finance Center	**Shimao International Plaza**	**Sinar Mas Center 1**	**Australia 108**
67% 31%	67% 29%	72% 26%	84% 14%
Ground floor	Low-rise floor	Low-rise floor	Typical floor

(*Continued*)

APPENDIX C (*Continued*)

Freeform Supertall Buildings' Floor Plan, Space Efficiency, and Core/GFA Ratio

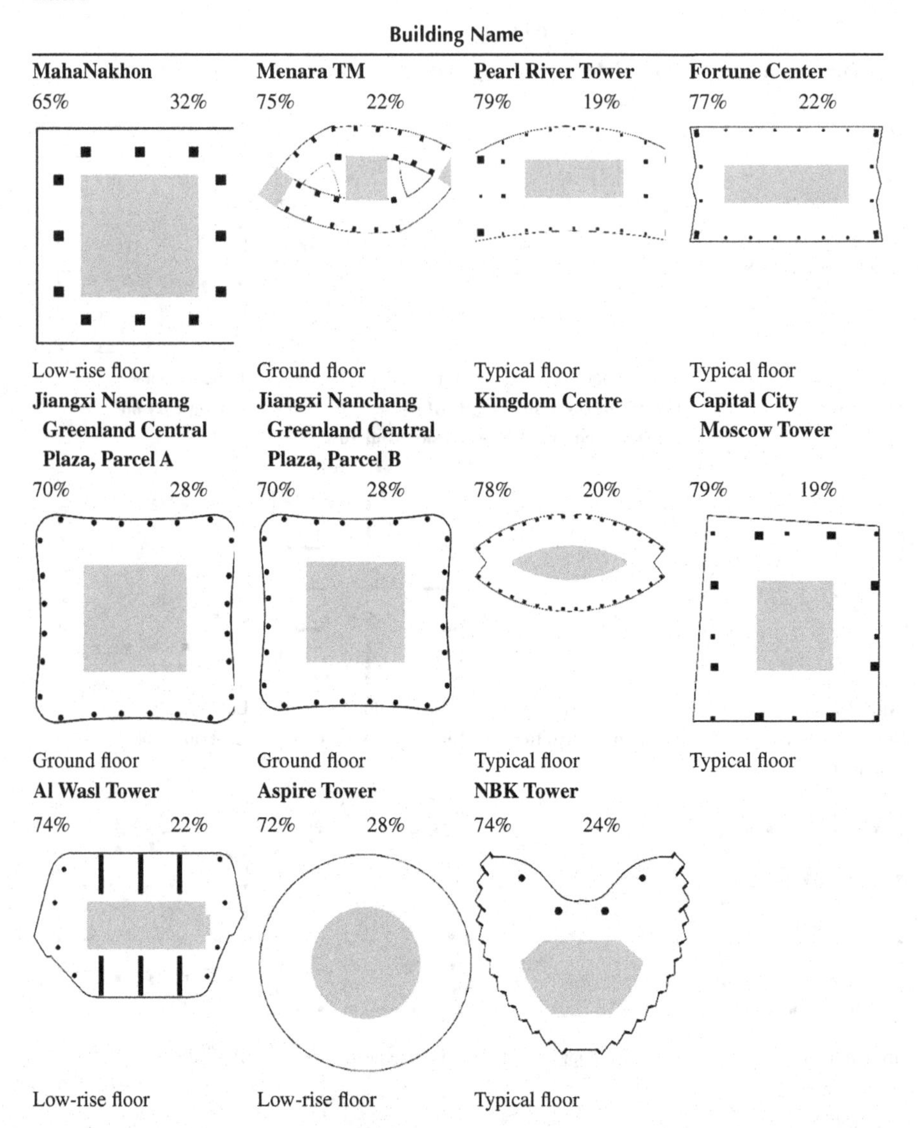

Building Name	Space efficiency[a]	Core/GFA[b]	Floor plan
MahaNakhon	65%	32%	Low-rise floor
Menara TM	75%	22%	Ground floor
Pearl River Tower	79%	19%	Typical floor
Fortune Center	77%	22%	Typical floor
Jiangxi Nanchang Greenland Central Plaza, Parcel A	70%	28%	Ground floor
Jiangxi Nanchang Greenland Central Plaza, Parcel B	70%	28%	Ground floor
Kingdom Centre	78%	20%	Typical floor
Capital City Moscow Tower	79%	19%	Typical floor
Al Wasl Tower	74%	22%	Low-rise floor
Aspire Tower	72%	28%	Low-rise floor
NBK Tower	74%	24%	Typical floor

[a] Space efficiency is calculated as the ratio of net floor area [obtained by subtracting service core (gray area on the floor plan) and structural components from GFA] to GFA.

[b] Core/GFA is calculated as the ratio of the service core (gray area on the floor plan) to GFA.

6 Linking between Renewables Development and Energy Security

A Scoping Review

Olena Chygryn and Liliia Khomenko
Sumy State University

6.1 INTRODUCTION

In recent years, academic community has been significantly interested in renewables development and energy security trends and processes. Immanent and constant fluctuations in global energy policy and global and local energy markets are the huge incentive to the sustainable energy transition, provoking scientific studies and research activity throughout the world. Renewable development has great importance for the future of humanity, since combustible minerals and fossil fuels, which were the basis of energy production at the beginning of the 21st century, have limited reserves that will sooner or later be exhausted. Researchers note that the ideal for human survival would be sustainable development, as a concept in which production and consumption in society are balanced and do not depend on resources that are only temporarily available (Rosokhata et al., 2021). Countries, government institutions, and academic circles face ambitious and complicated challenges for the Sustainable Development Goals 2030 implementation, the requirements to mitigate and adapt to climate changes, and reducing resource extraction, environmental pollution, and greenhouse gas emissions (Dobrowolski et al., 2022; Ziabina, & Navickas, 2022). The environmental problems are a complex and integral part of global sustainable development and the destructive influence of which is greatly spread due to the exhaustion of fossil resources and insufficient and unsystematic nature protection.

Today in most developed countries of the world, renewable energy sources are one of the main development priorities of energy security, and simultaneously the need to eliminate energy instability in countries related to wars, energy crises, and increasing volumes of harmful emissions generated in the process of using traditional energy. Globally, active actions are being taken to reduce pollution in the environment, primarily due to the minimization of greenhouse gasses. Agreement within the United Nations Framework Convention on Climate Change (UNFCCC) regarding the regulation of measures to reduce emissions of carbon dioxide since 2020 was signed in Paris in 2015 and aims to direct efforts to limit temperature rise from pre-industrial indicators level. Unlike the Kyoto Protocol, the Paris Climate Agreement stipulates

DOI: 10.1201/9781003403456-6

that reduction commitments harmful emissions into the atmosphere are assumed by all states, regardless of the degree of their economic development. It is envisaged that such goals also will be achieved through subsidies to developing countries for the implementation of low-carbon technologies, renewable energy development, and reducing production and use of fossil fuels (Chaparro-Banegas et al., 2023; Omer & Capaldo, 2023).

At the same time, the task of energy security strengthening is considered through the possibility of using renewable energy potential. The researchers emphasize the link between renewable energy development and energy security and socioeconomic development of the countries (Wu et al., 2023). However, not all regions and countries have been able to implement widely renewables and, as a result, achieve decreasing CO_2 emissions, increase energy efficiency targets, and support national economies' energy security. Many scholars noticed a significant gap between potential and real energy generation from renewables and assessed the efficiency of renewable energy development on a global and local scale. A potential solution, which will support energy security, will be spreading green technologies and innovations, green transition, smart grid construction, and developing energy storage approaches. The development in communities of the consumption processes greening will be the base for renewables stimulating. To accelerate the transition to renewables, people in the world are joining energy communities, which are designed to provide environmental (reduction of greenhouse gas emissions), economic (achieve energy independence of the community), or social (creation of new jobs) benefits for the community. The involvement of stakeholders in the energy communities strengthens the capacity of the community itself, increases the pace of transition to renewable energy, and the implementation of planned measures for adaptation to climate change. Therefore, new scientific topics and spheres are rapidly appearing, accelerated by relevant changes and approaches in the energy sector. New topics require a detailed investigation, emphasizing the key determinants, drivers, and directions of the transition to sustainable technologies in the energy sector, and identifying the main forces to stimulate green energy breakthrough, which creates the aim of this study.

This paper includes five parts: Introduction, Research Methodology, Literature Review, Results, and Conclusion.

The work examines the main stages of scoping review, the research results of new scientific areas and clusters, scientists and institutions in the subject area "Development of renewable energy and energy security" for 2000–2022.

6.2 RESEARCH METHODOLOGY

Based on information obtained from the Scopus database, publications on renewables development & energy security were collected. The search was conducted by article title, abstract, and keywords. More than 4,000 publications were found in total. Only articles, reviews, and conference papers published in English up to 2022 were selected for filtering.

The article analysed 3,323 publications. The work used a bibliometric approach for the analysis and visual display of articles published in the Scopus database during this period.

The document examined the number of publications by year, subject areas, the most prolific institutions and countries, the finding sponsors, prolific authors, prolific journals, general and annual citations, and keywords.

Excel was used to construct graphs and tables. In addition, VOSviewer software was used to visualize keywords and themes, allowing the generation and identification of relationships between key terms in the research field.

6.3 LITERATURE REVIEW

Based on the information presented in the Scopus database, 498 review articles were found. Most of them are devoted to energy security, energy policy, sustainable development, renewable energies, renewable energy resources, fossil fuels, and climate change. The considered issues in the most cited reviews are listed below.

The most cited review is Ellabban et al. (2014). It examines how renewable energy sources are used, developments to improve their use, and the impact of power electronics and smart grid technologies on renewable energy sources.

A review article (Owusu & Asumadu-Sarkodie, 2016) examines the opportunities associated with renewable energy sources, including energy security, energy access, social and economic development, climate change mitigation, and reducing environmental and health impacts. Challenges hindering the sustainability of renewable energy sources to mitigate the effects of climate change are also addressed. The paper provides recommendations on how to achieve the renewable energy goal to reduce emissions, mitigate climate change, and ensure a clean environment and clean energy for all and future generations.

The publication (Asif & Muneer, 2007) contains an overview of the current and projected energy situation using the examples of China, India, Russia, Great Britain, and the United States. The paper defines the period of depletion of coal, oil, gas, and nuclear fissile material and presents the forecasted demand for energy. The paper also estimates the size of the wind and solar farms that could meet the energy needs of these five countries by 2020.

According to the authors of the paper (Ahmad et al., 2011), of the three generations of biodiesel feedstocks, only microalgae can be sustainably developed in the future. The authors found that microalgae are a more sustainable source of biodiesel in terms of food safety and environmental impact compared to palm oil. The paper also describes the advantages of using microalgae for biodiesel production compared to other available raw materials, primarily palm oil.

The paper (Joselin Herbert et al., 2007) examines wind as a source of renewable energy. The authors considered wind resource assessment models, site selection models, and aerodynamic models; compared different efficiency and reliability assessment models and problems related to wind turbine components (blades, gearbox, generator, and transformer) and wind power system network. The work also considers various methods and loads for the design, control systems, and economics of the wind energy conversion system.

Other most cited reviews connected with the catalytic process for the conversion of syngas to ethanol, opportunities and challenges for biodiesel fuel, big data-driven smart energy management and others.

6.4 RESULTS

According to the information obtained from the Scopus database, 3,323 publications were published during 1980–2022 (Figure 6.1). Of them, 1,811 articles, 498 reviews, and 10,575 conference papers.

As can be seen from Figure 6.1, starting in 2003, the number of publications began to increase, which indicates interest in this topic. At the same time, almost half of the works were published during 2018–2022. This may indicate the relevance of this topic in recent years.

Research on renewables development and energy security is of interest to scientists in various fields of activity (Figure 6.2).

There are studies in energy, engineering, environmental science, computer sciences, social sciences, mathematics, earth and planetary sciences, business and economics, chemistry, physics and astronomy, medicine, veterinary medicine, psychology, etc. This indicates the application of renewables development & energy security achievements in many areas of human life. As can be seen from Figure 6.2, most research are carried out in the field of energy, engineering, and environmental science. These three fields alone account for 60% of all publications. At the same time, a small part of economic studies: business, management, accounting only 2.7%, economics, econometrics, and finance – 2.6%.

Researchers from many organizations and countries of the world are interested in this topic (Table 6.1).

As can be seen from Table 6.1, most research is organized by Chinese institutes, in particular, North China Electric Power University, Tsinghua University, Chinese Academy of Sciences, Ministry of Education China, and State Grid Corporation of China. Also, scientists from China took part in the largest number of studies, which accounted for 15.3% of all publications.

The most prolific institutions list includes two Malaysian organizations (University Malaya and University Technology Malaysia), and Malaysian scientists participated in 4% of the research.

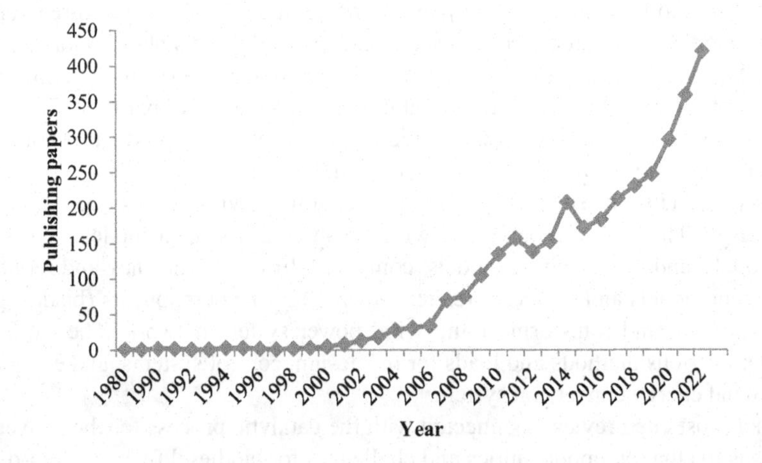

FIGURE 6.1 The dynamic of published articles by year (1980–2022). (Build on the base of Scopus.)

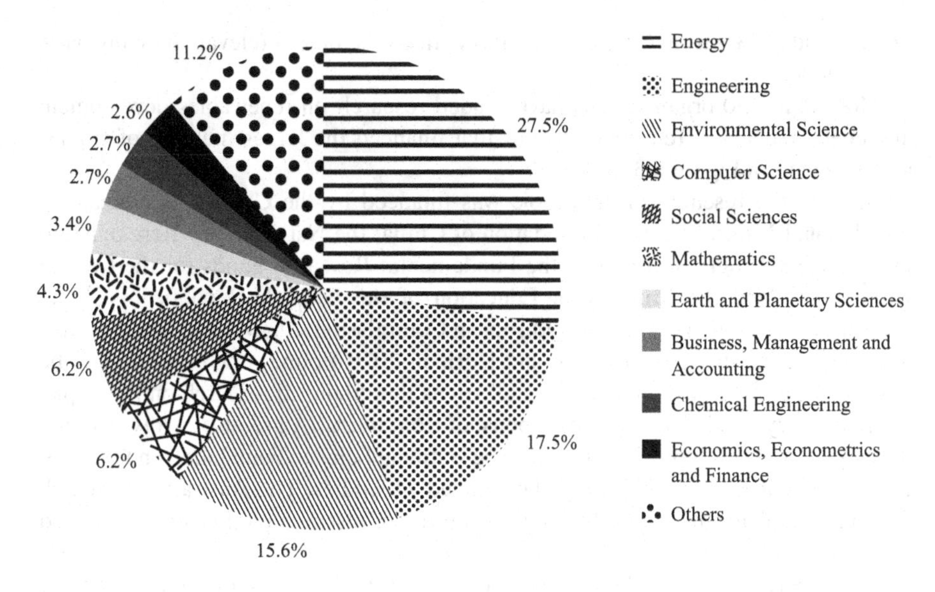

FIGURE 6.2 Subject area. (Build on the base of Scopus.)

TABLE 6.1
The Most Prolific Institutions and Countries (1980–2022)

Institutes	No. of Articles	Country	No. of Articles
North China Electric Power University	44	China	509
Tsinghua University (China)	35	United States	482
Chinese Academy of Sciences	31	United Kingdom	269
Universiti Malaya (Malaysia)	25	India	245
Ministry of Education China	24	Germany	189
Aalborg University (Denmark)	23	Italy	139
State Grid Corporation of China	22	Malaysia	135
Technical University of Denmark	21	Australia	133
University Technology Malaysia	21	Poland	113
The University of Queensland (Australia)	21	Canada	96

Source: Build on the base of Scopus.

Two institutions of Denmark (Aalborg University and Technical University of Denmark) are represented in the most prolific institutions list, but the scientists of this country did not make it to the most prolific countries.

The University of Queensland (Australia) completes the most prolific institutions list, and scientists from Australia published 4% of all works on this topic.

14.5% of the publications involved scientists from the USA, 8% – from the United Kingdom, 7.4% – from India, 5.7% – from Germany, 4.2% – from Italy, 3.4% – from

Poland, and 2.9% –from Canada. This shows that this topic is relevant in many parts of the globe.

More than 150 organizations have funded research on renewables development and energy security. Ten organizations that financed the largest number of studies are presented in Figure 6.3.

Most of the research on this topic was financed by the Chinese organizations the National Natural Science Foundation of China, the National Key Research and Development Program of China, the Fundamental Research Funds for the Central Universities, and the Ministry of Education of the People's Republic of China. Together, they funded 219 studies. European Commission, Engineering and Physical Sciences Research Council (UK), and Horizon 2020 Framework Program can be singled out among the European organizations that financed research on this topic. Together, they funded 111 studies. Sponsors from the USA are also involved in the development of this topic, in particular, the National Science Foundation and the U.S. Department of Energy, which together funded 74 studies. The Australian Research Council completes the list of the largest sponsors of research on this topic, it funded 18 studies.

Thus, there is a group of sponsors who fund research on this topic on an ongoing basis.

More than 150 authors write about renewables development and energy security. The authors with the largest number of publications are presented in Table 6.2.

At least five publications have more than 10 scientists–researchers renewables development and energy security. As can be seen from Table 6.2, Huang has the most publications, with 12 works cited 922 times. Tan has published only six papers, but they have been cited 670 times (an average of 112 times – most of the authors). Seven works by author Lin were cited an average of 55 times. This testifies to the impact of these authors on the chosen subject.

Works on renewables development and energy security have been published in more than 120 journals. Journal titles in which the most works were published are presented in Table 6.3.

Most of the works were published in *Renewable and Sustainable Energy Reviews*, *Energies*, and *Energy Policy*. Together, they published 490 works, which is 14% of the sample. This indicates the impact of these journals. The journals *Renewable*

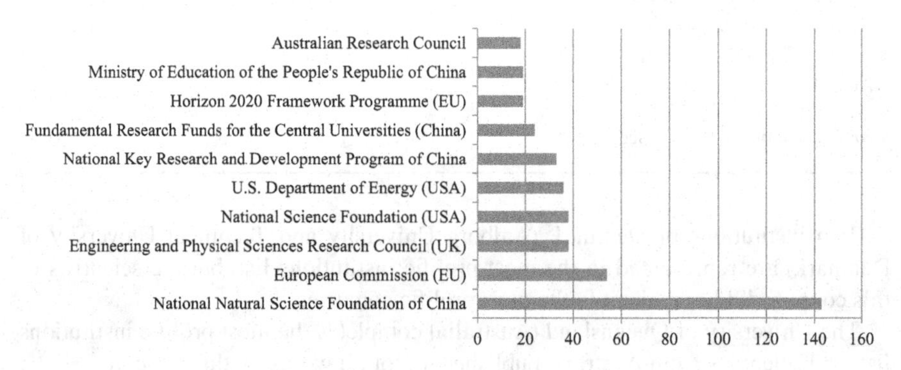

FIGURE 6.3 Funding sponsors (1980–2022). (Build on the basis of Scopus.)

TABLE 6.2
The Most Prolific Authors (1980–2022)

Authors	Publication Papers	Citations	Average Number of Citations Per Publications
Huang	12	922	77
Sovacool	11	449	41
Carlini	10	66	7
Mondal	9	220	24
Lin	7	385	55
Streimikiene	7	230	33
Aslani	7	224	32
Tan	6	670	112
Lund	6	257	43
Mathiesen	6	242	40

Source: Build on the base of Scopus.

TABLE 6.3
The Ten Most Prolific and Cited Journals (1980–2022)

Journal Title	Publication Papers	H-index	Quartiles
Renewable and Sustainable Energy Reviews	194	378	Q1
Energies	142	132	Q2
Energy Policy	124	254	Q1
IOP Conference Series Earth and Environmental Science	65	41	-
Energy	61	232	Q1
Renewable Energy	61	232	Q1
Journal of Cleaner Production	57	268	Q1
Sustainability Switzerland	50	136	Q2
Applied Energy	45	264	Q1
Energy Procedia	39	107	-

Source: Build on the base of Scopus.

and Sustainable Energy Reviews (H-index is 378), *Journal of Cleaner Production* (H-index is 268), and *Applied Energy* (H-index is 264) have the highest H-indexes. Most journals belong to the first quartile (Q1). The list also includes collections of two conferences: *IOP Conference Series Earth and Environmental Science* and *Energy Procedia*. This indicates the influence of these journals.

The most cited publications are listed in Table 6.4. The most cited publications were analysed according to the search criteria of the keywords "Development of renewable energy" and "Energy Security" in the articles' title, introduction, and keywords.

TABLE 6.4
The Ten Most Cited Articles (1980–2022)

Rank	Title	Author(s)/Year	Journal Title	Total Citations	Annual Citation
1	Renewable energy resources: Current status, future prospects and their enabling technology	Ellabban, Abu-Rub & Blaabjerg, 2014	*Renewable and Sustainable Energy Reviews*	1713	190.3
2	A review of renewable energy sources, sustainability issues and climate change mitigation	Owusu & Asumadu-Sarkodie, 2016	*Cogent Engineering*	1304	186.3
3	Hydrogen futures: Toward a sustainable energy system	Dunn, 2002	*International Journal of Hydrogen Energy*	1158	55.1
4	Biofuels sources, biofuel policy, biofuel economy and global biofuel projections	Demirbas, 2008	*Energy Conversion and Management*	881	58.7
5	Virtual synchronous machine	Beck & Hesse, 2007	*2007 9th International Conference on Electrical Power Quality and Utilization*	865	54.1
6	Energy supply, its demand and security issues for developed and emerging economies	Asif & Muneer, 2007	*Renewable and Sustainable Energy Reviews*	815	50.9
7	Microalgae as a sustainable energy source for biodiesel production: A review	Ahmad, Yasin, Derek & Lim, 2011	*Renewable and Sustainable Energy Reviews*	796	66.3
8	A review of wind energy technologies	Joselin Herbert, Iniyan, Sreevalsan & Rajapandian, 2007	*Renewable and Sustainable Energy Reviews*	786	49.1
9	The hydrogen economy in the 21st century: A sustainable development scenario	Barreto, Makihira & Riahi, 2003	*International Journal of Hydrogen Energy*	698	34.9
10	Towards global phosphorus security: A systems framework for phosphorus recovery and reuse options	Cordell, Rosemarin, Schröder & Smit, 2011	*Chemosphere*	640	53.3

Source: Build on the base of Scopus.

As can be seen from Table 6.4, all articles have been cited more than 50 times, indicating significant interest. Most of them were published between 2007 and 2011. The most cited article (Ellabban, Abu-Rub & Blaabjerg, 2014) was published in *Renewable and Sustainable Energy Reviews* in 2014. It has 1,713 citations, on average it is cited 190 times a year. In second place is the article (Owusu & Asumadu-Sarkodie, 2016), which was published in the journal *Cogent Engineering*, which is not presented in Table 6.3. The article (Ahmad, Yasin, Derek & Lim, 2011) is also published in *Renewable and Sustainable Energy Reviews* and has a high Annual citation – 66.3. This also confirms the influence of this publication.

Two articles, which were included in the top 10 most cited publications on the selected topic, were published in the International Journal of Hydrogen Energy (H-index 248, Q1). 20 articles on this topic were published in it. This may indicate the influence of this publication on *Renewable and Sustainable Energy Reviews.*

The analysis of the most cited publications emphasized that the chosen topic is interdisciplinary. Leading works of the authors are presented in the fields of engineering, sustainable development, energy, management, economics, etc.

The represented results of citation analysis provided using the VOSviewer 1.6.13 tools outlined the most significant clusters of scholars' teams, which greatly influence the theory of green energy development (Figure 6.4).

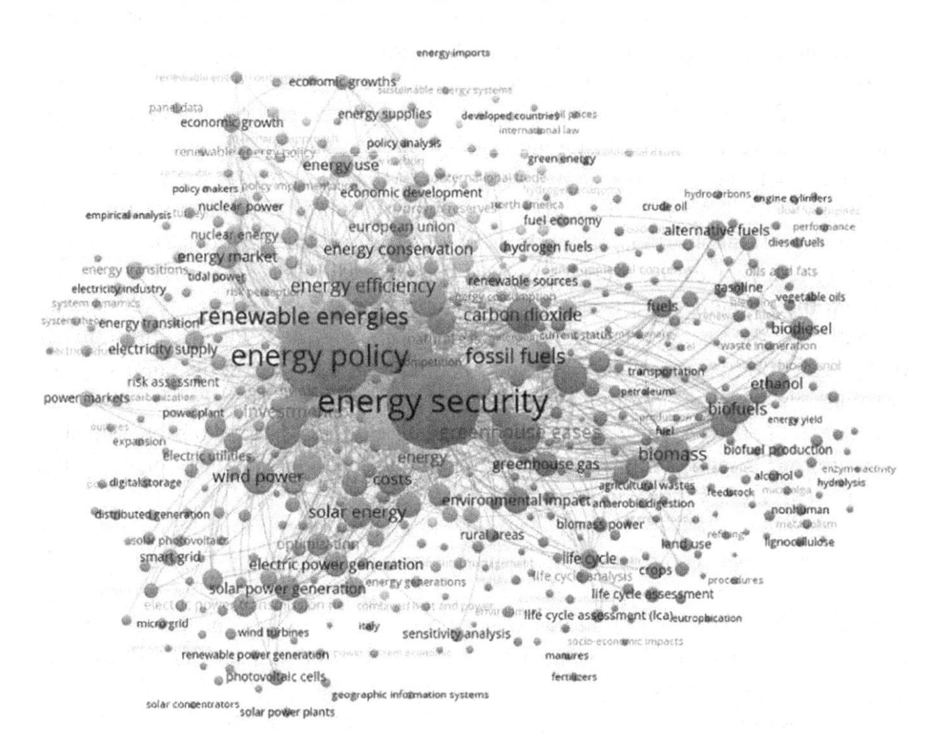

FIGURE 6.4 The co-citation analysis map of the categories "renewables development" and "energy security".

Sources: build on the base of Scopus and VOSviewer tools

As shown in Figure 6.4, a significant number of publications targeted green energy practices, energy policy, and energy security, including sustainable development, renewable energy, development of wind, solar, bioenergy, energy transition, and smart technologies in the energy sector. Simultaneously, a considerable amount of paper is devoted to ensuring economic growth and energy security through greening energy demand, incenting green energy consumption supporting green investments, etc.

The analysis of a co-citation described the most meaningful scholars, as well as the seven clusters of research unions, who had a significant influence on the theory of renewable energy development as a driver of ensuring countries' energy security (Figure 6.3). The biggest blue cluster contains the categories (more than 200 items) related to the energy security processes and their development and acceleration through the implementation of a wide range of alternative sources. It includes such categories as energy, alternative fuel, bioenergy, bioconversion, biogas, biomass energy, conservation of energy resources, energy independence, energy balance, environmental assessment, environmental benefits, environmental impact, global climate changes, global energy, optimization, energy storage, solar energy, and wind power. The next green cluster (more than 130 items) mirrors regulatory processes in the energy sector and includes the next categories: energy market, energy supply, economic growth, investments, energy planning, costs, power generation, carbon emission, efficiency, economic impact, energy infrastructure, energy intensity, energy budget, energy policy, energy sustainability, environmental regulations, mitigation policy, pollution control, renewable energy development, renewable policy, sustainability assessment, sustainable energy systems, etc. The described cluster structure emphasizes the relevance of the processes of energy security supported through the implementation of renewable energy practices and sustainable transformation of the energy sector. The described cluster structure emphasizes the relevance of the processes of energy security supported through the implementation of renewable energy practices and sustainable transformation of the energy sector. Simultaneously, the well-developed yellow cluster (more than 100 items) also outlined the sustainable tendencies in the energy sector development because contains categories of the renewable industry: decarbonization, green energy, ecosystem, electric power generation, electric power transmission, microgrid, photovoltaic systems, smart power grid, solar power plants, wind power, clean energy, emission reduction, etc.

Representation of terminological visualization allows exploring the interconnection of citation between researchers in the relevant scientific fields. Figure 6.5 identifies the 32 clusters of scholars.

Results of co-citation illustrate the cooperation between scholars and their co-citation number. Co-citation analysis showed the interconnection between authors and scientific sources where the papers were published. Note that scholars mostly prefer high-ranked journals as follows: *Energy Policy* (Q1, H-index 254, United Kingdom), *Renewable and Sustainable Energy Reviews* (Q1, H-index 354, United Kingdom), *International Journal of Hydrogen Energy* (Q1, H-index 248, United Kingdom), *Energy for Sustainable Development* (Q1, H-index 76, the Netherlands), and *Energy* (Q1, H-index 232, United Kingdom).

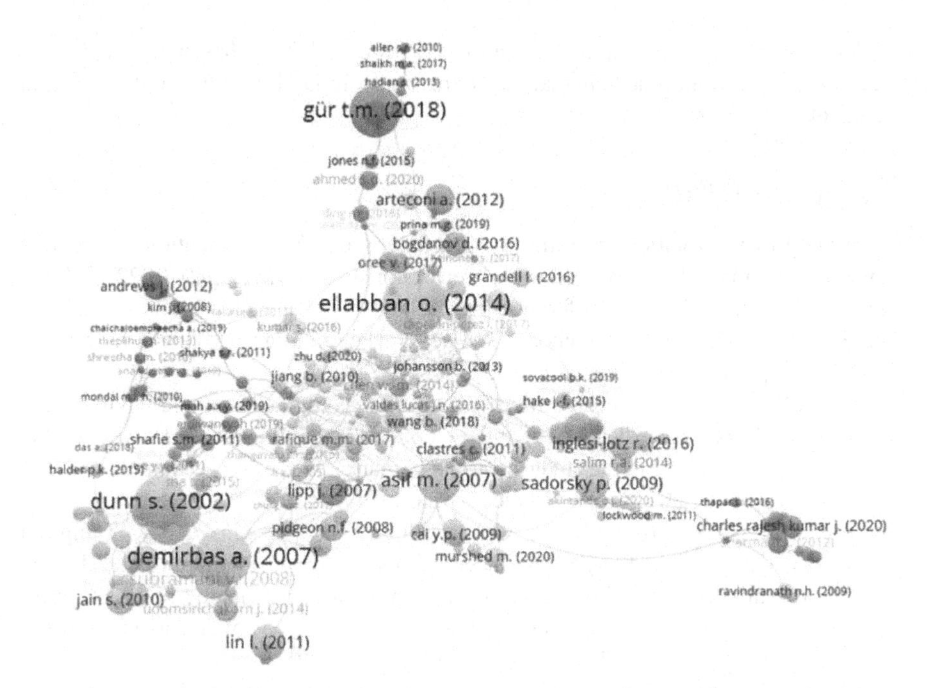

FIGURE 6.5 Results of co-citation analysis by publications in the field of "renewables development" and "energy security". (Build on the base of Scopus and VOSviewer tools.)

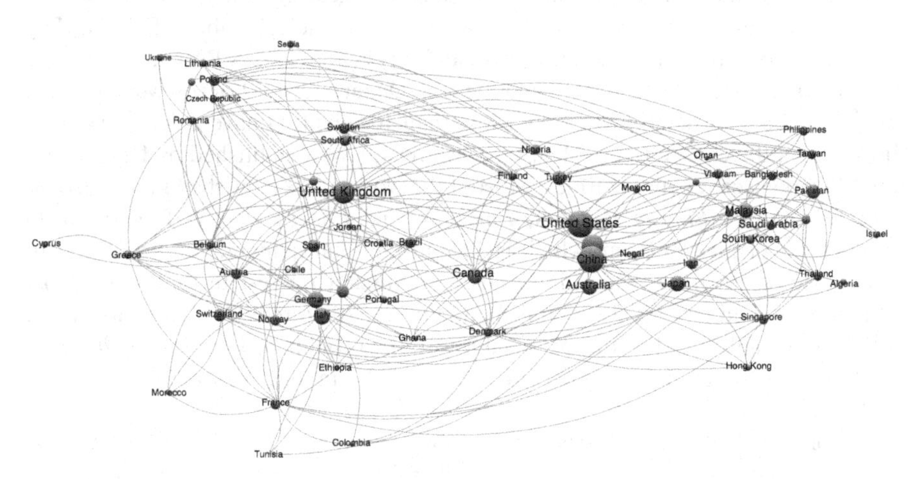

FIGURE 6.6 The countries' scientific collaboration in the field of "renewables development" and "energy security". (Build on the base of Scopus and VOSviewer tools.)

The bibliometric map in Figure 6.6 represents the countries' networks in the sustainable energy security field. The United States, the United Kingdom, Germany, and Malaysia are significant scientific clusters.

The United States was found to have more scientific collaborations with scholars from Turkey, China, Israel, United Emirates, and Taiwan. China has also a huge global research network, and they built scientific collaborations with the United States, Canada, Denmark, Canada, the Netherlands, Ireland, Sweden, and the United Kingdom.

6.5 CONCLUSION

According to information from the Scopus database, 3,323 publications on renewables development & energy security were published during 1980–2022, 54% of them are articles, 32% are conference papers, and 14% are reviews. Since 2003, the number of publications on the chosen subject has begun to grow. Almost half of the works were published during 2018–2022.

Most of the research concerns the fields of energy, engineering, and environmental science. These three fields account for 60% of all publications. However, only 5.3% of them are in economic studies (business, management, and accounting – 2.7%, economics, econometrics, and finance – 2.6%). Generally, they were organized by the Chinese institutes North China Electric Power University, Tsinghua University, Chinese Academy of Sciences, Ministry of Education China, and State Grid Corporation of China. Also, scientists from China participated in 15.3% of all publications. Also, a large number of studies are organized by scientists from the USA, the United Kingdom, India, and Germany.

More than 200 studies on this topic were financed by the Chinese organizations National Natural Science Foundation of China, the National Key Research and Development Program of China, Fundamental Research Funds for the Central Universities, and the Ministry of Education of the People's Republic of China. The European Commission, the Engineering and Physical Sciences Research Council, the National Science Foundation, and the U.S. Department of Energy were also the most frequent sponsors of research. More than 10 scientists have at least five publications on the chosen topic. Most research results were published by Huang, and his 12 works were cited 922 times. Most of the works were published in *Renewable and Sustainable Energy Reviews*, *Energies*, and *Energy Policy*. Together, 490 works were printed in them, which is 14% of the total number. Ten articles were cited at least 50 times, which indicates a significant interest in this topic. Most of the most cited articles were published between 2007 and 2011. The most cited article is Ellabban, Abu-Rub & Blaabjerg (2014), published in *Renewable and Sustainable Energy Reviews* in 2014. It has 1,713 citations, on average it is cited 190 times a year.

Unfortunately, the study does not take into account publications published during 2023, since the year was not completed at the time of writing. Given the rapid increase in the number of publications, they could reflect the most modern research topics. The work also does not include information from other databases, including Web of Science, Google Scholar, ResearchGate, and others due to the difficulty of comparing information obtained from different databases.

The large-scale provision of low-carbon energy should take place through constant interaction with various groups of stakeholders who are able to introduce environmental innovations into the relevant energy sectors of the economy. This also includes supporting the state policy of countries aimed at achieving zero emissions by 2050. The relevant conclusions should strengthen the basis for the formation of a sustainable energy sector, with zero emissions, supporting energy security, following the progress of society in the implementation of the Paris Agreement.

The study of the influence and interconnection of renewable energy development and the energy security of the countries can be carried out not only by involving in the study of other scientometric bases but also in the section of separate nations, international academic institutions, the levels of funding of relevant scientific research, etc.

The research results can be used by scientists interested in renewable energy issues and looking for co-authors, influential journals for publication, sponsors, etc.

REFERENCES

Ahmad, A. L., Yasin, N. H. M., Derek, C. J. C., & Lim, J. K. (2011). Microalgae as a sustainable energy source for biodiesel production: A review. *Renewable and Sustainable Energy Reviews*, 15(1), 584–593. https://doi.org/10.1016/j.rser.2010.09.018.

Asif, M., & Muneer, T. (2007). Energy supply, its demand and security issues for developed and emerging economies. *Renewable and Sustainable Energy Reviews*, 11(7), 1388–1413. https://doi.org/10.1016/j.rser.2005.12.004.

Barreto, L., Makihira, A., & Riahi, K. (2003). The hydrogen economy in the 21st century: a sustainable development scenario. *International Journal of Hydrogen Energy*, 28(3), 267–284. https://doi.org/10.1016/s0360-3199(02)00074-5.

Beck, H.-P., & Hesse, R. (2007). Virtual synchronous machine. In *2007 9th International Conference on Electrical Power Quality and Utilisation*. IEEE. https://doi.org/10.1109/epqu.2007.4424220.

Chaparro-Banegas, N., Mas-Tur, A., & Roig-Tierno, N. (2023). Driving research on eco-innovation systems: Crossing the boundaries of innovation systems. *International Journal of Innovation Studies*, 7(3), 218–229. https://doi.org/10.1016/j.ijis.2023.04.004.

Cordell, D., Rosemarin, A., Schröder, J. J., & Smit, A. L. (2011). Towards global phosphorus security: A systems framework for phosphorus recovery and reuse options. *Chemosphere*, 84(6), 747–758. https://doi.org/10.1016/j.chemosphere.2011.02.032.

Demirbas, A. (2008). Biofuels sources, biofuel policy, biofuel economy and global biofuel projections. *Energy Conversion and Management*, 49(8), 2106–2116.

Dobrowolski, Z., Sulkowski, L., & Adamisin, P (2022). Innovative ecosystem: the role of lean management auditing *Marketing and Management of Innovations*, 3, 9–20. https://doi.org/10.21272/mmi.2022.3-01.

Dunn, S. (2002). Hydrogen futures: Toward a sustainable energy system. *International Journal of Hydrogen Energy*, 27(3), 235–264.

Ellabban, O., Abu-Rub, H., & Blaabjerg, F. (2014). Renewable energy resources: Current status, future prospects and their enabling technology. *Renewable and Sustainable Energy Reviews*, 39, 748–764. https://doi.org/10.1016/j.rser.2014.07.113.

Joselin Herbert, G. M., Iniyan, S., Sreevalsan, E., & Rajapandian, S. (2007). A review of wind energy technologies. *Renewable and Sustainable Energy Reviews*, 11(6), 1117–1145.

Omer, O., & Capaldo, J. (2023). The risks of the wrong climate policy for developing countries: Scenarios for South Africa. *Ecological Economics*, 211. https://doi.org/10.1016/j.ecolecon.2023.107874.

Owusu, P. A., & Asumadu-Sarkodie, S. (2016). A review of renewable energy sources, sustainability issues and climate change mitigation. *Cogent Engineering*, 3(1). https://doi.org/10.1080/23311916.2016.1167990.

Rosokhata, A., Minchenko, M., Khomenko, L., Chygryn, O. (2021). Renewable energy: A bibliometric analysis. In *1st Conference on Traditional and Renewable Energy Sources: Perspectives and Paradigms for the 21st Century (TRESP 2021): Conference Paper*, 11 p. https://doi.org/10.1051/e3sconf/202125003002.

Wu, F., Wang, X., & Liu, T. (2023). Sustainable development goals, natural resources and economic growth: Evidence from China. *Resources Policy*, 83. https://doi.org/10.1016/j.resourpol.2023.103520.

Ziabina, Y., & Navickas, V (2022). Innovations in energy efficiency management: Role of public governance. *Marketing and Management of Innovations*, 4, 218–227. https://doi.org/10.21272/mmi.2022.4-20.

7 Energy Policy
Formulation, Monitoring, and Adaptation for Moving Towards a Low Carbon Economy

Amit R. Patel
The Maharaja Sayajirao University of Baroda

Prabir Sarkar, Harpreet Singh, and Himanshu Tyagi
Indian Institute of Technology Ropar

7.1 INTRODUCTION

The UN has given significant attention to sustainable energy, as seen by the designation of 2012 as the year of sustainable energy for all and the Sustainable Energy for All (SE4All) project, which called for global access to modern energy (Cozzi, n.d.). There are three crucial aspects to sustainable energy. The first is energy security, which is defined as the production and distribution of primary energy derived from domestic and imported fuel sources. The second is energy equity, which is concerned with ensuring that customers have access to a reasonably priced energy supply. Third, environmental sustainability, which deals with using low-carbon energy and energy-efficient sources from both the supply and demand sides (Piere Gadonneix, World Energy Council, 2012). Any energy strategy should aim to achieve these requirements while pursuing the goal of sustainable energy. Although their various energy systems, infrastructure, financial strength, etc., are in varying states of development, the majority of countries around the world are working towards the single goal of establishing sustainable energy systems. Based on its unique mix of natural resources, geographic location, economic development, population growth, consumption habits, environmental responsibilities, and other factors, each nation has a unique understanding of sustainability. The way they go about achieving their sustainable energy goals differs, which has an impact on their energy policy. The responsibility of policymakers is to create a plan that makes an accessible, safe, and environmentally responsible energy system a reality (Piere Gadonneix, World Energy Council, 2012).

DOI: 10.1201/9781003403456-7

A country's energy system may be in a different stage of development than another, and there may be other factors that influence this, so the concept of sustainability will vary accordingly. Using only renewable energy is what Germany considers sustainable [3], whereas the sustainable level of 20% is a convenient target for a developing nation like India. Other nations, such as the United Kingdom (UK), which is ahead of India in terms of development, are concentrating on using 15% of renewable energy by 2020 and becoming carbon-free by 2030 (*UK – Bioenergy Strategy*, 2012). On the other hand for nations like China and India, achieving carbon neutrality is practically unachievable. However, compared to other countries, China had the highest capacity for renewable energy at the end of 2011 (Ei-Ashry, 2012). China has also committed to reaching its carbon dioxide (CO_2) emission peak by 2030 and increasing its reliance on non-emission sources from the current 10% to 20% (*U.S.-China Joint Announcement on Climate Change*, n.d.). By designating low-carbon inclusive growth as the primary tenet of its XIIth five-year plans and committing to cut its Gross Domestic Product (GDP)'s emissions intensity by 20%–25% above 2005 levels by 2020 (Parikh, 2011), India has demonstrated its strong commitment to the fight against global warming.

7.2 RECENT TRENDS IN ENERGY POLICY

7.2.1 Energy Trilimma Index

The World Energy Council has identified three key components that make up the Energy Trilemma Index: environmental sustainability, energy security, and equity in terms of affordability and accessibility (Oliver Wyman & Marsh & McLennan, 2022). For a nation, it is essential to adopt a sustainable energy policy. For this, it must take into account, in the analytical process, all pertinent aspects. As a result, developing an energy policy is a complex and difficult process. These elements typically span a wide range, including technological, infrastructural, financial, and even socio-economic ones. Classifying the elements into three independent heads is a modern strategy for dealing with these intricate processes. The three pillars are (a) energy security, (b) energy equality, and (c) environmental sustainability (Grigoryev & Medzhidova, 2020). The selection of appropriate indicators is the next step in index development. The next step is to give each of these indicators, which are components of each category, the appropriate waits. Finally, the analysis is conducted using a multicriteria assessment framework in this way. There are several case studies like this one can be found in the literature (Zafeiratou & Spataru, 2018). Employing such a framework will assist the assessor in producing valuable data that will aid in evaluating the nation's current situation. Such an annual assessment's findings will aid in understanding a country's performance and ability to address important thrust areas. Furthermore, it is crucial to understand that there is no one specific way to reform the current energy strategy. The goal of reducing the carbon footprint of energy, for instance, can be achieved in a number of different ways. One is by increasing the use of renewable energy sources. Another option is to adopt technologies that increase energy efficiency. A third option is to perform load management and demand control. Another option is to invest in less polluting fossil fuel technologies. A fourth option is to consider nuclear power. The acceptance of such solutions and their potential repercussions are entirely dependent on the current energy usage and economic structure of a country. Since two ecologies cannot be similarly configured or exposed to resources and natural strengths. As a result, the creation of national policy varies from nation to nation and is an art rather than a science. However, when approached scientifically, a multicireria energy trilemma can provide a good indication of its potential impact on a nation's energy policy (Pliousis et al., 2019).

Despite the fact that the trilemma concept is multidimensional, one of its weaknesses is that each of its three pillars by definition has spillover effects and does not operate rigidly within a single compartment. For instance, minimising energy dependence is one way that sustainable variables that increase energy intensity and efficiency automatically support energy security. Similar to this, using energy sources that are renewable when they are natural encourages environmental sustainability by reducing emissions as well as energy security. One such example is the use of solar power in India, where the majority of solar photovoltaic (PV) panels are currently domestically produced. All three aspects will be simultaneously affected by the decision. As a result, it might be challenging to define each dimension precisely and to consider it in isolation from other dimensions (Pliousis et al., 2019).

The following are succinct explanations of the three pillars. When a country has the financial means to purchase energy and makes its energy supply available to the broader population, it is said to have achieved energy security. In other words, it relates to the ability of an individual or a group to pay for the energy that has been gained. The second is sustainability; to be sustainable, we must discover ways to meet our energy needs now without jeopardising our capacity to produce energy in the future. The third is energy equity, which is also known as inequality in access to energy. It refers to the lack of access to energy that is brought on by a variety of barriers, including inadequate infrastructure and, most crucially, excessive energy costs relative to available funds, etc. The traditional trilemma notion, as depicted in Figure 7.1, is improved and extended by the inclusion of the government policy dimension in this. Evaluation of a country's progress towards a low-carbon economy at an affordable price depends on policies that are either mandated or recommended in nature, supported by appropriate legislative framework and market regulation. Such an energy policy could review a nation's position and have significant economic repercussions. With the aid of such policies, sustainable technologies (change) can be implemented, and technological innovation can boost competitiveness.

The successful implementation of the appropriate strategy under each pillar and the realisation of an overall beneficial effect depend on the establishment of the appropriate assessment parameters, which are sometimes referred to as indicators, despite the lack of a well-defined set of precise measures or book-specific markers. However, Table 7.1 provides a preliminary list of suggested indicators.

7.2.2 Energy Transition

Energy transitions have existed throughout history. Past era-defining changes include the switch from using wood to utilising coal in the 19th century, the switch from coal to oil in the 20th century, and the recent switch from oil to gas. The urgency of defending

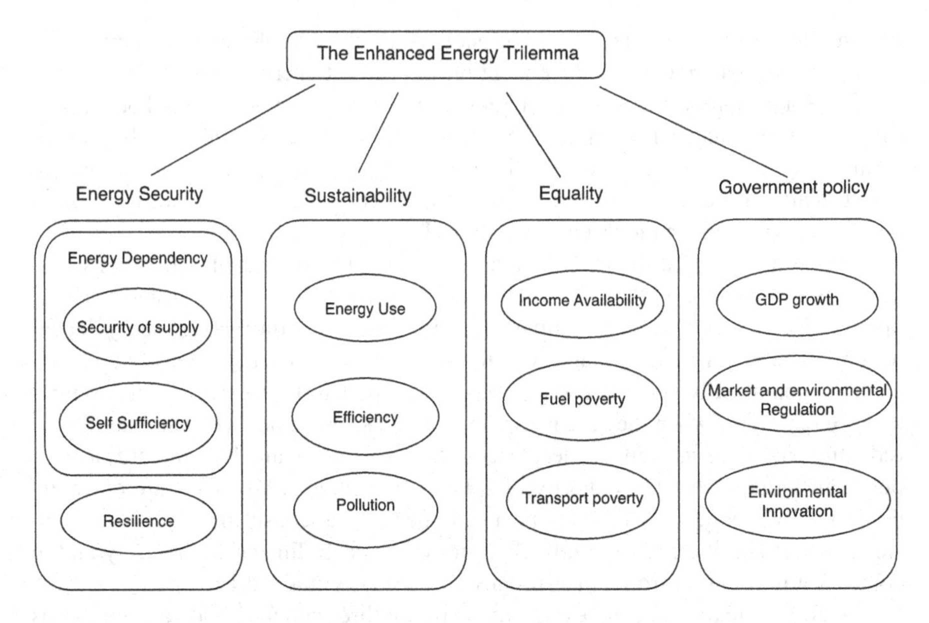

FIGURE 7.1 The enhanced trilemma framework (Pliousis et al., 2019).

TABLE 7.1
Proposed Indicators on Each Pillar of Energy Trilemma (Pliousis et al., 2019)

Pilar	Proposed Indicators
Energy Security	Fraction of energy imports, fraction of fossil fuel energy use, transmission and distribution losses of electricity, per capital energy consumption, change in electricity mix, or change in grid emission.
Sustainability	Energy consumption as a fraction of GDP, emission as a fraction of GDP, emission index of electricity, per capital emission of CO_2, particulate measurement in air pollution.
Equality-Poverty	Fraction of cost of procuring energy to household income, fraction of cost of transportation to household income, fraction of cost of total energy to household income, fraction of disposable income to household income.
Government Policy	GDP adjustment for pollution abatement, environmentally adjusted multifcator productivity growth, enviromental policy stringency, environmentally related R&D expenditure, relative advantage in environment-related technology.

the world from the gravest threat it has ever faced, and of doing it as soon as feasible, sets this shift apart from its forerunners. This motivation has hastened the developments in the energy sector: in just 10 years (2010–2019), the cost of renewable technology has decreased by 60% for on-shore wind power and 80% for solar PV. In the discussion, it is crucial to keep in mind that the energy transition and the ensuing shift to a low-carbon economy are causing changes in energy legislation. There are many challenges that prevent a country from making a smooth energy transition. There are various forms of these barriers. Taking this into consideration, one must keep in mind that any policy must be dynamic in nature and so allow for changes (Heffron et al., 2021). A notable example here in this discussion is of UK. The UK has had success in a transition to low-carbon power generation. The country undertook a massive expansion of offshore wind and a steep decline in coal-fired power generation and the carbon emission from the power generation is well controlled. However, even today a sizeable portion of the UK's power generation still comes from unabated carbon-emitting sources. Other sectors including transport, heating of buildings, and certain industries are major carbon emitters and need a dedicated policy to control the same. It shows that a total reduction in carbon emissions from power generation alone is not enough to achieve green targets (Heffron et al., 2021).

7.2.3 ESG

The primary trends in the global economy are determined by decarbonisation, Environmental, Social, and (corporate) Governance (ESG) principles, modern climate policies of various nations, and sustainable development, which leads to a new perspective on the issue of energy conservation. The ESG (World Energy Trilemma indicator) indicator is used to assess both industrialised nations and businesses in the industrial sector of the economy. According to Ernst & Young Global Limited, a British audit and consulting firm, 97% of investors now base their investment decisions on the ESG index. The index's high values are important for all businesses operating in the industrial sector of the economy. It is quite important to improve environmental indicators that are directly tied to any certain activity (Verstina et al., 2022). From the perspective of the investor, ESG investments are those that take into account each of the ESG pillars and aspects while making investment decisions. This will aid in risk management, sustainable investing, and long-term return generation. The understanding of what ESG covers is made clearer by describing each pillar. The "Environment" pillar deals with concerns like energy and electricity usage, pollution, climate change, global warming, and other elements pertaining to protecting our world. The "Social" indicator gauges how businesses handle the welfare and security of their workers. It also takes into account philanthropic and corporate citizenship facets. The "governance" indicator takes into account board independence, executive compensation, corporate culture, ethics, and accounting structure (Brune et al., 2023).

7.3 FACTORS AFFECTING THE ENERGY POLICY

To interpret a country's energy policy, one must first comprehend the variables that affect or contribute to the policy. Various such elements are briefly explained in this section. Numerous factors, both internal and external to the country as well as region-specific, might have an impact on the design of a policy. Prudent policymaking strikes a balance between the two and meets the sustainability goals. By exposing it to a critical evaluation based on the pertinent already developed or newly generated indicators, the performance of the framed policy is evaluated. These aspects are thoroughly covered in this section.

7.3.1 Internal Factors

The evaluation of internal strengths and weaknesses is the first phase in the policy-making process. It is essential that the decision-makers thoroughly assess the current internal situation or constraints. The commitment to energy security and environmental protection heavily depends on factors like this and can be evaluated by looking at things like energy diversity, energy efficiency, and flexibility within the energy sector, etc. It is important to take note of things like how much free and open markets predominate in the home country. It needs to be followed by an evaluation of the capacity to handle emergencies, such as those involving the oil supply or the use of energy in a way that is environmentally responsible (the "polluter pays principle") (Metz, 2000). Next is the creation of sustainable energy sources, which can be accomplished by the creation of affordable non-fossil energy sources, the clean and efficient use of fossil fuels, increased energy efficiency, cost-effective energy security, strong government commitment, robust Research and Development (R&D) programmes, undistorted energy prices, the accurate calculation of the environmental costs of energy production, and its proper implementation as directed by procedures along with liberal trade and a secured investment framework (*Energy Policies of IEA Countries JAPAN 2008 Review*, 2008).

7.3.2 External Factors

Since policy formulation is an iterative process and the initial steps are frequently taken with implicit aims, these goals inevitably lack complete clarity. Reduced nuclear energy in the energy mix (EM) and an increase in the proportion of bioenergy are a few good examples (Harmelink, 2006). Noticing global cue and demonstrating readiness to act by incorporating it into the policy is a good first step in the policy-making process. A national policy is the turning of a worldwide cause, such as the Kyoto Protocol, into an international initiative. Global issues that all countries must address include petroleum depletion, the energy crisis, greenhouse gas (GHG) emissions, etc. While some of the obligations are discretionary, others are necessary. The initial description of the policy design process typically begins with global initiative and countries' commitment to such initiatives. The global indicators are suitably modified based on how committed a nation is to sustainability and energy security. Canada, for instance, has pledged to reduce its overall GHG emissions by 60% by 2020 and to 70% by 2050. In addition, it has promised to work towards the objective of generating 90% of its electricity from non-carbon emitting sources by 2020 (*Energy Policies of IEA Countries - Canada 2009 Review*, 2010). The actions taken by Ukraine have committed to limiting its GHG emissions to levels that are 20% below those of 1990 and have set a long-term objective of a 50% reduction by 2050 (*Energy Policies beyond IEA Countries - Ukraine 2012*, 2012). However, Sweden has committed to achieving net zero GHG emissions by 2050 as part of the Climate Roadmap 2050. The scenario incorporates high industrial and transportation efficiency as well as the use of carbon capture and storage (CCS) and biogenic CCS technology (IEA, n.d.). The New Zealand government established an emissions reduction goal of 20% by 2020 and 50% by 2050 (*Energy Policies of IEA Countries - New Zealand 2010 Review*, 2010). However, a country's capacity to phase out fossil fuels for transportation, energy with no net GHG emissions, and heating sources is constrained, necessitating difficult measures to achieve the same (IEA, n.d.). In certain cases, a previous study or a report from an expert group's interim committee many times serves as the starting point for a policy design. For instance, low-carbon strategies for inclusive growth are a useful starting point when determining India's policy for the upcoming five-year plan (Parikh, 2011). In doing so, the policy must adequately reflect consideration of the international obligation.

7.3.3 Developing Right Indicators

The performance of the energy policy must be evaluated, and corrective measures must be made on a regular basis. Since indicators will be a numerical window to evaluate performance, developing accurate indicators is one of the most difficult tasks. Indicators play a crucial role in improving and comprehending the policy planning process. A single indicator will not always be effective because each country has varied energy resources, growth priorities, and environmental commitments. The establishment of appropriate indicators is crucial for accurate result interpretation and assessment. Energy intensity indicators serve as a guide for policymakers when drafting energy efficiency laws and regulations, in addition to aiding in monitoring (Nanduri, 1998; Metz, 2000). The Energy Development Index (EDI) is used to gauge the development of the energy system. The country's success in supplying fuels and providing energy services is gauged by the EDI, which also serves to quantify the contribution of energy efficiency to human development (*Measuring Progress towards Energy for All*, 2012).

The characteristics of the indicators must also be well defined; for example, when using the term "energy," it must include both power and fuel for cooking facilities. Further, the energy must be assessed holistically, taking into account its availability, price affordability, supply sufficiency, simplicity of use, and convenience in terms of time and distance. It is also crucial to consider the supply's quality, including the voltage level and reliability in the case of electricity. Finally, all of these factors work together to support the achievement of institutional and policy goals (*Measuring Progress towards Energy for All*, 2012). The success of such energy assessment exercises depends on adequate, consistent, trustworthy, and robust data, yet these can occasionally be quite scarce (*Measuring Progress towards Energy for All*, 2012). If these benchmark data do not already exist, developing them is the first and most crucial step. It is imperative to focus on implicit aspects that are difficult to quantify in this difficult undertaking, such as the inconsistent electrical supply that causes power outages and an unmet demand for electricity. Although this is mostly handled by the use of quite pricey diesel generators, etc. (*Measuring Progress towards Energy for All*, 2012), in the case of Iraq, where daily power outages are very common, Iraq needs 70% more power to make up for this, and as of right now, there is no systematic data for replenishment of alternate power arrangements (*Iraq Energy Outlook - World Energy Outlook Special Report*, 2012). Indicators must be customised to meet the needs and goals of a policy-making process. The importance of renewables should be raised along with evolving policies that largely forgo the use of fossil fuels in order to attain "low carbon" EDI (*Measuring Progress towards Energy for All*, 2012).

7.4 PROCESS OF POLICY FORMATION

Even if there is not a set formula for creating an energy policy for a certain nation, this article discusses a basic approach that should be taken in this regard. Policy formation steps given hereunder are a tentative path followed by the policymakers showing various parameters and its interactions and the general flow of the process. The basic process comprises the following steps. Note that these steps remain identical to almost any policymaking, although the outcome of the policy may be different and unique in each case. Figure 7.2 shows a basic design process (Ei-Ashry, 2012).

The process starts with initial characterisation, setting goals and intended targets, defining end-use areas and targeted technologies as well as identifying the barriers that must be overcome, determining stakeholders, choosing executives to carry out the policy, allocating available working capital, etc. The active period for policy instrument, the policy context, and deciding stakeholders are all steps in the process. The next step is to create a development strategy for a policy area, design and implement the first policy tool, document all presumptions, map cause-and-effect relationships, and assess the impact of the policy on other policies. Create pertinent indicators or change current indicators to reflect elements that may cause success or failure. The following step, which involves verification, feedback, and monitoring, is an iterative process. In essence, this is a highly important phase, as it is during this time that the policy is shaped based on the data gathered. The next step is to gather data, examine every angle, and determine the most important signs and causes. The final stage is to calculate the overall effect and target success.

7.5 ISSUES AND ITS INFLUENCE ON POLICYMAKING

There are several factors affecting the selection of a particular option while formulating an energy policy. These factors are many times overlapping in effect. Additionally as discussed there are several indicators developed to monitor the change or study the

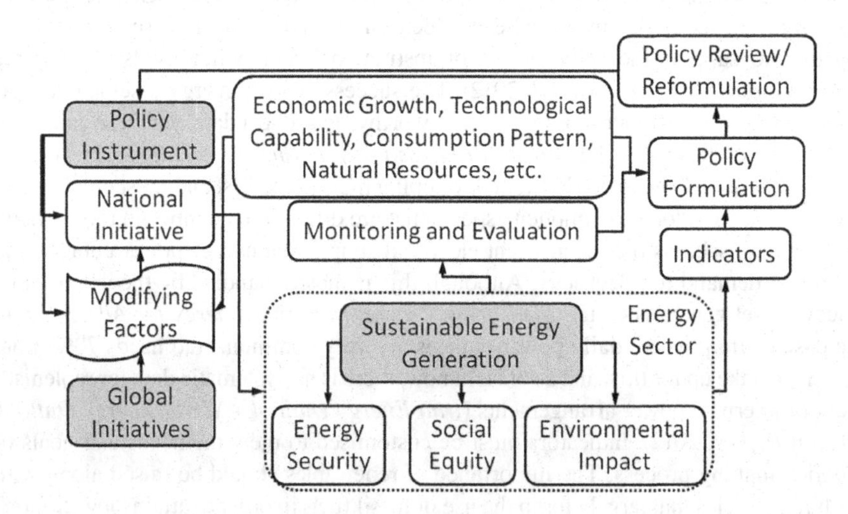

FIGURE 7.2 Policy design process (Adapted from Ei-Ashry, 2012.)

performance of an energy policy. It is critical to comprehend the various influences of these factors on energy policy. As suggested in the literature, these components can be logically categorised under a number of topics, including those that deal with issues with the environment, energy security, international agreements, the development of technology, restrictions due to infrastructure, economic and financial concerns, subsidies or freebies, public preference, built-in capabilities, Grey zones which are relatively unaddressed, etc. (Heffron & Talus, 2016).

7.5.1 Environment Concern

Under this use of renewable will be the first option to explore. The energy policy clearly places a high premium on renewable energy. This is brought on by both a decrease in fossil fuel imports and owing to obligations to other nations. It is clear that industrialised countries are best equipped for renewable energy options due to their high capital costs. This has caused the industrialised world to converge quickly on sustainability objectives. Sweden, for example, reduced its carbon intensity from its 1990 level by 40.5% in 2010. Its renewable energy made up 33% of the total in 1990, but by 2011 it had climbed to 48%. This is primarily due to Sweden's high proportion of nuclear energy and renewable energy sources in the country's EM (IEA, n.d.). Sweden has taken more steps to reduce its reliance on oil. Oil is by far the most important source of CO_2 emissions when seen in terms of fuel, accounting for 68.9% of CO_2 emissions in 2010, while coal provided 16.9%. The country's energy-related CO_2 emissions in the transport sector, which accounted for 45.3% of all emissions, were the highest in 2010 (IEA, n.d.). By 2030, it wants to phase out the use of fossil fuels for heating and develop a fleet of non-fossil fuel vehicles. Due to the use of solid biofuels for heating and the widespread use of heat pumps, Sweden's heating and cooling sector (including industrial) contributes a high 65% of renewable energy (IEA, n.d.). Up to 44% of the nation's total electricity is produced from hydroelectricity (IEA, n.d.). In addition to hydropower and existing nuclear power, Sweden has committed to developing wind energy as the third pillar of its electricity supply, with the dual goals of enhancing energy security and lowering vulnerability. The goal of Sweden's 2050 vision is to have an energy supply that is sustainable, resource-efficient, and emits no net GHGs (IEA, n.d.). Those nations who already have a significant amount of renewable energy in their EM can use it to gradually phase out thermal heating. For instance, hydropower accounts for up to 58% of Canada's total electricity production, including 61% renewable energy. In 2008, 27% of export revenues, primarily to the US, came from the energy sector (*Energy Policies of IEA Countries - Canada 2009 Review*, 2010). In future once fossil fuel-based subsidies are eliminated, solar water heating replacing the fossil fuel emission can be another practical alternative for making buildings more environmentally friendly. In the same way, Ghana has abundant access to renewable energy sources, including biomass, solar, and wind power. Ghana's contribution is renewable to the tune of (76.8%) (REP, 2013). Ghana's energy security would be taken care of together with the detrimental impacts of climate change if renewable energy sources were developed and used, as well as waste-to-energy resources used on a commercial basis. Additionally, the national sanitation programme can benefit greatly from the waste-to-energy resources (*Country Energy Profile: Ghana - Clean Energy Information Portal - Reegle*, 2013). The "energy economy" plan, which the Ghanaian government established, aims to boost the production of renewable energy with a particular emphasis on electrifying rural areas by 2020 (*Country Energy Profile: Ghana - Clean Energy Information Portal - Reegle*, 2013). The legislature is on the approach of passing the nation's Renewable Energy Bill, which includes a feed-in-tariff provision to make the investment attractive for independent power supplies (*Country Energy Profile: Ghana - Clean Energy Information Portal - Reegle*, 2013).

Some nations have access to a specific renewable energy resource, which helps them move up the global energy trilemma index. The wind energy resources in New Zealand are among the most effective in the world; at some locations, installed wind farms operate at net capacity factors between 45% and 50%, making them commercially viable enough to function without public assistance. The plant load factor for wind park in case of India is very low, often in the single digits (Samantaray & Patnaik, 2010). Another area that shows promise is geothermal energy, and New Zealand is a global pioneer in this field. The government is actively supporting the industry by conducting feasibility studies for its development and reviewing relevant technology at the policy level. Thirdly, hydel electricity makes a significant contribution by producing over 55% of its capacity and is still growing as planned capacity increases are under way. Even bioenergy resources have promise, but they have not been fully utilised yet (*Energy Policies of IEA Countries - New Zealand 2010 Review*, 2010). All things considered, New Zealand has abundant renewable resources and therefore it has a solid policy foundation.

On the other hand, there is relatively little room for integrating renewable energy in some nations. Iraq serves as a typical illustration of this group. The wind speed in Iraq is quite low, the biomass resource is minimal, and the hydropower potential in Iraq is very limited (*Iraq Energy Outlook - World Energy Outlook Special Report*, 2012). However, widespread access to fossil fuels requires the country to remain in the category of underperformer on the global energy trilemma index. Another similar case is Japan. Japan's ability to develop renewable energy technology is constrained by its resources. Despite this, they overcame this limitation. Japan is the world leader in the production of solar panels and has the second-largest installed solar PV capacity in the world (*Energy Policies of IEA Countries JAPAN 2008 Review*, 2008). This resulted in around very significant component of Japan's electricity being produced using renewable energy sources. The majority of renewable electricity is produced by hydro, followed by biomass, geothermal, and wind and solar PV (*Energy Policies of IEA Countries JAPAN 2008 Review*, 2008). Another example is of Ireland. Ireland has a proactive energy policy with a legally enforceable goal to use 40% renewable energy by 2020 and cut GHG emissions by 20% by that time (*Energy Policies of IEA Countries - Ireland 2012*, 2012). Ireland has some of the best wind and ocean resources in all of Europe thanks to its location at the edge of the Atlantic Ocean, and the government is promoting the production of power from renewable sources (*Energy Policies of IEA Countries - Ireland 2012*, 2012).

Another option to explore renewable energy is by use of biofuel. Although there are several experimental biofuels' projects running all throughout India, there is no large-scale biofuel usage in the transportation sector. Utilising biofuels can significantly reduce the usage of fossil fuels, particularly diesel fuel, for transportation and rural power generation. 20% of the world's CO_2 emissions come from transportation (Weiss et al., 2000), and there are very few options for reducing these emissions, such as solar PV, biofuel, fuel cells, and hydrogen as a fuel. Biofuel has lately come to light as one of the most promising methods for lowering GHG. For a nation like India with a sizable agrarian base, the biofuel alternative works great. Governments all across the world support the use of biofuel in a variety of ways, including; by providing tax breaks, capping the percentage of conventional gasoline, and providing specific vehicles that can only run on biofuel. The Renewable Fuel Standard mandates that the US use 36 billion gallons of biofuels by 2022 (Kesan & Ohyama, 2011). States with a 10% mix requirement include Minnesota, while Kansas has a tax deduction. Tax incentives exist in addition to Alaska, Illinois, Kansas, Michigan, Minnesota, and Oregon (Kesan & Ohyama, 2011). In Illinois, there is an 80% automobile refund offered. Credits or subsidies for alternative fuel vehicles, which promoted the use of ethanol in Brazil, are another example. There is a push to adopt alternative fuel for school buses (Kesan & Ohyama, 2011). Additionally, it is more crucial to promote cellulosic biofuel feedstocks like switchgrass, Miscanthus, maize stover, and wheat straw in order to limit the use of food as fuel than it is to forbid the production of ethanol from grains (Brown, 1980).

It is well recognised that the energy industry plays a significant role in the environmental problem. Examples include the contamination of water, soil, and gas during the production of oil and gas, as well as the air pollution caused by the use of low-quality leaded fuels by an ageing fleet of vehicles. The extremely high emission demands prompt attention from policymakers. The "Cool Earth 50" programme was introduced by the Japanese government in May 2007. It calls for a "global consensus" on the aim and proposes a global target to cut GHG emissions in half by 2050 (*Energy Policies of IEA Countries JAPAN 2008 Review*, 2008). It is interesting to note that Iraq's energy-related emissions in 2010 totalled 11 billion cubic metres of gas flared and around 100 million tonnes of CO_2 (*Iraq Energy Outlook - World Energy Outlook Special Report*, 2012). Due to the reduced population in that country, per capita emissions are relatively high. Emissions are therefore very important for that nation. Australia ranks among the nations with the highest per capita emission rates despite having low overall emissions (Garnaut, 2011). In 2000, the fourth-highest emission rate per capita was 25.6 tonnes CO_2 equivalent (Data, 2005). Despite producing only 2% of the world's total GHG emissions, Canada ranks among the top emitters per person globally due to its small population (*Energy Policies of IEA Countries - Canada 2009 Review*, 2010). It emitted 22.1 tonnes of CO_2 equivalent with a seventh-place ranking according to statistics from the year 2000 (Data, 2005). Poor energy conversion practices also contribute to high GHG emissions. India's rural areas provide a suitable case study; there, 86% of people (REP, 2013) predominantly rely on solid fuels for cooking, which is usually always an inefficient technique. Coal is the most carbon-intensive fossil fuel among those that are often used, followed by oil and gas (Garnaut, 2011). Given how much coal dominates its energy supply, India is predicted to see the greatest rate of increase in carbon intensity (Piere Gadonneix, World Energy Council, 2012). For the same reasons, China's carbon intensity dramatically increased from 2005 to 2009; however, it is anticipated that they would plateau at such levels for the ensuing 20 years (Garnaut, 2011). This necessitates that the emission be given a more prominent role in energy strategy.

7.5.2 Technology Related Constrain

There are certain options for policy formation, but they are limited by technological limitations. This indicates that even while the advantages of such actions are broadly acknowledged, they are not technologically ingested locally. CCS, or carbon capture and storage, is one such measure. Currently, CCS is one such significant technology that can be used to minimise CO_2 emissions in the power production industry. The lack of cost-competitive technology is preventing the European Union (EU) from supporting the initiative, which necessitates an urgent policy reaction to reaffirm both the technical and economic sustainability of such programmes (*On the Future of Carbon Capture and Storage in Europe*, 2013). CCS is the only other option to offer base load power in areas where renewable sources are not available, and it is equal to nuclear power in terms of low-carbon emission (Institute & London, 2012). Unchecked coal consumption results in significant CO_2 emissions into the atmosphere, which supports the deployment of CCS systems. The wealthy countries must be the ones to use CCS first, and the developing countries must only follow (Gibbins & Chalmers, 2008). Canada provided evidence that a policy instrument may support CCS technology for commercial applications through specific incentives, and it also has plans for large-scale investment (*Energy Policies of IEA Countries - Canada 2009 Review*, 2010). The New Zealand Ministry of Economic Development is funding a research and demonstration programme with the goal of purchasing 100,000 tonnes of CO_2 (*Energy Policies of IEA Countries - New Zealand 2010 Review*, 2010). Coal mining businesses must make payments under the Climate Change Response Act of 2002 for the carbon emissions of coal. Despite the fact that there are no fossil fuel exploration opportunities in New Zealand, CCS wants to further development and the gift is having development plans in place (*Energy Policies of IEA Countries - New Zealand 2010 Review*, 2010). Sweden has plans for both biogenic and conventional CCS technologies (IEA, n.d.). Around the world, there are at least eight commercial-scale projects in operation (*On the Future of Carbon Capture and Storage in Europe*, 2013). An appealing solution for enhanced oil recovery (EOR) is CCS. Positive carbon balance is seen in the life-cycle inventory of CO_2 consumption in EOR. The same, however, do not currently receive any policy support, not even Clean Development Mechanism (CDM) benefit. EOR can play a significant role in nations with exhausted hydrocarbon reserves and salty aquifers, such as Mexico and Indonesia (Mikunda & Kober, 2012). Such alternatives should definitely be taken into account when developing policy actions (Mikunda & Kober, 2012).

Access to advanced technology and enhanced systems is linked to the option of improving the current system through the usage of system components that operate more efficiently. The choice is extremely pertinent to systems that have been in operation for a long time or have not undergone modernisation. More economic growth, better energy security, and decreased pollution are just a few advantages of higher energy efficiency. However, the current initiatives in many countries are insufficient to realise their true economic potential (*World Energy Outlook 2012 Factsheet*, 2012). Ireland set a target to reduce its energy consumption by 33% in the public sector and 20% nationally by 2020 by implementing energy-saving measures (*Energy Policies of IEA Countries - Ireland 2012*, 2012). In Canada, energy efficiency programmes are a priority, and they promise to boost energy efficiency by about 20% (*Energy Policies of IEA Countries - Canada 2009 Review*, 2010). The areas that are being improved upon include regulations covering a wider range of energy efficiency, aid with retrofits, control of energy-consuming products, green building codes, and improvements to energy audits. It is laudable that Canada increased its energy efficiency utilisation by about 15% between 1990 and 2005 (*Energy Policies of IEA Countries - Canada 2009 Review*, 2010). The "ecoACTION" programme in Canada promotes the use of energy-efficient technology like biofuels and intends to implement new energy-saving technologies (*Energy Policies of IEA Countries - Canada 2009 Review*, 2010). Sweden's energy efficiency programme (2010–2014) aims to support green public procurement, strengthen local and regional climate and energy initiatives, and encourage Small and Medium-sized businesses Entities (SMEs) to monitor and audit their energy usage and purchase energy-efficient technology (IEA, n.d.). In this aspect, New Zealand's situation is an intriguing one. It has efficient energy conservation policies, such as minimal requirements for energy ratings based on set performance standards. Additionally, it features a database on energy usage created by the Energy Efficiency and Conservation Authority that displays precise end-use energy consumption (*Energy Policies of IEA Countries - New Zealand 2010 Review*, 2010). The goal of the Energy Efficiency and Conservation Act of 2000 in New Zealand is to generate 90% of the country's electricity from renewable sources by 2025 (*Energy Policies of IEA Countries - New Zealand 2010 Review*, 2010). This act also promotes energy efficiency, energy conservation, and renewable energy. The removal of inefficient products from the market is handled through a revolutionary concept known as Minimum Energy Performance Standards (*Energy Policies of IEA Countries - New Zealand 2010 Review*, 2010). To increase energy efficiency, India introduced the Perform, Achieve, and Trade (PAT) scheme (Singh, 2013). The PAT scheme establishes emission reduction targets for a facility rather than a sector, and after the first year of the compliance period, 2012, the qualified facilities receive credits annually for energy savings (Singh, 2013). One must examine the impact of energy-efficient measures on the units' Break Even Point (BEP) while contemplating efficiency improvements for the production unit. Production costs will rise as BEP rises, reducing the competitive edge in cost (Alam, 2016).

The gathering of accurate field data both before and after the policy's implementation is necessary for the realisation of sound policy. This will make it easier to gauge the increase in efficiency. The potential for electricity in New Zealand's residential, commercial, and industrial sectors was evaluated by the New Zealand Electric Energy Efficiency Potentials Study in September 2007. The study's objectives were to identify the levels of economic power efficiency potentials and to calculate the cost-effectiveness of projects to realise these potentials. According to the report, lighting has significant economic potential in the residential and commercial sectors, and the Electricity Commission's plans to date have placed a lot of emphasis on this topic. In New Zealand's residential and commercial sectors, KEMA Inc. of NZ projects that efficient lighting will generate an annual economic potential of 1,750 GWh, or 40% of all other energy-saving initiatives, by 2016. The survey also noted how lighting affects peak demand; in the residential sector, 50% of lighting electricity use is thought to be used during peak hours (*Energy Policies of IEA Countries - New Zealand 2010 Review*, 2010). New Zealand's energy-efficient construction codes are performance-based. Instead of defining the architecture and construction of the building, it specifies how the building and its parts must function (*Energy Policies of IEA Countries - New Zealand 2010 Review*, 2010). The housing sector presents one of the biggest prospects for efficiency gains and GHG emission reductions, even though some advantages of stricter building rules take longer to manifest due to the slow turnover of the building stock (*Energy Policies of IEA Countries - Canada 2009 Review*, 2010). Energy efficiency is also given great importance in Japan, which aids in achieving the dual objectives of environmental conservation and energy security. In order to implement its climate and efficiency policies, the nation mainly relies on voluntary agreements with business (*Energy Policies of IEA Countries JAPAN 2008 Review*, 2008).

It is a difficult undertaking to make improvements in the energy efficiency of the transportation industry. Countries are unaware of how much energy has actually been gained by the transport sector. For instance, in New Zealand, it is unclear how the sector's shown potential for energy efficiency benefits would be released under the current policy framework. Weak policies that change regularly further add to the complexity (*Energy Policies of IEA Countries - New Zealand 2010 Review*, 2010). Other government initiatives, such as taxing less energy-efficient cars and lowering the acquisition taxes for more energy-efficient cars, have also been implemented in the transportation sector (*Energy Policies of IEA Countries - New Zealand 2010 Review*, 2010). In this context, Japan has implemented various novel strategies, such as the "Top Runner" programme for product efficiency. Japan demonstrated its dedication to energy efficiency by establishing goals for improvements of 30% by 2020 (*Energy Policies of IEA Countries JAPAN 2008 Review*, 2008). The steel, paper, cement, and chemical industries saw performance improvements between 1973 and 2005 of 20%, 24%, 29%, and 52%, respectively (*Energy Policies of IEA Countries JAPAN 2008 Review*, 2008). A further intriguing development in Japan is the increased emphasis on energy efficiency improvements in the sector of SMEs, where the Japanese government has attempted to support SMEs by establishing a number of fiscal schemes to do so. The 'Biomass Nippon Strategy' was launched by the government in 2006 with the goal of promoting the use of biomass fuels in transportation. The government set a target for 2010 of 0.5 million kL (crude oil equivalent) of biofuels in the transport sector with the intention of reducing CO_2 emissions by 1.3 Mt (*Energy Policies of IEA Countries JAPAN 2008 Review*, 2008). Another fascinating feature is the fully computerised Household Energy Management System (HEMS) and Business Energy Management System (BEMS), which give the offices real-time data on energy usage (Amoo et al., 2016).

7.5.3 Lack of Infrastructure

Due to a lack of corresponding infrastructure, a nation could not adopt a number of technical possibilities. Financially restricted policies, not a technical difficulty, are the main cause of the deployment of necessary infrastructure. If sufficient electrification work is done in the area, especially in remote and hilly rural areas, users from rural areas will be able to obtain electricity. The country and the specifics of the grid levelised cost of electricity determine whether a region is electrified. For instance, a typical approach to decentralising electrification requirements is the use of diesel (and, more recently, solar PV). Diesel generators are more cost-effective for off-grid applications, although they are less so due to rising fuel prices and the high expense of delivering fuel to remote locations (Szabó et al., 2011). When transportation costs are included in, along with additional expenses like fuel subsidies, the cost of importing crude, the cost of local air pollution, etc., a modern diesel generator of a lesser size will need roughly 0.4 L of fuel for every kWh of energy, with these prices rising 10–15 times the effective cost of electricity (*Iraq Energy Outlook - World Energy Outlook Special Report*, 2012). For the concerned state agencies, connecting off-grid investments in grid extension projects is a crucial area of concern (Szabó et al., 2011). To get a high proportion of electrification, energy policy must explicitly concentrate on energy availability. Rolling blackouts and limited access to energy hurt Ghana's ability to expand its economy. Energy Development and Access Project, a focused initiative with a 2007–2013 period, was introduced. The program's goal was to reduce the energy disparity between rural and urban areas by increasing power access, efficiency, availability, and reliability (*Country Energy Profile: Ghana - Clean Energy Information Portal - Reegle*, 2013). The programme assisted in raising the 60.5% (REP, 2013) current power access rate to a very high level. In another similar program in Brazil, a similar initiative called "Luz Para Todos" (Light for all) ran from 2003 to 2014 with the goal of providing energy to everyone by that year. The initiative made it possible to electrify over 99% of the population. There is also a programme to teach people how to set up solar and biogas power systems, which are community-based, decentralised renewable energy sources. The program generated a huge infrastructure for the operation of these systems. Due to increased access to power, it also created close to 300,000 additional jobs and income opportunities. Additionally, it encouraged a number of social initiatives involving the provision of services in the fields of health, sanitation, and education (*Measuring Progress towards Energy for All*, 2012). India's rural population lacks access to electricity in 66%, 94%, or 404 million households ("India Boasts a Growing Economy, and Is Increasingly a Significant Consumer of Oil and Natural Gas," 2011). The total achievement in rural areas between April 2005 and January 2010 was the electricity of the tune of 71,793 unconnected or un-electrified communities (Das, 2011).

7.5.4 Built-In Capabilities

The inherent resources that a country possesses will determine how it responds to the global push for environmental issues. Its overall emission will be determined by these resources. The fuel mix, which will then change into the EM, will determine the energy emission, also known as grid emission. The predominant fuel mix varies significantly between nations. The fuel mix varies from one country to another depending on a combination of the price and accessibility of different energy sources. One must include other sectors, such as transportation and housing, in addition to the energy-intensive industries when developing a widely approved policy. As a result of the steady replacement of a significant stock of durable energy infrastructure, changes in the fuel mix typically occur slowly. However, in addition to energy policies, such as those aimed at reducing dependency on imports, climate change policies also have a significant impact on a nation's long-term fuel mix (Garnaut, 2011). The combination of sources that currently make up the energy supply mix also affects policy formation. For instance, Ghana has abundant biomass resources, which between 2000 and 2008 accounted for 81% of primary energy and 76% of final energy. Petroleum makes up 17% of total energy consumption and 12% of the primary energy supply. For cooking, rural households use wood (charcoal, firewood, and crop/sawmill residue), whereas around 60% of urban households use charcoal predominantly. This is demonstrated by the fact that about 60% of all energy is derived from biomass resources (*Country Energy Profile: Ghana - Clean Energy Information Portal - Reegle*, 2013). The current EM is primarily made up of renewable sources of energy; but, as the nation moves towards industrialisation, its fossil fuel component will increase and distort the current EM. Another example is Ukraine, which is the world's thirteenth-largest producer of coal and is therefore well-endowed with energy resources, particularly coal. Despite the fact that most of Ukraine's domestic demand is met by coal production, 39% of its total energy needs are met by imports of gas and oil. Natural gas, which made up 40% of Ukraine's EM in 2010 but dropped from 47% in 2004, is the primary source of energy. In 2010, coal accounted for 31%, up from 23.6% in 2004. Hydropower (2%) and nuclear (17%) were the other energy sources in 2010 (*Energy Policies beyond IEA Countries - Ukraine 2012*, 2012). The cost of gas rises when coal is used, and any further increase in the demand for energy is highly emission sensitive, which creates a barrier to further energy expansion. Sweden contributed 4.1% of the world's coal and 2.4% of its natural gas. The majority of the fossil fuels supplied to Sweden come in the form of oil. In 2011, Sweden generated 40.5% of its power using nuclear energy, which is the second-highest percentage of nuclear energy in the world behind France. Sweden is third in terms of the percentage of renewable energy, with little over 35% in 2011. This is due to the country's abundant natural resources and effective, long-term policies that incentivise renewable energy. This is mainly due to both the second-highest share of biofuels and waste (22.7%) after Finland and the fourth highest share of hydro (11.7%) (IEA, n.d.). It is observed that Sweden is having a good balance of the energy resources and can expand the energy generation in a sustainable way if required.

In case of Japan, four fuels make up the majority of the total, which has a fuel mix that is rather well diversified. The largest portion comes from oil, followed by coal, which makes up one-fifth, nuclear energy, and natural gas, which each accounts for 15%. When industrial and non-renewable municipal garbage are excluded, renewables account for 3.2% of the total (*Energy Policies of IEA Countries JAPAN 2008 Review*, 2008). It is challenging to increase the proportion of renewable sources in the country since it lacks a land basis and relies heavily on fuel imports. Additionally, the nation is extremely sensitive to emissions. In case of New Zealand, the oil and natural gas account for 57% of energy supply, which is equally distributed and dominated by both fuels. Coal makes up 10%, while geothermal and hydro sources contribute 15% and 11.3%, respectively. The proportion of geothermal energy in New Zealand is substantial. Large, high-quality coal reserves are also present in New Zealand (*Energy Policies of IEA Countries - New Zealand 2010 Review*, 2010). The majority of the electricity was produced by hydropower (51%) with the remainder coming from geothermal energy (9.6%), natural gas (24.3%), and coal (11%) sources. Wind (2.4%), biomass (1.3%), and oil (0.3%) contributed less. 64.2% of the electricity produced comes from renewable energy sources. Notably, among IEA member nations, geothermal energy accounts for the biggest share of electricity in case of New Zealand (*Energy Policies of IEA Countries - New Zealand 2010 Review*, 2010). As a result, it is less sensitive to the addition of new energy sources in terms of emission.

Changing the fuel option in the current EM can save costs and emissions simultaneously. Iraq is currently the third-largest oil exporter in the world (*Iraq Energy Outlook - World Energy Outlook Special Report*, 2012). Iraq's economy is based on energy, with oil exports making about 95% of government income and more than 70% of GDP in 2011 (*Iraq Energy Outlook - World Energy Outlook Special Report*, 2012). For Iraq, natural gas is strategically significant since it can lessen the dominance of oil in the country's energy balance. It will be crucial to collect and process the associated gas from Iraq, much of which is being vented. The expected demand for gas in Iraq, which will exceed 70 Billion Cubic Metres (BCM) in 2035, will not be met by gas alone (*Iraq Energy Outlook - World Energy Outlook Special Report*, 2012). Without making the necessary transformation, Iraq is expected to lose $520 billion in revenue from oil exports (*Iraq Energy Outlook - World Energy Outlook Special Report*, 2012). Iraq needs to add roughly 70 GW of production capacity and shift from a power mix that relies mostly on oil to one that relies more on efficient gas-fired generation. China has a highly carbon-intensive energy profile with 79% thermal coal energy, 16% hydro, 1% other renewable energy, and 2% nuclear energy. India has an energy profile with 82% thermal coal energy, 13% hydro, 2% nuclear, and 2% other renewable energy, and both countries are emission-sensitive and dependent on coal imports, which results in poor energy security (Piere Gadonneix, World Energy Council, 2012). Due to their small size and reliance on coal for all of their energy needs, nations like Israel and Jordan have higher emission sensitivity (Piere Gadonneix, World Energy Council, 2012). Some nations, like Korea (31%), France (76%), and others, have a good energy supply due to improved nuclear technology. Some nations, including Ghana (hydro 78%), Iceland (74%), and Switzerland (55% hydro, 40% nuclear) (Piere Gadonneix, World Energy Council, 2012), as well as Turkey (95% hydro in 2009 (Saygın & Çetin, 2011) have a large renewable component. It has been noted that the price of coal has the greatest impact on the cost of power. According to simulation studies conducted on a German power market, the decline in coal and petrol prices is responsible for almost 50% of the wholesale power price decline in Germany. While subsidies for new renewable energy sources like wind and solar only lead to a 30% price decrease (Everts et al., 2016).

7.5.5 Energy Security

The most important issue in designing the energy policy for a developing economy is energy security. Therefore, it is reasonable to predict that while developing an energy strategy, energy security will take precedence over market reform and environmental preservation. Despite Turkey's low hydrocarbon reserves, fossil fuels account for the majority of the country's energy use. Renewable resources, with the exception of hydro, have been minimal. In these conditions, fast-rising energy demand entails rapidly rising reliance on foreign energy imports. Imports supply more than 70% of the nation's overall primary energy needs (Saygın & Çetin, 2011). In contrast to this, Ghana has 76.3% more energy self-sufficiency (REP, 2013) and is well-positioned to meet its energy growth. Since Japan is an island country that imports a significant portion of its fossil fuels and a huge portion of its total energy supply, it only makes sense that the government places energy security at the top of its list of policy objectives. It is the world's largest importer of Liquefied Natural Gas (LNG) and a pioneer in the LNG trade due to its high energy needs. The island nation uses nuclear power extensively and will likely continue to hold the top spot in the nuclear EM in the future due to its restricted resource base (it produces just 20% of the energy it consumes) (*Energy Policies of IEA Countries JAPAN 2008 Review*, 2008). The data will undoubtedly be more affected if the Fukushima Daiichi facility is included. Energy security is crucial for the country's continued development because of its remote position and distance from major energy providers (*Energy Policies of IEA Countries JAPAN 2008 Review*, 2008).

Ireland's gas market is known for its extreme reliance on imports, and 93% of its gas supplies pass through just one transit point in Scotland, Moffat, making it extremely susceptible to a supply-related issue. The existence of many transit points tends to significantly reduce this danger. The development of upstream gas reserves like Corrib and the proposed construction of an LNG terminal on the Shannon Estuary would both tend to significantly improve Ireland's security of supply (*Energy Policies of IEA Countries JAPAN 2008 Review*, 2008). Inadequate energy security may result from technological limitations as well. As compared to the rest of West Asia, where oil makes up around 50% of the whole EM, Iraq's primary energy source is oil, which accounts for about 80% of its total EM. Although it is well known that using natural gas instead of liquid fuels to generate power is more cost-effective, in 2011 approximately 60% of the nation's gas production was flared (*Iraq Energy Outlook - World Energy Outlook Special Report*, 2012). In the case of New Zealand, inadequate energy security is seen as a result of inadequate gas storage. All emergency efforts would, at most, be to regulate demand and supply in the nation because there is no gas storage in the nation and no infrastructure in place to import gas supplies from overseas (*Energy Policies of IEA Countries - New Zealand 2010 Review*, 2010). With a vast network of pipelines, Canada is a sizable net exporter of oil. Canada is the largest supplier of natural gas to the United States and a net exporter of gas. Both the massive system of gas pipelines and the highly integrated electrical transmission network shared by Canada and the United States contribute to the energy security of each nation (*Energy Policies of IEA Countries - Canada 2009 Review*, 2010). It is recognised that there are substantial energy security issues caused by usage patterns, weather, and geographic position, like in the case of Iraq. Iraq lacks 6 GW of the necessary energy demands since its net capacity to supply electricity at the peak demand of 15 GW in 2011 is just 9 GW. Add to this the fact that the production that is typically possible is decreased at high ambient temperatures. The country's theoretically available energy supply must be cut to account for the offsetting due to high temperature because Iraq's peak energy consumption occurs on summer days, when the temperature is around 45°C (*Iraq Energy Outlook - World Energy Outlook Special Report*, 2012). The problem of energy availability is strongly impacted by this.

7.5.6 Infrastructure-Related Issue

Infrastructure issues may also have an impact on policy objectives. In the case of New Zealand, the grid is heavily loaded, and structural restrictions make planned repairs very challenging. In order to achieve the government's 90% renewable electricity target, this necessitates the development of a smart grid (*Energy Policies of IEA Countries - New Zealand 2010 Review*, 2010). Large saving opportunities are wasted as a result of technical shortcomings or technological innovation. Due to a lack of modern refineries, Iraq's petroleum production falls short of domestic demand and is inefficient. The oil's quality, which produces only 15% of the petrol and 45% of heavy fuel oil, contributes to the shortfall. The petroleum deficit would have been contained by a more efficient production, and the current reliance on imports could have been avoided and had the technology employed in Iraq's refineries been similar to that utilised in United States (US) refineries (*Iraq Energy Outlook - World Energy Outlook Special Report*, 2012). The majority of heavy industries, such as those producing iron and steel, fertiliser, chemicals, petrochemicals, cement, glass, and ceramics, as well as light industries, such as those producing textiles, leather, furniture, and dairy products, are all operated by the public sector, which is reliant on antiquated technology (*Iraq Energy Outlook - World Energy Outlook Special Report*, 2012). Gas and electrical infrastructure are also in development. In 2010, Ukraine committed to creating an energy community and reforming its energy industry, and since then, it has made a number of promises in that direction. Some of the significant steps include encouraging energy trade, creating an enabling environment for energy efficiency and renewable energy projects, creating a stable market and responsible regulatory framework to attract investment in power generation, transmission, and distribution networks, and offering incentives for energy efficiency improvements by the end of 2012 (*Energy Policies beyond IEA Countries - Ukraine 2012*, 2012). Iraq also faces significant infrastructure development challenges and issues, which contribute to erratic electrical supplies. Despite the power plants producing more electricity, there is still a shortage, which causes daily power outages on a regular basis. This results in an overreliance on diesel power plants. The construction of a modern electrical system is urgently required because the power distribution network in Iraq, which provides electricity to 98% of all households, is deteriorated (*Iraq Energy Outlook - World Energy Outlook Special Report*, 2012). Particularly for gas-fired plants, efficiency is now at 31%, whereas under ideal operating conditions, modern combined cycle gas turbines can achieve efficiency levels of up to 55%. However, for financial reasons, their broad use is not permitted (*Iraq Energy Outlook - World Energy Outlook Special Report*, 2012).

7.5.7 Policy Measures

Policy is the requirement that is supplemented by resources and supported by the government's legal and financial authority. The wheel is ultimately turned by the policy. A policy tool needs to be applied differently in various nations. A New Zealand Bioenergy Strategy was produced in 2010 by the New Zealand Forest Owners Association and the Bioenergy Association of New Zealand. According to the Strategy, biomass accounts for about 8.5% of energy consumption, with the heating sector using the majority of the energy to heat both households and businesses. By 2040, bioenergy will provide more than 25% of New Zealand's overall energy needs, including 30% of the country's transport fuels, according to the strategy (*Energy Policies of IEA Countries - New Zealand 2010 Review*, 2010). The New Zealand government's "The Engine Fuel Specification Regulations" require diesel engines to be blended with 5% biodiesel. An analysis of a research undertaken in 2007 with the aim of developing an electric car policy that would remove obstacles to their use and promote their adoption of the potential of electric vehicles was also published by the New Zealand Ministry of Transport (*Energy Policies of IEA Countries - New Zealand 2010 Review*, 2010). It is crucial to remember that thorough market research is crucial and can be the foundation for developing policies. The Bill 2009/10:41, which focuses on energy and the environment as well as taxation for the years 2010, 2011, 2013, and 2015, was approved by the Swedish Parliament. A significant increase in CO_2 taxes on non-ETS sectors (agricultural, forestry, and some businesses) and an increased CO_2 factor in the car tax were other significant components of the act. The change improved the efficiency of the energy and CO_2 taxation by reducing the exemptions given to domestic industries (IEA, n.d.). The Turning the Corner framework, which Canada created in response to its inability to fulfil its obligations under the Kyoto Protocol, was released in 2007. The policy document "Turning the corner" included an action plan with multiple parts, including a regulatory framework, new energy efficiency performance standards, steps to enhance indoor air quality, and mandated fuel economy standards for cars. This worked well to control the nation's GHG emissions (*Energy Policies of IEA Countries - Canada 2009 Review*, 2010). In order to give Japan's future energy policy direction, the Japanese government passed The Basic Act on Energy Policy in June 2002. As essential policy directions, it particularly names ensuring a stable supply, environmental suitability, and use of market processes. Additionally, this law mandated that the relevant branch of government "formulate a basic plan on energy supply and demand in order to promote measures on energy supply and demand on a long-term, comprehensive, and systematic basis" (*Energy Policies of IEA Countries JAPAN 2008 Review*, 2008). The Resource Management Act (RMA) of 1991 (RMA) is administered by the Ministry for the Environment of New Zealand, which is also the agency tasked with creating and upholding the country's national policy statements and regulations (also known as national environmental standards) under that

Act (*Energy Policies of IEA Countries - New Zealand 2010 Review*, 2010). In order to promote the use of light electric vehicles, the New Zealand Ministry of Transport stated in 2009 that the government will waive road user fees for these vehicles for a period of 4 years, from 2010 to 2013. In July 2009, a 3-year Biodiesel Grant Scheme went into effect to promote domestic biofuel production (*Energy Policies of IEA Countries - New Zealand 2010 Review*, 2010).

7.5.8 Price Disparity and Subsidy

Despite the fact that the global subsidy for the consumption of fossil fuels increased by 30% in 2011 compared to 2010, the fossil fuel industry essentially distorts the global energy market. In 2011, financial support for renewable energy amounts to over $88 billion (*World Energy Outlook 2012 Factsheet*, 2012). However, the trend is away from subsidy increases for the majority of the countries. Except for transportation, diesel fuel used as heating oil and in generators is tax exempt in Canada (*Energy Policies of IEA Countries - Canada 2009 Review*, 2010). Iraqi petrol and diesel are 28% and 11% less expensive than imports (*Iraq Energy Outlook - World Energy Outlook Special Report*, 2012). In 2012, Bolivia gradually reduced its petrol and diesel subsidies. In 2012, Egypt lowered its natural gas and energy subsidies. El Salvador offered cash rewards for households using less than 300 kWh of energy each month when it changed its LPG subsidies in 2011. In 2011–2012, India lowers the subsidy by 2% of GDP. In order to raise energy prices to their full cost in 2010, Iran abolished fossil fuel subsidies. Malaysia announced in 2012 that the current administration would continue to provide subsidies for petrol, diesel, and cooking gas. Mexico gradually raised the cost of petrol, diesel, and LPG in 2011 to do rid of subsidies. In 2012, Morocco increased the cost of petrol and diesel by 20% and 10%, respectively. Nigeria staged a state-wide strike in response to the complete elimination of petrol subsidies in 2012, which caused prices to treble. Qatar raised the cost of petrol, diesel, and kerosene in 2011. 2013 will see a 15% increase in natural gas prices in Russia. Through 2015, South Africa will increase electricity costs by 20% annually. Sudan likewise started a programme to cut subsidies in 2012. In 2010, the United Arab Emirates (UAE) raised the price of petrol. According to the terms of International Monitory Fund financing, Ukraine would gradually raise its natural gas prices (*Recent Developments in Energy Subsidies*, 2012). Due to the cheap cost of electricity in some countries—for example, Canada's hydropower—natural resources are readily available. The cheapest electricity in North America may be found in Canada. The hydro-dependent provinces of Manitoba, Québec, and British Columbia have the lowest electricity costs (*Energy Policies of IEA Countries - Canada 2009 Review*, 2010). Taxes are not applied to diesel fuel used for heating. Except for diesel fuel used principally to operate a vehicle, diesel fuel utilised in the production of electricity is similarly excluded (*Energy Policies of IEA Countries - Canada 2009 Review*, 2010). Some emissions are concealed, and the correspondingly subsidised energy has also shown to be expensive. The irregular public energy supply in Iraq is handled by private diesel gensets, which are used by about 90% of domestic houses. Private generators make it simple to reach isolated rural communities in this way. Diesel generators are expensive, misuse subsidies, and also add to local air pollution (*Iraq Energy Outlook - World Energy Outlook Special Report*, 2012).

7.5.9 Gray Zones

A number of pockets of policy-related measures exist, and if they are put into place, they will undoubtedly have a significant impact on the country's energy strategy. The following section explains these elements.

7.5.9.1 Renewable Energy Certificate (REC)

REC now in use provides a strong base for the economically advantageous expansion of renewable electricity. The REC quotas are distinctive in that they do not define technology-specific goals, demonstrating sufficient flexibility to reach the goals. Japan's clean energy sector can benefit from this flexibility by creating effective ways to meet the renewable energy requirements. The system also has the advantage of enabling credit banking, which enables investors to spread the expenses of compliance and reduce the cost of implementation. A large trade volume will stimulate the production of renewable energy where it is most affordable. However, feed-in tariffs are beneficial for the advancement of more advanced technologies like wind power (*Energy Policies of IEA Countries JAPAN 2008 Review*, 2008). One of the concerns with the implementation of REC is few of its features overlap with the CDM projects. The issues can be addressed by considering the price of carbon credit based on the cost of emission and not on an arbitrary basis. Both the price of CDM and the price of carbon credits will remain stable as a result of this. Additionally, there will be no possibility for an energy project to simultaneously qualify for CDM and REC benefits. However, no energy strategy in place today takes such shifts into account. The literature elsewhere demonstrates an in-depth methodology with an example of intermediate pyrolysis technology (Patel et al., n.d.). The prime objective of the REC mechanism is to offer the fullest utilisation of the RE and bring down the operating cost for renewable energy to its most competitive level. However, the very purpose of the REC cannot be fully utilised if intrastate energy transfer is not allowed or possible.

7.5.9.2 Single Grid

Having a shared grid for the energy transfer can overcome the limitation mentioned in the previous section. The EU's energy strategy placed a strong emphasis on creating the infrastructure necessary for the simple exchange of power across its most diverse member nations, which are diverse in terms of their energy sources, rates of technical and economic advancement, infrastructure, etc. There is currently international trade in electricity. The real environmental cost of emissions, however, is not reflected in the price of power. It is crucial to take into account the equivalent emission component in it, i.e., the cost of electricity must take into account both the production cost and the environmental cost of the emissions that the production generates. In the electrical industry, there is currently no equivalent "energy to emission factor" accounting method. Lack of such a system might lead to unreasonably cheap electricity coming from nations with greater emission factors (EFs), such as China (Institute & London, 2012) or India (GoI, 2011), who have a higher proportion of coal in their EM than nations like Germany (which has a higher proportion of renewable energy). Taking into consideration the cost of transmitting energy from one state to another is the grid EF (Lewis, 1997). This can deter states that pollute the environment by generating energy at a higher EF and will encourage member states to incorporate more renewable energy sources into their state EM. Given that some Indian states, such as Manipur and Meghalaya, currently lack enough energy infrastructure, implementing such a plan is challenging. In terms of kg of CO_2/kWh, no technology-specific EF is currently known. The idea of a single grid will also help with cross-border product chain optimisation. Given that the US grid's average EF is larger than Germany's (750 vs. 680 CO_2/kWh), every process-line electricity reduction strategy results in a steeper reduction of GHG emissions in the US than in Germany (Fthenakis et al., 2009). This suggests that the US will profit more from the same energy conservation efforts than Germany. If so, energy conservation will become less important as GHG levels fall (which is counterproductive), and GHG cannot be decreased below a certain level. Such inconsistencies can be resolved with the use of an integrated climate policy that can optimise global supply chains. The concept of single grid will also benefit in optimising product chain across boundaries. An integrated climate policy which can optimise international supply chain can help resolve such discrepancies. A single such policy is more effective than several policies created for the domestic or regional emission reduction criterion (Quirk, n.d.). Additionally, this will support the development of comparative risk and sustainability evaluations, source-switching scenarios, and capacity growth scenarios (Lenzen & Munksgaard, 2002). Energy cannot be stored; therefore, a single grid will result in an unlimited demand, as well as benefiting underutilised or wasted resources, such as related gas waste at an Iraqi oil extraction site (*Iraq Energy Outlook - World Energy Outlook Special Report*, 2012) or the potential for coal gasification at mines in Bihar, India, etc.

Having a calculation-based technique for grid emissions with a provision for the penalty for using energy components in the EM containing higher EFs is one way to address this imbalance. Figure 7.3 shows a proposed intra-state energy transfer model. The model allows the transfer of energy considering the cost of the energy as well as the cost of emission into the account. A sustainable grid must have clearly defined requirements for the simple transfer and sale of power among individual member states in a nation like India, which has several states. For this, assume that two states each have a unique EM and that both transfer their energy production into a single grid with a known amount of grid emissions (based on statistics on national

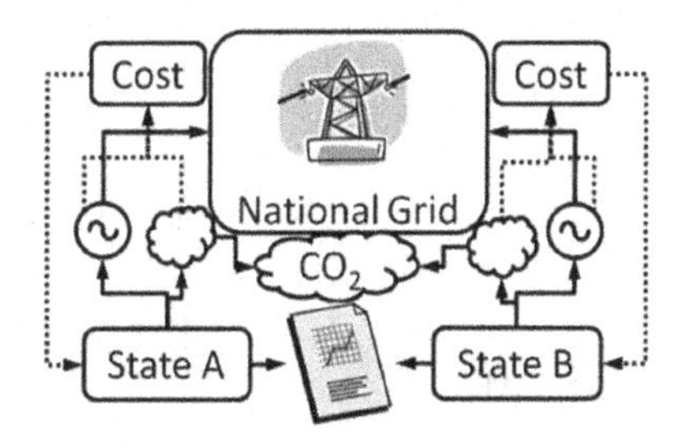

FIGURE 7.3 The proposed intrastate energy transfer model.

grid emissions from (Data, 2005)). Additionally, it is presumable that each state's emission indexes for different energy-producing kinds are known. Based on this, a state's overall EF can be calculated, and the difference in emissions between the state and the common grid can be expressed in kg of CO_2/kWh. Using the price of carbon credits, which is expressed in Rs/t of CO_2, it is simple to calculate the cost of such emission differential in Rs/kWh. The net difference in cost of emission is the cost difference between the two states exchanging electricity. The price of power for the state producing emissions less than the emission produced by the national grid and exchanging energy with the state producing emissions more than the national grid may be credited for the difference in total emissions. Such treatment usually makes green energy a pricey product to sell. States will be encouraged to move to cleaner energy sources as a result of this. It is interesting to note that such energy transfer costs depend only on the emission intensity of the two states exchanging the energy and are independent of national grid emissions (GoI, 2011).

In Table 7.2, one such potential energy transaction is shown. Table 7.2 displays the standard national EF (kg of CO_2/kWh) and the EM of States A and B. A multiplier called the Availability Factor (AF) is used to take a source's access to carbon credit benefits into consideration. Its value can range from 0 to 1; 0 otherwise, or 1 if the carbon credit advantage is used. For instance, if a source of power only qualifies for carbon credits for 30% of its output, its value can be rounded to 0.3. The table shows that electricity from state A emits more CO_2 since a large portion of it comes from coal, up to 70%, and because its hydro component does not support carbon credits. Due to the same factors, energy from state B emits less CO_2 because a significant portion of it is generated using hydroelectricity. This means that whereas the EF for the nation as a whole is 0.84 kg of CO_2/kWh, the EF for States A and B is 1.5 and 0.8 kg of CO_2/kWh, respectively. State A's emissions above the national grid result

TABLE 7.2
Energy Mix of States A and B

EF of National Grid		0.84		kg of CO_2/kWh	
Cost of Electricity		4.00		Rs/kWh	
Credit Price		720		Rs/t of CO_2	
	Category	**State A**		**State B**	
Energy Source	**EF**	**EM**	**AF**	**EM**	**AF**
Coal	2	70%	1	30%	1
Hydro	0.45	20%	0	40%	1
Nuclear	0.05	5%	0	15%	0
Solar	0.255	3%	1	5%	0
Wind	0.01	2%	1	10%	0
	1.50	100%	94%	100%	97%
	EF	**Diff**	**Diff**	**Price**	
State	**kg of CO_2/kWh**			**Rs/kWh**	
A	1.50	−0.66	−0.475	3.497	(A → B)
B	0.80	0.039	0.028	(B → A)	

in a net increase in EF of −0.66 whereas State B receives credit for a difference of 0.039. Taking into account the carbon credit at 720 Rs/t of CO_2, the state A's net penalty for emitting emissions in excess of the national grid is −0.475 Rs/kWh, whereas the state B's net credit is 0.028 Rs/kWh. According to the assumption that the base price of electricity is 4 Rs/kWh, the net payable price for energy transfer from state A to state B is 3.497 Rs/kWh (less than the base price of electricity), while the price for energy transfer from state B to state A is 4.503 Rs/kWh (more than the base price because the energy is greener).

7.5.9.3 Nuclear Justification

On whether nuclear energy should be regarded as renewable or non-renewable, there is a heated discussion. Around 14% of the electricity produced globally comes from nuclear energy; thus, the argument is worth addressing. Nuclear energy is a profitable choice for nations due to its greater output potential, efficiency, and comparative affordability. Numerous nations' decisions to pursue civil nuclear programmes have been motivated primarily by the relatively low cost of nuclear energy (Ernest and Young, 2012). The catastrophic impact of nuclear power plant accidents is not openly acknowledged or accounted for. On the other hand, the nuclear option is sometimes viewed as green and even renewable in extreme cases. Nuclear power generation is a desirable alternative for countries with limited resources because it emits less CO_2 than other ways of generating. For instance, Japan tends to use nuclear energy more because it has very limited hydropower (*Energy Policies of IEA Countries JAPAN 2008 Review*, 2008). According to estimates, the cost of producing nuclear electricity is 5.3 yen/kWh, compared to 11.9 for hydropower and 10.7 for thermal oil-fired power (Matusuo, 2012). The trend shows that nuclear power has the highest potential for mitigation and the lowest average cost in the energy supply sector (*Climate Change and Nuclear Power 2012*, 2012). In terms of g CO_2-eq./kWh, nuclear power, along with hydropower and wind-based electricity, is also one of the lowest emitters of GHGs (*Climate Change and Nuclear Power 2012*, 2012). Nuclear power generation does not release CO_2 throughout the power generation process, so it has an exceptionally important position with regard to the promotion of global warming countermeasures, according to the Kyoto Protocol Target Achievement Plan adopted by Japan (*Energy Policies of IEA Countries JAPAN 2008 Review*, 2008). However, following the Fukushima Daiichi nuclear power plant accident, the Japanese government said in October 2011 that it would drastically cut the nation's reliance on nuclear energy (World Nuclear Association, 2013). It is interesting to note that major US, Canadian, Russian, French, and South Korean vendors reported no drop in demand following the accident. Additionally, six European nations—the UK, France, Finland, the Czech Republic, the Netherlands, and Poland—have confirmed their plans to build new nuclear power plants, and other nations are still considering extending the lives of their existing nuclear plants (Ernest and Young, 2012). Germany is an exception, as it wants to totally phase out the usage of nuclear energy by 2020 and now gets 23% of its electricity from nuclear sources (Institute & London, 2012). However, it is predicted that the cost of nuclear electricity will increase by around 50% over what was predicted 7 years ago as a result of stricter safety rules following the Fukushima event (*Even Higher Costs and More Headaches Ahead for Nuclear Power in 2012*, n.d.).

However, a crucial query to pose is, "Should nuclear energy be taken into account as a renewable energy option?" In the Energy Sustainability Index published by the World Energy Council, UK (Piere Gadonneix, World Energy Council, 2012), nuclear energy is ranked on par with renewable energy or granted the status of carbon-free energy. Nuclear energy-using nations have very high energy indices, for example. The substantial proportion of nuclear energy to the Swedish power mix—40% in 2010, 40.5% in 2011, and 37% in 2012 (IEA, n.d.)—aided Sweden's performance on the Energy Sustainability Index (Piere Gadonneix, World Energy Council, 2012). Due to its zero CO_2 emissions, several nations want to enhance the nuclear energy alternative by holding it to the greatest safety standards (*Energy Policies of IEA Countries JAPAN 2008 Review*, 2008). However, following Fukushima, the mindset towards nuclear power has shifted (World Nuclear Association, 2013). The incident brought to light a significant flaw in how nuclear energy costs are calculated and how the Energy Sustainability Index is established (Piere Gadonneix, World Energy Council, 2012). The denial of risk's existence is harming the nuclear sector (*Even Higher Costs and More Headaches Ahead for Nuclear Power in 2012*, n.d.). The price, or more precisely the risk of an accident occurring, was severely understated. So much so that even Japan is getting ready to forget about this incident (*Japan Prime Minister Shinzo Abe Promises to Speed up Recovery on Second Anniversary of Deadly Tsunami, Fukushima Nuclear Disaster*, 2013). Nuclear power can be substituted, and it is not impossible. If thermal power-producing utilities took the place of nuclear energy, studies conducted in Japan predict that the revised cost of power generation would increase by roughly 0.7% of GDP. Additionally, if the nuclear facilities stayed down, the cost of power would need to rise by up to 15% (*Climate Change and Nuclear Power 2012*, 2012). However, the question is whether we are willing to acknowledge that nuclear power is not a renewable source and that it is not sustainable or safe just because it produces no carbon. Is there a practical way to account for the risk posed by technology, specifically nuclear energy? Excluding the expense of new power, the clean up after the Three Mile Island tragedy is expected to cost $1 billion. The full clean-up effort might take an unbelievable 40 years (Joskow & Parsons, 2012). The operator of the facility and its insurers allegedly paid residents at least $82 million in publicly disclosed compensation for "loss of business revenue, evacuation expenses, and health claims" ("Three Mile Island Accident," 2013), according to Eric Epstein, chair of Three Mile Island Alert. On the other hand, it is claimed that nuclear energy has a life-cycle cost of only 17 kg of CO_2/kWh (Miere, 2002). If the social cost of power is taken into consideration, the true cost of nuclear power can be accurately calculated. The biggest obstacle at this point is the need to develop a social cost model for power generation that is clearly defined and accepted globally, taking into account not only nuclear power but also a number of other competitive energy sources.

7.5.9.4 Convenient Definition of Cost

Due to the statistics' discrepancy, a more realistic and long-term costing technique needs to be updated. In a number of situations, the true cost of operation is either not taken into account or is completely unknown. The thermal power plant's extensive water demand is not properly accounted for. Cooling tower resource loss must be taken into consideration when calculating the societal cost. Water utilised for oil exploration must also be considered a resource and certainly not a free resource. There is no provision for quantifying the impact of aerosols emitted during open field burning, such as burning firewood for cooking or burning agricultural waste in the open in fields, on human life. Aerosol release significantly lowers the potential for global warming of GHG. Including Cryosphere forcing, the radiative forcing from open-burning emissions is $+0.20 \pm 0.08$ Wm^2 for black carbon (Bond et al., 2011). However, the impact of toxic gases that result in lung ailments and other negative effects is not taken into account. There is no record of deaths caused by hazardous gas inhalation. Burning biomass and utilising firewood for cooking are unsafe practices that can lead to smoke and aerosol inhalation. For instance, at least 570,000 deaths per year in India are attributable to exposure to indoor air pollution from household solid fuel burning (such as residential cooking). Additionally, it is estimated that outdoor air pollution kills nearly 490,000 people annually throughout all of Asia (*Reducing Black Carbon Emissions in South Asia - Low Cost Opportunity*, n.d.). There are no established metrics or statistics to gauge the aerosol impact of open field burning. Some sort of model accounting is urgently required.

Biomass is transformed into a gas in an anaerobic digester, and the biogas gas can be burned to produce CO_2. The same organic material will naturally decompose, if not processed by anaerobic digestion, can also produce Methane (CH_4). This CH_4 will reach the environment directly and may contaminate it. It is worth noting that CH_4 has a stronger potential for global warming than CO_2 (Dessus et al., 2008). The true cost of the harm caused by global warming or its financial implications produced by GHGs cannot yet be predicted by a single model. In this case, it might be argued that the social cost of anaerobic digesters is not fully appreciated or assessed. Other environmental welfare initiatives carried out by corporations under the Corporate Social Responsibility (CSR) umbrella must also be evaluated using this costing approach. Instead of just falling under an unaccounted welfare act, any CSR or action must be properly costed utilising social costing techniques.

Correct cost estimation is required for renewable energy sources as well since renewable energy certificates for solar and non-solar sources have different pricing in India. In the case of renewable, there are no global standards or logical principles in use. An arbitrary attempt to make up for environmental deterioration is the collection of the coal utilisation fund in India, from Rs 10 per tonne to Rs 50 per tonne (Kakkar, 2013) and the provision of financial aid for solar PV installations (*Coal Tax to Fund India's 750 MW Solar Power Capacity Addition Plan in 2013*, n.d.) is an arbitrary attempt to compensate for environmental degradation. The issue of disposing of nuclear waste, the effects of radiation on human life, and the impact of extreme events like Three Mile Island, Chernobyl, or Fukushima on the lives of human beings all call for the development of an effective social accounting model to forecast the social cost of nuclear energy (and several other technologies). If the higher risk to human life were properly factored into the price of nuclear energy, it might aid in correctly evaluating the nuclear option in comparison to alternative carbon-free energy sources. This issue might be resolved by taking into account and assessing the social cost component, which is not currently taken into account in the accounting approach.

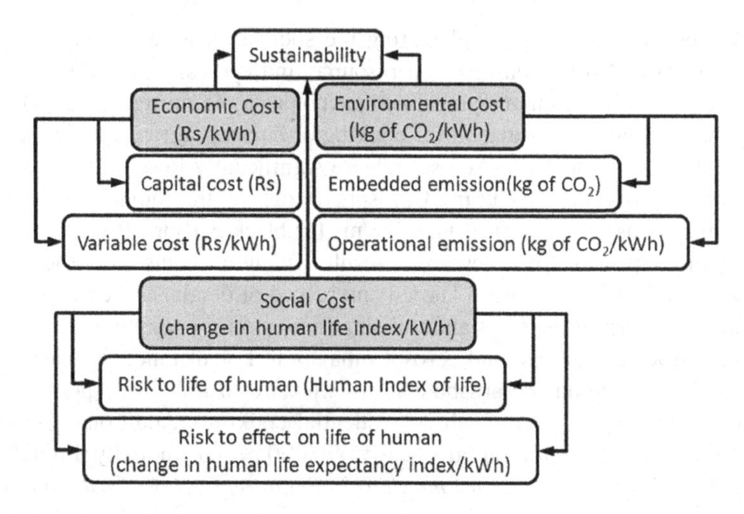

FIGURE 7.4 Sustainability models for different pillars.

An outline of this sustainable approach is shown in Figure 7.4. Noting that the social cost can also be evaluated in terms of risk to human life, for installation and operation of the plant, in addition to the economic cost being in terms of rupees and the environmental cost being in terms of kg of CO_2. A single phase diagram can represent multiple technologies, and careful waits can aid in accurately ranking them.

7.5.9.5 Inclusion of Emission in Policy

The non-energy related emission mitigation factors were ignored or underdeveloped when formulating an energy policy. Such an approach has a tendency to assume that the only cause of the global emission issue is the energy boom, which is scarcely accurate. Consider the use of air ventilators, which provide fresh air to the plant and offices without the need of electricity. In contrast to the moderating impact, policy does not encourage such actions. Solar water heaters and solar cookers are two other examples that lessen the demand for gas and energy, respectively. A nation would be better equipped to achieve its low-carbon objectives if an integrated energy and emission plan were developed in place of an energy policy (without taking into account the benefits of emission replacement).

7.5.9.6 Energy and Emission in Transport Sector

More than 50% of the world's oil is consumed by the transportation sector, and 20% of CO_2 emissions worldwide (Weiss et al., 2000). One of the fastest increasing modes of transport, aviation and maritime transport account for about 5% of worldwide GHG emissions and 3% of GHG emissions in the US combined (McCollum et al., 2009). There are not many alternatives available to reduce emissions from the transportation industry. Biofuels, fuel cells, hydrogen vehicles, and electric cars (provided the electricity is produced sustainably) are among the available technological possibilities. In contrast to emissions from energy generation, it is therefore more difficult to replace emissions from transportation. Additionally, much fewer trucks have been adopted that meet the fuel economy criteria (*World Energy Outlook 2012 Factsheet*, 2012). Heavy trucks, in particular, have extremely low efficiency, especially since there are rarely any emission inspections because most nations lack strict regulations or efficiency criteria. Recently, the US government issued the first-ever efficiency guidelines for medium- and heavy-duty trucks. The first-ever standard is anticipated to progressively phase in between the model years of 2014 and 2018 and is anticipated to increase fuel economy by 7%–20% while lowering demand for about 500 million barrels of oil or 250 million Metric Tonnes of GHG (*Is No One Against the New Truck Fuel Efficiency Standards?*, 2010). In contrast to the statistics on energy generation, there are no systematic data on the amount of fuel used by a vehicle, the distance it travels, its efficiency, the type of emissions it produces from its tail pipe, etc. As a result, it is impossible to track how and how much fuel is used on the road, making it difficult for policymakers to suggest meaningful solutions. According to the study, CO_2 emissions from highways in India and Vietnam are greater by 144% and 138% than those from railways, which emit between 377 and 1,180 tonnes of CO_2/km annually (Satish Rao, 2010). Since public transport significantly reduces GHG emissions by multiple orders of magnitude, there is absolutely no policy-driven push for it in terms of earning tax breaks or other benefits for the users. There is no comprehensive study that takes into consideration the emissions produced by trains or buses in terms of CO_2 emissions per person per kilometre. Although it is a commendable effort on the part of air carriers to provide emission statistics for a specific route, there should be a means by which to charge for emissions incurred during a particular route.

Since it is more challenging to substitute emissions from transportation than from power generation, there are now no adequate alternatives for energy consumption from these two separate routes. It is not appropriate to treat the additional energy released per kilogram of fuel for transportation and for power generation equally. More consideration needs to be given to any solution that replaces emissions from transportation. In order to produce heat or power, crop waste may be compressed into a pellet and put into a commercial boiler. The pellet itself would not be a biofuel in this case, right? The US Department of Agriculture's (USDA) interpretation of the Biomass Crop Assistance Program (BCAP) final regulation, which oversees the biomass crop support programme, leads us to the conclusion that the answer is indeed yes. If implemented in policy, this understanding will broaden the scope of uses for biomass. The "sustainability" and "renewability" standards that constitute the definition of "biofuel" must, however, be continually revised from a policy perspective. The issue might be resolved if the need for electricity generation is larger than the requirement for transportation fuel substitution (Kesan, n.d.).

7.5.9.7 No Legal Provision for *RE*

The German Renewable Energy Act of 2000 ("German Renewable Energy Act," 2013) grants renewable energy a distinct legal position. There are no significant energy laws in any of the African nations. Under their energy laws, Australia, the Philippines, Saudi Arabia, and India all cover renewable energy. There are numerous energy laws in Canada. Iraq does not yet have an energy law, but one is being considered. Instead of having an energy law, Russia, Turkey, and Malaysia do have an energy policy ("Energy Law," 2013). Additionally, some of the benefits of renewable energy do require more legal definition (Platner & Frenkil, 2011), for example. Can someone sell a wind farm with no book value in the event of a resale after claiming 100% depreciation? If so, does it make sense? Laws governing energy or renewable energy should be in excellent harmony with other legal matters. There should be rules for selling RE divides that have depreciated but are still functional. There must be some circumstances for 100% depreciation. Additionally, there is no legal support for Restriction of Hazardous Substances (RoHS), Waste from Electrical and Electronic Equipment (WEEE), etc. enforcement in Brazil, Russia, India, China and South Africa (BRICS) nations. The promotion of Renewable Energy (RE) is not uniform. Some nations support the use of REC and Feed-in-Tariff (FiT) (Lauber & Mez, 2004). FiT is popular in Germany, while REC is a more sensible strategy for promoting renewable energy; both include restrictions on how the cost of energy is determined and widely used in US. In fact, the price should be determined by a genuine market-driven free-floating process, like the one that exists today in the market for oil or gold. The policy should either be in line with the market for carbon credits or be linked to it so as to enhance or strengthen the market's stability over time.

7.5.9.8 Data for Cost of *CCS* Technology

Utilising biochar created by pyrolysis technology is one potential method of biological CCS. Is there data available that assesses the cost of carbon sequestration using various processes, including biochar? A practical substitute, biochar can be created and used while generating carbon credits and benefiting a farmer by increasing the yields of crops. There are not any concrete statistics on biological CCS yet, though. The biofuel project must include biochar as a possibility, and its role in blend fuel usage must be made explicit. The extensive field trial ought to be used to defend the application of biochar. Pyrolysis technology has not yet been referenced in Indonesian and Indian bioenergy policies (Hutapea, 2011).

7.5.9.9 Hydro Is Conditionally Renewable

Whether hydropower qualifies as a conditionally renewable energy source, similar to nuclear energy, is a topic of intense controversy at the moment. If hydropower is generated in an area where dense woods have persisted for a number of years, it cannot be fully accounted for as sustainable. The ecosystem, which would ordinarily be able to support the growth of forests and serve as a natural carbon sink, as well as infrastructure and human settlement, could be harmed by the reservoir flooding these places. These activities also contribute to localised water and air pollution, biodiversity loss, infrastructure destruction, landscape changes, habitation destruction, loss of livelihood, and loss of cultural identity (Kaunda et al., 2012). Furthermore, it has been shown that hydroelectric facilities with significant water catchment area storage capacity may release some GHGs, mainly CO_2 and CH_4 as a result of buried organic waste decaying in the absence of sufficient oxygen (Howarth et al., 2011). Relatively high water temperatures may encourage this activity when dissolved gases in the water are turbulently degassed and escape to the atmosphere through surface bubbling in reservoirs. It can be suggested that the reservoir's top surface be covered using renewable energy options like growing algae, raft or marine farming, or floating solar PV in order to reduce evaporative water loss and temperature rise. Such decisions could further reduce the hydropower technologies' life-cycle GHG EFs, which vary from 15 to 25 g CO_2-eq./kWh (Howarth et al., 2011).

7.5.9.10 Data Assessment Study

The evaluation of a nation's data is crucial; for instance, India lacks a study on solar data assessment but has a reliable assessment of its wind data. At the moment, no policy makes it necessary to report country-specific energy data. The issue of the scarcity of up-to-date, correct, comparable, and reliable data on the performance of renewable energy systems globally needs to be addressed at the highest levels (Gielen, 2012).

7.5.9.11 Social Responsibilities

For small and medium-sized firms, processing the paperwork for carbon credits can be difficult. Many times, manufacturers, himself, fail to take the financial benefits of carbon credits into account due to ignorance and lack of awareness. This disadvantages renewable energy sources. For example, the benefit of carbon credits for community solar cookers is not taken into consideration and is not even provided to the customers by the manufacturer. Manufacturers of air ventilators are not eligible for carbon credits due to electricity replacement for the same reasons. In order for qualified consumers to acquire CDM advantages and the goods to become cost-competitive, assistance with document processing and other formalities is required on a civic level. In the same way as anti-tobacco campaigns, alcohol regulations, etc. do this, such messages must be displayed in technical specifications or brochures. Additionally, socially responsible initiatives in this area include marketing and awareness building. Certain uses are excluded from the CDM benefit preview, such as the carbon sequestration activity of using trees for furniture, doors, and windows. Users who use coal ash from power plants are also eligible for carbon sequestration, but they must be eligible for the CDM benefit in these situations.

7.5.9.12 Green Labelling Must Include Emission

The effort to green-label products must take into account the energy equipment's energy consumption as well as an EF in kg of CO_2/kWh. Low electricity consumption might be caused by high embedded emission responsibility. To distinguish items, for instance, between air coolers and air conditioners, the life-cycle EF must be used. The same labelling requirements must apply to food goods. The usage of fertiliser should be accurately recorded, and any offered subsidies must be incorporated and noted on the label. The use of excessive fertiliser and free electricity results in agricultural products that are inflated as inexpensive and emit a lot of unreported emissions. To lower a higher emission level, a particular percentage of biofertiliser must be made mandatory.

7.5.9.13 Disparity in the Consumption Pattern

In general, advanced countries appear to consume a lot of energy per person, whether for domestic or commercial uses. As a result of resource availability, it has been observed that the cost of energy is highly subsidised in some economies. For instance, hydel power accounts for 58% of Canada's total gross energy output, with some of the lowest hydel power tariffs in the world being found in the hydro-dominant provinces of British Columbia, Manitoba, and Québec. Although the cost of electricity is higher in other provinces than it is in these three provinces, they are still low in contrast to many International Energy Agency (IEA) member nations (*Energy Policies of IEA Countries - Canada 2009 Review*, 2010). The use of diesel fuel for heating is excluded. The justification given for low tariffs in hydroelectric power plants is to generate enough demand for electricity to generate enough base load for the hydroelectric plant to operate efficiently. Nevertheless, putting in a common grid and connecting it to the national grid are good substitutes. Similarly in Iraq for example, where the domestic price of petrol is 28% of the price paid for imports and 11% for diesel (*Iraq Energy Outlook - World Energy Outlook Special Report*, 2012). Advanced nations are less susceptible to energy conservation initiatives due to their high per capita energy use. Also for the nations that explore for oil, the same holds true. The conditions of the policy must specify that people cannot be given energy at unreasonably low prices that force an inefficient use. People should not be permitted to waste energy just because it is affordable or abundant. For the same reason, gas flare-ups at extraction sites (*Iraq Energy Outlook - World Energy Outlook Special Report*, 2012) should not be allowed.

India serves as a good illustration of the wide range of electricity demand. It is important to note that the majority of the nation's coal reserves are found in the east. The country's western half is where the electricity consumption centres are located. Thanks to extensive industrialisation, which has flourished in western states like Gujarat and Maharashtra, among others. This results in the conventional eastward location of several power plants. In these conditions, there is a daily huge transfer of many tonnes of coal, primarily by special trains, from eastern coal mines to western power facilities. The best course of action to fix the problem is to wisely utilise the common grid that is currently in place. It is observed that most of the transportation of coal is currently carried out by trains, which operate on DC electric current. This means that transportation of coal effectively consumes more electricity compared to the transmission and distribution (T&D) losses that might have occurred had electricity transported via a common grid. A logical approach must be developed to assess (a) the amount of electricity lost through T&D losses from the power plant to the point of use. As well as (b) the actual cost per kilometre and per tonne of coal transported. A significant activity of transporting fuel would have been avoided if such a scientific accounting system had been implemented. The nation will see less congestion on its already congested train network as an additional benefit. Such a concept will enable the relocation of power plants to the eastern parts of the nation and will aid in creating jobs, which is essential for these states given the area's high poverty rate. The excessive and unregulated use of costly off-grid power generation for farming or other purposes is a further problem. For instance, rural India has high levels of diesel consumption. Diesel usage that is not related to transportation needs to be closely controlled. Off-grid electricity production with diesel is actually 3–4 times more expensive per unit than electricity purchased commercially, hence it must be avoided. However, because of such uncontrolled diesel fuel waste, the country currently faces enormous import obligations. To stop such misuse, urgent action is needed in energy policy.

7.5.9.14 Consideration of Embedded Energy or Emission

There is no policy that incorporates sustainable criteria into the accounting process, while estimating emission cost. Embedded energy and emission are that quantity that sunk in to providing manufacturing facility of any product. It may be plant, machinery, transportation, mining, and processing of metal. Currently, there are no standard methods for accounting embedded energy and emissions for the plant and manufacturing facility. There are no region-specific embedded energy and emissions data available. Accurate data on emission levels are hard to come by because no nation has implemented energy controls (Khusru & Noor, n.d.). Partly as these quantities are not even considered in the analysis. It was noted that when it comes to some items, consideration of embedded energy is essential, such as Compact Florescent Lamp (CFL) and Light-Emitting Diodes (LEDs), which are known for operating more efficiently (four times) than incandescent light bulbs. However, 10%–30% of lifetime energy comes from embedded energy (Quirk, n.d.). The life-cycle analysis study issued a warning that the impact of manufacture and disposal plays a significant role in light of an obligatory end-of-life scenario as per WEEE recycling law (Welz et al., 2011). This reality is ignored by the current policy framework (Quirk, n.d.). Additional embedded emissions, which can affect 10%–30% of life-cycle emissions, depend on the EM of the manufacturing nation (Welz et al., 2011). An intriguing study of embedded energy and emission data for building materials, including brick, cement, sand, steel, stone chips, timber, and concrete, which account for 40% of global emissions and have a significant impact on natural resources, proposed an empirical equation of life-cycle CO_2 emission for these materials (Khusru & Noor, n.d.). A useful method in the case of fuel-efficient cars is to increase manufacturing embedded emission to minimise emission at the time of use. Since the product is dominated by emissions during operation, it was desirable to employ advanced materials, high-efficiency heat transfer surfaces, eliminate gear shifts, adopt a new design with safety features, reduce noise emissions, etc. (Pehnt, 2002). In some research, the impact of embedded energy can be extremely important. According to the study, using energy-efficient CFLs saves consumers (around 10%) more money than the initial production costs of the energy and resources required to make CFLs. In contrast, the payback for refrigerators is only around 50% of their useful lifetime, and to a great extent, energy-efficient air conditioners also have a negative payback (Quirk, n.d.). Large disparities are seen in the case of items like wind turbines because of the poor embedded energy and emission data (Lenzen & Munksgaard, 2002). The information about embedded energy and emission might alter how they are disposed of. Due to the presence of large metallic components, decommissioning or end of use actually produces energy and needs to compensate or credited back into embedded energy quantities (Rahman, 2003).

7.6 CONCLUSION

The aim of the chapter is to explain to the reader how energy policy is formed, and more specifically, the author aims to explain why different countries' energy policies differ despite though they all wish to pursue similar goals. What are their restrictions and limitations, and how did each one come up with a special strategy to get around them? The chapter also wishes to emphasise the strategies or plans each nation is doing to accomplish the same objective. The chapter also discusses the current criteria and index that have emerged in the process of analysis through time and how this is accomplished with the use of meaningful indicators created by various nations. The chapter also identifies and discusses some areas where there may be room for improvement to make energy policy more logical and dynamic so that a nation may successfully transition to a low-carbon economy.

ACKNOWLEDGEMENT

The authors are thankful for the support received from School of Mechanical, Materials & Energy Engineering at IIT Ropar and it is gratefully acknowledged.

REFERENCES

Alam, S. (2016). In Support of a Market Mechanism for Energy Efficiency to Address Energy Trilemma: Bangladesh Context.

Amoo, A., Flores, F. C., & Ranalkar, S. (2016). Home Energy Management System. https://doi.org/10.13140/RG.2.2.26165.27368.

Bond, T. C., Zarzycki, C., Flanner, M. G., & Koch, D. M. (2011). Quantifying immediate radiative forcing by black carbon and organic matter with the Specific Forcing Pulse. *Atmospheric Chemistry and Physics*, 11, 1505–1525. www.atmos-chem-phys.net/11/1505/2011/.

Brown, L. R. (1980). Food or fuel: New competition for the world's cropland. *Interciencia*, 5(6), 365–372.

Brune, J., Harder, D., & Klingenberger, L. (2023). Critical analysis of shareholder benefits from spin-offs and carve-outs of carbon-intensive businesses: A study of the energy industry. *Opportunities and Challenges in Sustainability*, 2(1), 1–17. https://doi.org/10.56578/ocs020101.

Climate Change and Nuclear Power 2012. (2012). IAEA, International Atomic Energy Agency.

Coal Tax to Fund India's 750 MW Solar Power Capacity Addition Plan in 2013. (n.d.). CleanTechnica. Retrieved June 2, 2013, from https://cleantechnica.com/2013/05/20/coal-tax-to-fund-indias-750-mw-solar-power-capacity-addition-plan-in-2013/.

Country Energy Profile: Ghana-Clean Energy Information Portal-Reegle. (2013, March 15). www.reegle.info/countries/nigeria-energy-profile/GH.

Cozzi, L. (n.d.). World Energy Outlook 2012.

Das, S. K. (2011). India Country Report 2010 Mid-Term Statistical Appraisal-SAARC Development Goals. Central Statistics Office Ministry of Statistics and Programme Implementation Government of India.

Data, G. G. (2005). Navigating the Numbers. https://www.consumieclima.org/download/navigating_numbers.pdf.

Dessus, B., Laponche, B., & Le Treut, H. (2008). Global Warming: The Significance of Methane. Unpublished Manuscript. https://global-chance.org/IMG/pdf/CH4march2008.pdf.

Ei-Ashry, M. (2012). REN21. 2012. Renewables 2012, Global Status Report. REN21 Secretariat.
Energy law. (2013). In Wikipedia, the free encyclopedia. https://en.wikipedia.org/w/index.php?title=Energy_law&oldid=555947294.
Energy Policies beyond IEA Countries-Ukraine 2012. (2012). OECD/IEA. www.iea.org/book
Energy Policies of IEA Countries JAPAN 2008 Review. (2008). OECD/IEA, 2008, International Energy Agency.
Energy Policies of IEA Countries-Canada 2009 Review. (2010). OECD/IEA.
Energy Policies of IEA Countries-Ireland 2012. (2012). International Energy Agency. https://www.iea.org/reports/energy-policies-of-iea-countries-ireland-2012-review
Energy Policies of IEA Countries-New Zealand 2010 Review. (2010). International Energy Agency. www.iea.org/publications/freepublications/.../NewZealand2010.pdf.
Ernest and Young. (2012). Benchmarking the Global Nuclear Industry 2012. EYGM Limited, 1205-1360198 EYG no. DX0137.
Even Higher Costs and More Headaches Ahead for Nuclear Power in 2012. (n.d.). Japan Today. Retrieved May 30, 2013, from https://www.japantoday.com/ category/opinions/ view/ even-higher-costs-and-more-headaches-ahead-for-nuclear-power-in-2012.
Everts, M., Huber, C., & Blume-Werry, E. (2016). Politics vs markets: How German power prices hit the floor. *The Journal of World Energy Law & Business*, 9(2), 116–123. https://doi.org/10.1093/jwelb/jww005.
Fthenakis, V., Kim, H. C., Held, M., Raugei, M., & Krones, J. (2009). Update of PV energy payback times and life-cycle greenhouse gas emissions. *24th European Photovoltaic Solar Energy Conference and Exhibition*, 21–25. https://elnostrefuturenergetic.cat/tutenslaparaula/images/pdf/ScientificPaperonPV.pdf
Garnaut, R. (2011). *The Garnaut Review 2011: Australia in the Global Response to Climate Change*. Cambridge University Press. https://books.google.com/books?hl=en&lr=&id=LR85_mD941cC&oi=fnd&pg=PR7&dq=%22of+Australia,+except+where+a+third+party+source+is%22+%22catalogue+record+for+this+publication+is+available+from+the+British%22+%22the+Commonwealth+of+Australia.+The+Commonwealth+of+Australia+and+all+persons+acting+for%22+&ots=4FnHawCBO1&sig=NAPwgJoHeltVZhi61hNXCQG8UAY.
German Renewable Energy Act. (2013). In Wikipedia, the free encyclopedia. https://en.wikipedia.org/w/index.php?title=German_Renewable_Energy_Act&oldid=547249376.
Gibbins, J., & Chalmers, H. (2008). Preparing for global rollout: A 'developed country first' demonstration programme for rapid CCS deployment. *Energy Policy*, 36(2), 501–507. https://doi.org/10.1016/j.enpol.2007.10.021.
Gielen, D. (2012). Renewable Energy Technologies: Cost Analysis Series. International Renewable Energy Agency.
GoI (2011). CO_2 Baseline Database for the Indian Power Sector, User Guide, Version 6.0 Government of India, Ministry of Power, New Delhi.
Grigoryev, L. M., & Medzhidova, D. D. (2020). Global energy trilemma. *Russian Journal of Economics*, 6(4), 437–462. https://doi.org/10.32609/j.ruje.6.58683.
Harmelink, M. (2006). Guidelines for the monitoring, evaluation and design of energy efficiency policies. "Active Implementation of the European Directive on Energy Efficiency" (AID-EE).
Heffron, R., Connor, R., Crossley, P., Mayor, V. L.-I., Talus, K., & Tomain, J. (2021). The identification and impact of justice risks to commercial risks in the energy sector: Post COVID-19 and for the energy transition. *Journal of Energy & Natural Resources Law*, 39(4), 439–468. https://doi.org/10.1080/02646811.2021.1874148.
Heffron, R. J., & Talus, K. (2016). The development of energy law in the 21st century: A paradigm shift? *The Journal of World Energy Law & Business*, 9(3), 189–202. https://doi.org/10.1093/jwelb/jww009.

Howarth, R. W., Santoro, R., & Ingraffea, A. (2011). Methane and the greenhouse-gas footprint of natural gas from shale formations. *Climatic Change*, 106(4), 679–690. https://doi.org/10.1007/s10584-011-0061-5.

Hutapea, M. (2011, March 2). Bioenergy Policy of Indonesia. Seminar on Bio Based Economy: Towards Sustainable Production and Application of Biomass, Jakarata, Indonesia.

IEA. (n.d.). Energy Policies of IEA Countries 2013 Review Sweden. OECD/IEA, 2013, International Energy Agency (IEA).

India Boasts a Growing economy, and Is Increasingly a Significant Consumer of Oil and Natural Gas. (2011, November 21). Energy Information Administration, Country Analysis Brief - India, www.eia.doe.gov.

Institute, G., & London, I. C. (2012, December 21). Is nuclear power necessary for solving climate change? *The Guardian*. https://www.guardian.co.uk/environment/2012/dec/21/nuclear-power-necessary-climate-change.

Iraq Energy Outlook-World Energy Outlook Special Report. (2012). OECD/IEA. www.iea.org.

Is No One Against the New Truck Fuel Efficiency Standards? (2010, November 10). Popular Mechanics. https://www.popularmechanics.com/cars/alternative-fuel/news/heavy-duty-truck-fuel-efficiency-standards.

Japan Prime Minister Shinzo Abe Promises to Speed Up Recovery on Second Anniversary of Deadly Tsunami, Fukushima Nuclear Disaster. (2013, March 11). *NY Daily News*. https://www.nydailynews.com/news/world/japan-pm-promises-speed-recovery-anniversary-deadly-tsunami-fukushima-nuclear-disaster-article-1.1285276.

Joskow, P., & Parsons, J. (2012). The Future of Nuclear Power after Fukushima. MIT Centewr for Energy and Environmental Policy Research.

Kakkar, J. (2013). Standing Committee Report Summary, The Coal Mines (Conservation and Development) Amendment Bill 2012. PRS Legislative Research, Institute of Research Policy Studies, New Delhi, www.prsindia.org.

Kaunda, C. S., Kimambo, C. Z., & Nielsen, T. K. (2012). Hydropower in the context of sustainable energy supply: A review of technologies and challenges. *ISRN Renewable Energy*, 2012, 1–15. https://doi.org/10.5402/2012/730631.

Kesan, J., & Ohyama, A. (2011). Understanding US ethanol consumption and its implications for policy: A study of the impact of state-level incentives. *University of Illinois Law Review*, Forthcoming, 10–31. https://papers.ssrn.com/sol3/papers.cfm?abstract_id=1792813.

Kesan, J. P. (n.d.). The Renewable Energy Policy Puzzle: Putting the Pieces Together. Retrieved April 11, 2013, from https://illinoislawreview.org/wp-content/ilr-content/articles/2011/2/Biofuel_Introduction.pdf.

Khusru, S., & Noor, M. A. (n.d.). Inventory of life cycle CO_2 emission of selected building materials of Bangladesh. Retrieved January 13, 2013, from https://www.iebconferences.info/370.pdf.

Lauber, V., & Mez, L. (2004). Three decades of renewable electricity policies in Germany, in: Lutz Mez (Guest editor), Green Power Markets - History and Perspectives. Special Issue of Energy & Environment.

Lenzen, M., & Munksgaard, J. (2002). Energy and CO_2 life-cycle analyses of wind turbines-Review and applications. *Renewable Energy*, 26(3), 339–362.

Lewis, C. A. (1997). Fuel and energy production emission factors. MEET Project: Methodologies for Estimating Air Pollutant Emissions from Transport. Retrieved July, 28, 2007.

Matusuo, Y. (2012). Summary and Evaluation of Cost Calculation for Nuclear Power Generation by the "Cost Estimation and Review Committee." *IEEJ*.

McCollum, D., Gould, G., & Greene, D. (2009). Greenhouse Gas Emissions from Aviation and Marine Transportation: Mitigation Potential and Policies. Sustainable Transportation Energy Pathways Program & Institute of Transportation Studies, University of California, Davis, USA.

Measuring Progress towards Energy for All: Vol. OECD/IEA. (2012). World Energy Outlook 2012- Special Topics. www.worldenergyoutlook.org.

Metz, B. (2000). International equity in climate change policy. Baltzer Science Publishers BV, Integrated Assessment 1, pp 111–126.

Mikunda, T., & Kober, T. (2012). CCS in Emerging Economies: Mexico and Indonesia. CATO2-WP2.3-D07, ECN-O--12-039.

Nanduri, M. (1998). An Assessment of Energy Intensity Indicators and Their Roles as Policy-Making Tools [Master of Natural Resource Management Thesis]. School of Resource and Environmental Management, Simon Fraser University.

Oliver Wyman & Marsh & McLennan. (2022). World Energy Trilemma Index 2022.pdf (Index No. 2022; World Energy Trilemma Index, p. 46). World Energy Council. https://www.worldenergy.org/publications/entry/world-energy-trilemma-index-2022.

On the Future of Carbon Capture and Storage in Europe. (2013). Brussels, COM(2013) 180 final, European Commission. ec.europa.eu/energy/coal/doc/com_2013_0180_ccs_en.pdf.

Miere, P. J. (2002). Life Cycle Assessment of Electricity Generation Systems and Applications for Climate Change Policy Analysis.

Parikh, K. (2011). Low Carbon Strategies for Inclusive Growth-An Interim Report. Planning Commission Government of India.

Patel, A., Sarkar, P., Tyagi, H., & Singh, H. (n.d.). System and Method for Tracking Details of Energy Associated with an Energy Generation Process (Patent No. 202011012620).

Pehnt, M. (2002). Life Cycle Assessment of Fuel Cell Systems. *Erscheint in Fuel Cell Handbook*, 3. https://gis.lrs.uoguelph.ca/AgriEnvArchives/bioenergy/download/fuel-cells_lifecycle_pehnt.pdf.

Piere Gadonneix, World Energy Council. (2012). Sustainability Index-2012. World Energy Council.

Platner, M., & Frenkil, J. (2011, April 12). IRS Guidance on 100% Bonus Depreciation Benefits Renewable Energy Sector. VanNess Feldman, Washington, DC.

Pliousis, A., Andriosopoulos, K., Doumpos, M., & Galariotis, E. (2019). A multicriteria assessment approach to the energy trilemma. *The Energy Journal*, 40(01). https://doi.org/10.5547/01956574.40.SI1.apli.

Quirk, I. (n.d.). Life-Cycle Assessment and Policy Implications of Energy Efficient Lighting Technologies. Retrieved January 13, 2013, from https://nature.berkeley.edu/classes/es196/projects/2009final/QuirkI_2009.pdf.

Rahman, S. (2003). Green power: What is it and where can we find it? *Power and Energy Magazine, IEEE*, 1(1), 30–37.

Recent Developments in Energy Subsidies. (2012). IEA.

Reducing Black Carbon Emissions in South Asia-Low Cost Opportunity. (n.d.). U.S. Environmental Protection Agency.

REP, I. (2013). International Standardisation in the Field of Renewable Energy. https://www.indiaenvironmentportal.org.in/files/file/International_Standardisation_%20in_the_Field_of_Renewable_Energy.pdf.

Samantaray, B., & Patnaik, K. (2010). A Study of Wind Power Potential in India [UG Project Thesis]. University of Rourkela.

Satish Rao, H. (2010). Reducing Carbon Emissions from Transport Projects. ADB Evaluation Study, EKB: REG 2010-16.

Saygın, H., & Çetin, F. (2011). Recent developments in renewable energy policies of Turkey. *Renewable Energy-Trends and Applications*, 25–40. https://www.intechopen.com/source/pdfs/22677/InTech-Recent_developments_in_renewable_energy_policies_of_turkey.pdf.

Singh, N. (2013). Creating market support for energy efficiency: India's Perform, Achieve and Trade scheme. *Climate & Development Knowledge Network*. https://www.ctc-n.org/resources/creating-market-support-energy-efficiency-india-s-perform-achieve-and-trade-scheme

Szabó, S., Bódis, K., Huld, T., & Moner-Girona, M. (2011). Energy solutions in rural Africa: Mapping electrification costs of distributed solar and diesel generation versus grid extension. *Environmental Research Letters*, 6(3), 034002. https://doi.org/10.1088/1748-9326/6/3/034002.

Three Mile Island Accident. (2013). In Wikipedia, the free encyclopedia. https://en.wikipedia.org/w/index.php?title=Three_Mile_Island_accident&oldid=556936689.

UK - Bioenergy Strategy. (2012). Dept. of Transport, Dept. of Energy and Climate Change, Dept of Food and Rural Affair. www.official-publications.gov.uk and www.decc.gov.uk.

U.S.-China Joint Announcement on Climate Change. (n.d.). Whitehouse.Gov. Retrieved October 9, 2015, from https://www.whitehouse.gov/the-press-office/2014/11/11/us-china-joint-announcement-climate-change.

Verstina, N., Solopova, N., Taskaeva, N., Meshcheryakova, T., & Shchepkina, N. (2022). A new approach to assessing the energy efficiency of industrial facilities. *Buildings*, 12(2), 191. https://doi.org/10.3390/buildings12020191.

Weiss, M. A., Heywood, J. B., Drake, E. M., Schafer, A., & AuYeung, F. F. (2000). On the Road in 2020. Report MIT EL 00-003, Laboratory for Energy and Environment, Massachussetts Institute of Technology, Cambridge MA. https://lees.mit.edu/public/In_the_News/el00-003.pdf.

Welz, T., Hischier, R., & Hilty, L. M. (2011). Environmental impacts of lighting technologies-Life cycle assessment and sensitivity analysis. *Environmental Impact Assessment Review*, 31(3), 334–343. https://doi.org/10.1016/j.eiar.2010.08.004.

World Energy Outlook 2012 Factsheet. (2012). www.worldenergyoutlook.org/media/weowebsite/2012/factsheets.pdf.

World Nuclear Association. (2013, April 17). Nuclear Power in Japan, Japanese Nuclear Energy, *World Nuclear News*. https://www.world-nuclear.org/info/Country-Profiles/Countries-G-N/Japan/#.UX1Y9bXviSo.

Zafeiratou, E., & Spataru, C. (2018). Sustainable island power system - Scenario analysis for Crete under the energy trilemma index. *Sustainable Cities and Society*, 41, 378–391. https://doi.org/10.1016/j.scs.2018.05.054.

8 Recent Developments in Large-Scale Solar Flat Plate Reflecting Systems

Optical Analysis Using Specialized Numerical and Analytical Tools

Jay Patel and Amit R. Patel
The Maharaja Sayajirao University of Baroda

NOMENCLATURE

Symbol	Description
A_a	Absorber area (m^2)
A_c	Total collector aperture area (m^2)
G_{sc}	Extra-terrestrial radiation incident outside of the earth (W/m^2)
$\hat{N}$	Unit vector normal to the plan (–)
n	Number of rays received by the basin (Nos.)
Q	Heat flux per unit area (W/m^2)
$\hat{S}$	Unit vector of ray (–)
T	Temperature (°C)
t	Time (s)
u_p, v_p, w_p	Coefficient in the wall (plane) equation (–)
u_s, v_s, w_s	Coefficient in the ray (line) equation (–)
v_{in}	Velocity of drying air (m/s)
W	Work output (kJ/s)
	Greek Letter
β	Angle made by surface to horizontal plane (°)
θ	Angle of incidence (°)
δ	Declination angle (°)
ρ	Density (kg/m^3)
ω	Hour angle (°)

DOI: 10.1201/9781003403456-8

φ	Latitude of the location (°)
γ	Surface azimuthal angle (°)
θ_Z	Zenith angle (°)
	Abbreviation
CPC	Compound Parabolic Concentrator
CTSRR	Curved-Type Solar Radiation Reflector
FTSRR	Flat Plate-Type Solar Radiation Reflector/Reflecting
HR	Heliostat Reflector
LFR	Linear Fresnel Reflector
MCOM	Monte Carlo Optimization Method
PTC	Parabolic Trough Concentrator
PV	Photovoltaic
RTS	Ray Tracing Simulation
SDC	Solar disk collector
STS	Solar Thermal System
STA	Solar Thermal Application
	Subscript
1,2,3	For the coordinate related to the first, second and third points chosen on wall
p	For the coordinate related to the plane
t	For the coordinate related to the point of intersection of ray and wall
o	For the coordinate related to the receiving point at basin
s	For the coordinate related to the sun ray

8.1 INTRODUCTION

Human evolution has been influenced by the availability and utilization of energy. Throughout history, energy has played a crucial role in advancing society, starting from the early use of fire and animal power to the modern world where electricity and cleaner sustainable fuels are utilized for various purposes (Hong et al., 2013; Hunt et al., 2018). Energy is indispensable in every sector of society, spanning from cooking, heating, cooling, lighting, transportation, and operating appliances to information technology, communication, and machinery. However, ensuring access to reliable and clean energy has become a challenging task for global development and the well-being of humanity (Dincer & Rosen, 2011; Manieniyan et al., 2009; Sharma & Savithri, 2021). Therefore, it is essential to comprehend the fundamental processes of energy circulation to optimize its utilization. The term "energy flow" is described as the production, import, export, storage changes, transformation, and consumption of energy by energy industries. The steps involved in the energy flow process are shown in Figure 8.1. It also illustrates the interrelationship between a country's energy demands (represented as the reference territory in Figure 8.1) and its import and export policies. Additionally, the diagram shows the energy losses that occur during the transformation process.

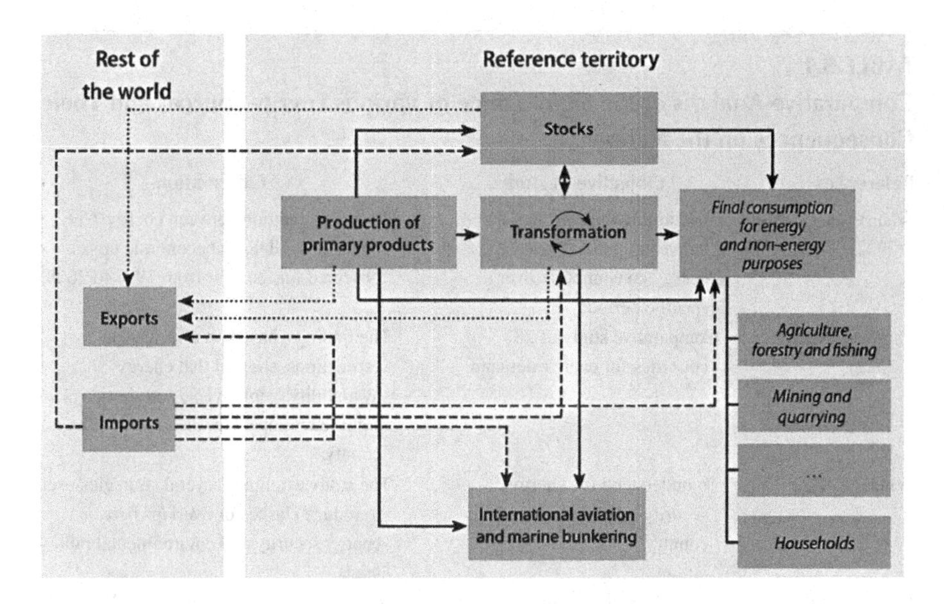

FIGURE 8.1 Diagram of energy flow (Sharma & Savithri, 2021).

The economic growth of a country is largely dependent on the availability of abundant and diverse energy sources (Amit, 2021). Table 8.1 provides a concise overview of the various energy sources required by a country and their impact on the financial situation of the country. From the open literature, it is found that energy sources are significant and have an impact on a nation's economic structure and there is a significant correlation between financial crises and energy demand. It is also noteworthy that the cost of production in different sectors is directly influenced by the availability and affordability of energy (Mannhardt et al., 2023; Paramati et al., 2022). An efficient energy supply that is also affordable helps enterprises run more productively and promotes economic competitiveness (Wang et al., 2022). Therefore, the country like India is essential to invest on the research, development, and use of various renewable energy sources for its long-term economic growth, energy security, and environmental sustainability (Li & Just, 2018).

A detailed examination of India's projected energy usage until the year 2020 is shown in Figure 8.2. It gives in-depth information on the patterns, trends, and percentages of energy consumption in the nation's different sectors or sources during the period of time. The potential for renewable energy production at the state level is shown in Figure 8.2b, whereas Figure 8.2a shows the potential for renewable energy generation at the national level. Figure 8.2c displays the output of petroleum products by its type during the 2019–2020 fiscal year. Figure 8.2d presents data on the power consumption of different industries. It can be seen that for the period of 2019–2020, there will be a total demand for electricity of 13×10^5 GWh, out of that around 67% of that demand coming from industries and home appliances. However, Figure 8.2e and f represents the year-wise installed capacity and energy generated from different sources

TABLE 8.1
Comparative Analysis of the Significance of Various Energy Sources and Their Consequences on the Nation

References	Objective of study	Observation
Wang et al. (2022)	To examine the correlation between financial years and energy consumption over a specific period	There is a relation between energy, CO_2 emissions, GDP, and economic crises observed for the year from 1997 to 2008.
Paramati et al. (2022)	Comparative study of 28 countries for energy demand.	The findings, based on a variety of estimations, showed that energy availability from diverse sources plays a significant impact on the growth of any county.
Winzer (2012)	To understand the energy security and reliability of any country.	The study examines several strategies used to reduce clashes or overlaps between energy security and environmental policy goals.
Li & Just (2018)	To examine the energy and climate policy for household application.	In the study article, the process behind households' short-term demand for electricity and natural gas is examined. It seeks to comprehend how families make decisions about their usage of different energy sources in a comparatively short amount of time.
Yanqin (2011)	To study the financial growth for the year from 1981 to 2010.	The study suggested a few policies based on the energy consumption and its relation to the import and export of goods for India.
Katircioglu et al. (2021)	To investigate the energy demand in developed countries for the period of 2000 to 2005.	Increasing or printing money within the markets has a negative impact on energy demand.
Alam et al. (2015)	To examine the relationship among energy consumption, and economic growth for the SAARC countries over a period of 1975–2011.	The SAARC countries face a choice between energy and economic growth, and it is important for them to work together and take coordinated actions to transition from a state of limited energy availability to a more efficient use of energy resources.
Aktar et al. (2021)	To study about the impacts of the COVID-19 pandemic on the world economy and world energy demand.	Nations must be ready to deal with these repercussions and create recovery plans. In order to maintain sustainable development, it is critical for countries to appreciate the value of natural resources and give their protection first priority.
Kusch et al. (2012)	To study the impact of solar and wind energy on the performance of the low-voltage grid of Germany.	It has been determined that raising the country's thermal storage capacity will aid in meeting its long-term energy needs.

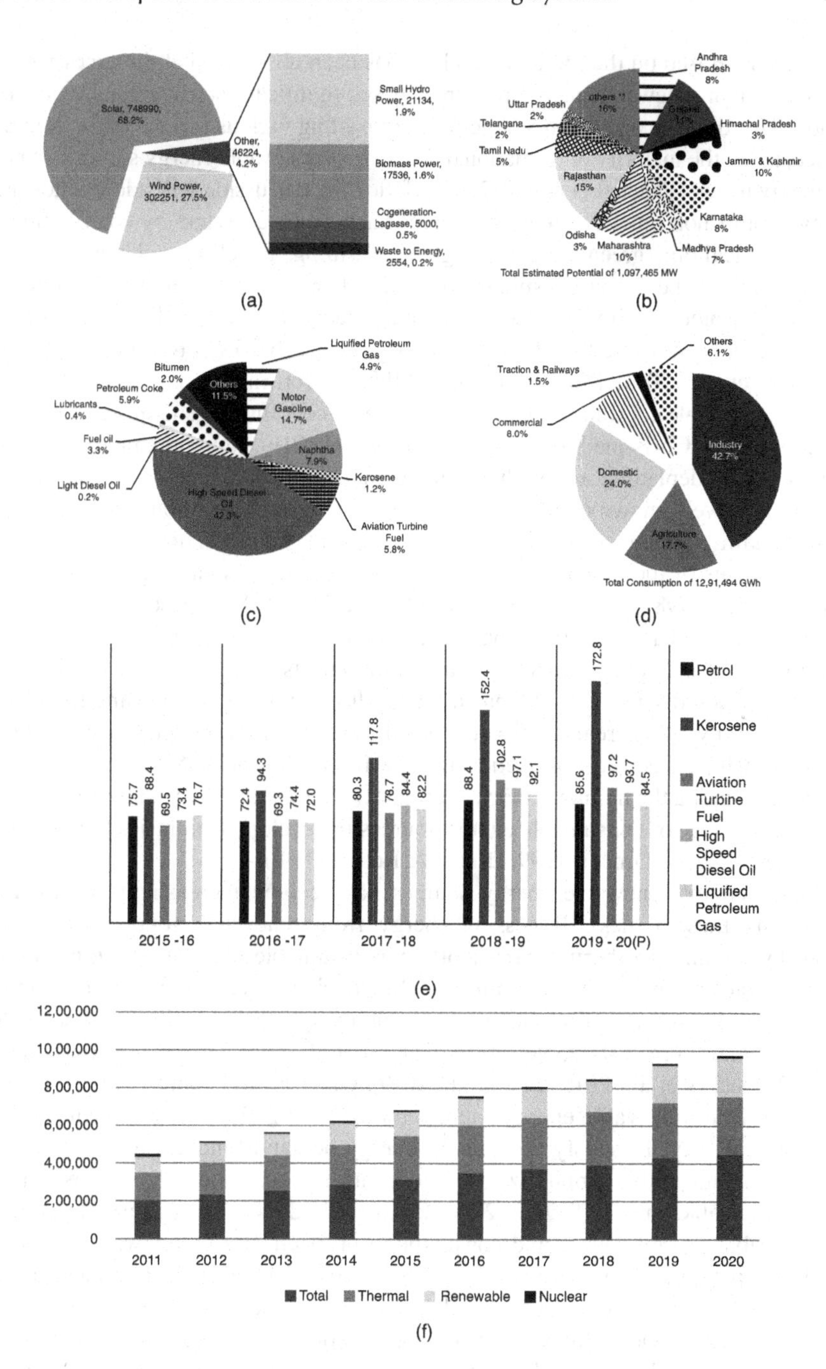

FIGURE 8.2 Specifics of India's energy usage. (a) Estimated potential of renewable energy as of March 31, 2020, (b) estimated potential of renewable energy as of March 31, 2020, (c) production of petroleum products by type of product, (d) consumption of electricity by sector during 2019–2020, (e) yearly wholesale price indices of selected energy commodities, and (f) yearly and state-wide installed capacity of generation of electricity in MW (Sharma & Savithri, 2021).

respectively. Based on the literature study, it has been observed that the excessive utilization of non-renewable energy resources poses significant detrimental effects on a nation. This damage occurs on a large scale, impacting various aspects of the country. Hence, today the majority of the nation relies on non-renewable energy sources to meet its energy needs (Panchal et al., 2017). The following discussion will address how utilizing a lot of non-renewable energy sources harms the ecosystem by contributing to issues like acid rain, greenhouse gases, global warming, and climate change.

On average, the globe consumes 76 million barrels of oil per day. Despite the well-documented detrimental environmental effects of fossil fuel burning, this is anticipated to rise to 123 million barrels per day by 2025 (Voronkin et al., 1995; Bhandari et al., 2014; Mohanakrishna, 2016). According to a growing body of research, if people continue to degrade the environment future generations' fates will be jeopardised (Christine Stork, 1965; Verma et al., 2016). Acid precipitation, stratospheric ozone depletion, and global climate change are three environmental issues that are now well-known on a global scale. Just a few years ago, conventional pollutants including carbon monoxide, nitrogen oxides, particulates, and sulphur dioxide were the focus of the majority of environmental evaluations and regulatory regulations (Dincer, 1998). Dincer provided a full overview of these gaseous and particle pollutants, as well as their effects on the environment and human life (Dincer, 2000). The impact of different concentrated acid rain and its damage to historical monuments has been discussed by Mohajan (2018). It is found that increasing the global temperature by +2°C creates a tremendous difference in the ice glacier of "Europe country," which leads to a reduction in 42% truism business. Moreover, to create awareness among the human society, a few articles have studied the damages done by the excessive use of consumable energy sources in developed countries (Amit, 2021; Damm et al., 2017; Dincer, 1999; Dincer & Rosen, 1998).

The need for renewable energy sources has become increasingly apparent in addressing these issues. Harnessing energy from renewable sources like solar, wind, hydro, and geothermal power offers a sustainable alternative. Transitioning to renewable energy is crucial to mitigate climate change, reduce pollution, and preserve natural resources. Embracing renewable energy technologies is essential for a greener and more sustainable future (Gandhi et al., 2017; Grodsky & Hernandez, 2020; Viviescas et al., 2019; Zhu et al., 2019). Only 9% of India's total power output came from renewable energy sources by the year 2020, which indicates a low installed capacity. To satisfy the nation's energy demands and advance sustainable power generation, this emphasizes the necessity of a considerable increase in the use of renewable energy (Sayigh, 2000; Zhao et al., 2020). The estimated potential of renewable energy source and consumption of electricity sector wise for Indian economic perspective has been depicted in Figure 8.2b and d. The greatest source of renewable energy is the sun and its direct as well as indirect impacts on the globe (solar radiation, wind, falling water, and various plants, i.e., biomass), gravitational forces (tides), and the heat of the earth's core are used to power the system of energy consumption (Akorede et al., 2010; Amit et al., 2018; Bellos, 2019; Jignesh et al., 2018; Khatri et al., 2021). These resources offer huge energy potential, but they are widely distributed and difficult to access. In the following section, the history of solar energy utilization for different applications is discussed.

8.1.1 History of Solar Energy and Its Evaluation

Solar energy is the world's most plentiful source of energy, and it has been utilized for thermal power generation, domestic lighting, and industrial heating for centuries (Chai et al., 2021; Yao et al., 2019). Because scientific research and technology have been generated throughout the previous many decades, it has been the most promising source of energy. India, Egypt, Morocco, and Mexico, for example, have significant amounts of solar radiation and are moving toward concentrating solar power for energy generation (Oluoch et al., 2020). A historical review of solar energy use is provided followed by a description of several types of collectors including flat-plate, compound parabolic, parabolic trough, Fresnel lens, parabolic dish, and heliostat field collectors (Kumar & Arora, 2020).

According to history, solar energy collectors or reflectors have been used to harness the sun's power since prehistoric times (Shukla et al., 2017). Archimedes, a Greek scholar and physicist, used concentrated sun rays to destroy the Roman fleet in 212 BC in the war, which is presented in Figure 8.3. Archimedes is claimed to have set fire to the Roman fleet attacking him by reflecting hundreds of polished shields on the same ship with a concave metallic mirror as mentioned in few of the literature (Liu et al., 2021; Shen et al., 2020; Spencer McDaniel, 2019; Wrighton & Society, 1978; Zahedi, 2011). Surprisingly, the early applications of solar energy entail the use of concentrating collectors, which are more "difficult" to accomplish owing to their nature (exact form fabrication) and the requirement to follow the sun (Apostolopoulos et al., 2020). In the 18th century, polished iron, glass lenses, and mirrors were used to create solar furnaces capable of melting iron, copper, and other metals. Moreover, the lenses used to concentrate solar energy can be claimed by giving the example of French chemist Lavoisier in 1862. Lavoisier has installed a large glass lens on intricate supporting structures to concentrate sun energy on the

FIGURE 8.3 Historical evident, the use of solar energy to destroy ship by Archimedes (McDaniel, 2019).

contents of distillation flasks (Ali et al., 2019). Later on, Becquerel discovered in 1839 the photovoltaic (PV) effect in selenium. Other PV materials, such as gallium arsenide (GaAS), were discovered during research in the 1960s. These were more expensive than silicon and could work at higher temperatures (Borah et al., 2014). It is worth indicating the global installed capacity of PV was close to 30 GW by the end of 2021 (International Energy Agency, 2021). However, several articles (Carmona et al., 2021; Giaconia & Grena, 2021; Herrando et al., 2023; Ramalingam & Indulkar, 2017; Panchal et al., 2017; Priyanshu et al., 2018; Wang et al., 2019) claimed that the amount of solar energy converted to PV is significantly lesser than the amount of solar energy converted to thermal energy.

In addition, there are a number of obstacles to solar energy's widespread adoption and integration into the energy system. Intermittency is one of the main problems. The ability to generate solar power depends on the availability of sunshine, which varies during the day and is affected by seasonal changes. This intermittency poses difficulties in meeting consistent energy demand, especially during periods of low sunlight. Effective energy storage solutions, such cutting-edge batteries or pumped hydro storage, are required to store excess energy for use during periods of poor solar power in order to solve this difficulty (Apaolaza-Pagoaga et al., 2023). Therefore, in the current study, the primary focus is on the various solar thermal applications (STAs) and various advanced methodologies used.

Over the past 50 years, numerous versions of STA have been developed and constructed. The performance of STA depends on mainly two factors: (a) geometry of reflector and (b) purpose of the use. However, there are two types of solar reflectors mainly used, (a) Curved-type Solar Radiation Reflector (CTSRR) and (b) Flat Plate-Type Solar Radiation Reflector (FTSRR) (Coventry, 2005). Additionally, the FTSRR offers a bigger surface area for solar energy collecting, enabling a higher amount of incident sunlight to be caught. Second, it directs sunlight onto the receiver with more uniformity. The equal distribution of solar energy on the flat surface minimizes the possibility of hot spots and improves thermal performance. Furthermore, the flat plate reflector is generally easier to manufacture, install, and maintain compared to the curved reflector. Its simpler design and construction process contribute to cost-effectiveness and practicality (Kalogirou, 2004).

The aim of this study is to analyse and explore mathematical models and numerical approaches for simulating and analysing FTSRR systems. The chapter's structure includes an introduction that addresses the total energy demand of a country, the environmental harm caused by non-renewable energy sources, and the potential of STA as a renewable energy solution. Section 8.2 focuses on various techniques for utilizing solar thermal energy. Section 8.3 provides a detailed explanation of the mathematical and computational methodologies used by relevant literature. In Section 8.4, various numerical modelling techniques for solar thermal systems (STS) are covered, along with their benefits and drawbacks. A summary of the study's findings and recommendations is provided in the concluding section, which is followed by a citation to the research article that served as the study's foundation.

8.2 SOLAR THERMAL ENERGY UTILIZATION TECHNIQUES

Solar energy usage is one of the most promising approaches to dealing with modern concerns such as global warming, fossil fuel depletion, and population expansion. Solar energy is a plentiful and sustainable source of energy with a large exergy potential due to the sun's high temperature (around 5,760 K) (Singer & Peterson, 2011). There are so many applications of solar energy which are later used for thermal applications for different temperature ranges. Table 8.2 lists the most common applications that use solar energy as the only source of energy and its different working temperature range as mentioned applications. It is observed that the average temperature range is from 50°C to 150°C for solar heating and solar refrigeration systems. For industrial applications, the operating temperature range is from 50°C to 400°C. The usage of various geometric constructions for various solar collector applications is explained in the section that follows.

8.2.1 Solar Reflectors (According to Geometric Construction)

In order to harness a sufficient amount of solar energy for various applications, it is important to capture the incoming sunlight in a specific arrangement or on a receiving surface. The design of the solar reflector depends on the geometric considerations as discussed previously in Section 8.1.1. The detailed classification of curved-type and flat plate-type reflectors is shown in Figure 8.4. CTSRR and FTSRR have been used for different STAs, i.e., baking of food with a solar oven, cooking of food with the help of a solar cooker and a few general uses such as solar water heating, solar still, solar furnace, solar air heating, space heating and cooling, and solar refrigeration (Selvakumar & Barshilia, 2012).

8.2.1.1 Curved-Type Solar Radiation Reflector

The curved-type reflecting surface reflects all received solar radiation at the focal point and to better efficiency it should be tracked very precisely and mounted very sturdy as

TABLE 8.2
Applications of Solar Energy in General and Temperature Information

	Applications	Temperature
Heating Application (50°C–70°C)	Domestic hot water production	~50°C
	Space heating applications	50°C–70°C
	Drying applications	50°C–70°C
	Cooking and baking	120°C–260°C
Cooling/refrigeration with sorption machines (80°C–150°C)	Single-stage machines	80°C
	Multistage machines	150°C
Industrial heat, for instance, steam production (50°C–400°C)	Washing procedures	90°C
	Chemicals production	200°C
	Methanol reforming	~300°C

Source: Adapted from Bellos & Tzivanidis (2019a) with permission from Springer Nature.

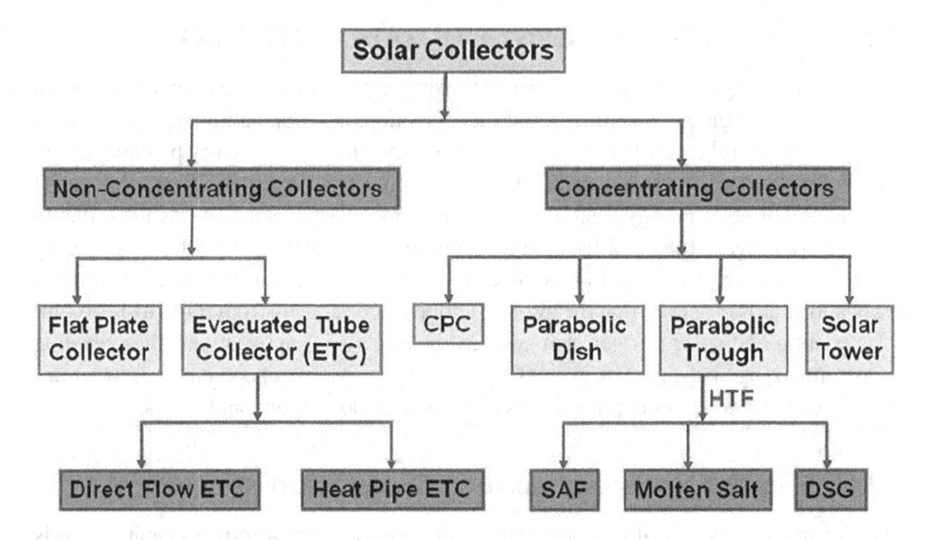

FIGURE 8.4 Classification of types of solar receiving surfaces according to geometric construction (Selvakumar & Barshilia, 2012).

they are exposed to different weather conditions (Donev, n.d.). The temperature level achieved with the use of the curved type of reflector is about 250°C–1,200°C and it mostly depended on the value of concentration ratio (*C*) which can be calculated with the help of Eq. (8.1) (Ruivo et al., 2022). Normally, the value of *C* is kept 10 and it is critical to note that a higher value of *C* allows for more efficient operation at higher temperatures. In the following discussion, the use of CTSRR for different STAs like solar parabolic trough collector (SPTC), compound parabolic collectors (CPCs), solar disk collector (SDC), parabolic dish reflector (PDR) (Liu et al., 2015) has been discussed. The SPTC is a device that receives solar radiation in discrete form and concentrates it to the focal point of the receiver. An absorber has been placed at the focal point of SPTC (Jebasingh & Herbert, 2016). However, the CPC is a non-imaging concentrating collector. It primarily uses sun beam irradiation, although a little amount of diffuse irradiation is also used (Bellos & Tzivanidis, 2019a). The SDC is also a focusing type concentrating device as shown in Figure 8.5a. It can handle medium and high temperatures between 230°C and 500°C and it concentrates the solar radiation on the receiver in the linear form (Blanco & Miller, 2017).

$$\text{Concentration Ratio}(C) = \frac{\text{Total concentrator aperture}}{\text{Receiver surface}} \tag{8.1}$$

The PDR is a point-focus system with a paraboloid geometry given by the revolution of one half of a parabola around its normal axis. Sunlight entering the collector aperture with a normal incidence is concentrated onto a heat receiver located at the focal point of the dish. Typically, the value of *C* is high compared to SPTC and SDC, hence it is used for achieving large temperature range of 130°C–580°C (Mlatho et al., 2010). Mostly, the CTSRR is used for achieving high temperature but to achieve it system must be aligned to the sun rays. Thus, it uses a solar tracking mechanism

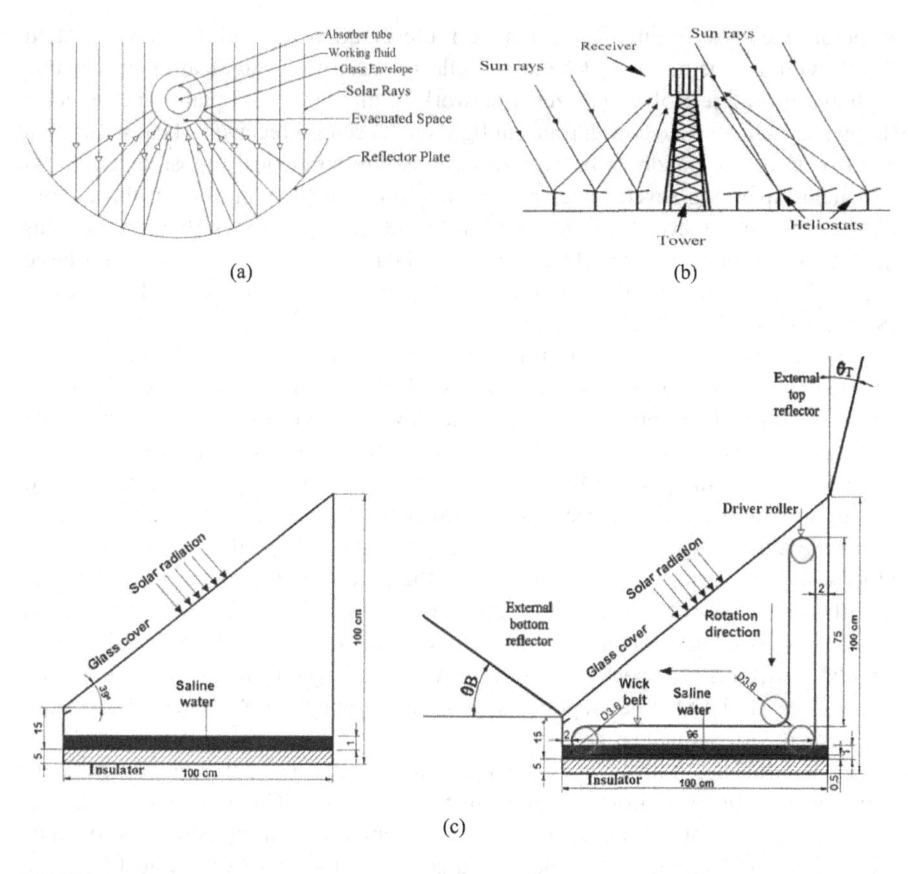

FIGURE 8.5 Schematic diagram of (a) SDC (adapted from Apostolopoulos et al., 2020 with permission from Elsevier), (b) CRS (adapted from Kalogirou, 2004 with permission from Elsevier), and (c) solar still with and without the flat plate reflector (adapted from Abdullah et al., 2021 with permission from Elsevier).

which itself is a costly system and led to an increase in the cost of overall energy generation. This limitation of CTSRR has been solved by using a flat plate-type reflecting surface (Masud, 2021).

8.2.1.2 Flat Plate-Type Solar Radiation Reflector

The flat plate type reflector is flat in nature and all solar radiation fall on it reflects back without concentrate to the receiving surfaces for further application. The prime objective to use of FTSRR is to reflect all incident radiation to deciding surface for different STAs. It received solar energy and reflects it back to the receiver without concentrate; hence, it does not require precise tracking (Patel & Patel, 2023). The manufacturing of this kind of reflector is very easy and more cost-effective due to its simple design, low cost, and relatively easier installation compared to its curved counterpart (*Alternative Energy Tutorials*, 2020). Heliostat Field Reflector and Linear Fresnel Reflector (LFR) are the two STAs that employ solar flat plate reflectors the most often. The schematic representation of central receiver system (CRS) with a large number of heliostat reflector used for solar steam power plants is shown in Figure 8.5b. It utilizes a network of mirrors or flat plate-type reflector (known as heliostats) to concentrate sunlight onto a central receiver. These heliostats, numbering in the hundreds or even thousands, are strategically positioned across vast fields spanning several hectares. By tracking the sun's movement throughout the day, it can achieve the value of *C* from 500 to 1,500 times the normal solar flux. This concentrated sunlight enables the plant to generate electricity on a large scale, with capacities ranging from tens of megawatts to several hundred megawatts (Sánchez-González, 2022).

The LFR is a linear concentrating device and it can able to handle medium and high temperatures between 230°C and 500°C. Additionally, it has less operating challenges, remains stable while in operation, has lower wind loads, and is less expensive compared to PTC (Wang et al., 2022). The solar still is also used a flat plate reflector to increase its overall performance as shown in Figure 8.5c. It has been demonstrated that improving the performance of solar stills through the use of flat plate reflectors is a successful strategy. The reflectors work by bouncing the sunlight that would have otherwise been lost or scattered back onto the still, thus increasing the solar heat absorbed by the system. Improved evaporation rates and increased distillate output are the results of this enhanced solar energy input (Reji et al., 2021). Additionally, flat plate reflectors have been utilized in STAs for baking and cooking. The following discussion goes into further depth on solar cooking, solar baking, and solar furnaces augmented with flat plate reflectors.

A solar furnace with heliostat reflectors is mostly used for large temperature difference applications in the range from 520°C to 1,460°C (Monreal et al., 2015). It is a structure that harnesses sun energy to generate high temperatures, primarily for industrial applications, and uses a series of flat mirrors or heliostat reflectors to focus and redirect the sun's rays towards the receiver area which results in extremely high temperatures (Costa et al., 2018; Duffi & Beckman, 1989; Singh & Chandra, 2018). Figure 8.6a depicts the image of the solar furnace at Odeillo (France) where the temperature at the receiver can be reached up to 3,500°C, and this heat can be used to generate electricity and melt steel. A solar cooker is a device that transforms sunlight into thermal energy for the purpose of cooking food as shown in Figure 8.6b. It consists of a reflective surface, often made of mirrors or reflective materials, which reflects solar rays onto a receiver (Singh et al., 2021). The solar-box type cooker, for instance, is very easy to build and is made of low-cost materials, but cooking is very slow and the maximum reachable temperature is relatively low, which limits the cooking options. A solar cooker is an environmental friendly and cost-effective device for harnessing solar energy (Wassie et al., 2022). The use of FTSRR for enhancing the performance of a solar cooker is shown in Figure 8.6b. The flat reflector helped to boost the system's overall performance by directing all solar radiation to the receiver, which would have been lost otherwise (Weldu et al., 2019).

On the other hand, the baking process accounts for over a third of the total energy used in bread production. For baking to be finished in the bakery industry, a maximum temperature of 240°C is required and this temperature can be easily achieved

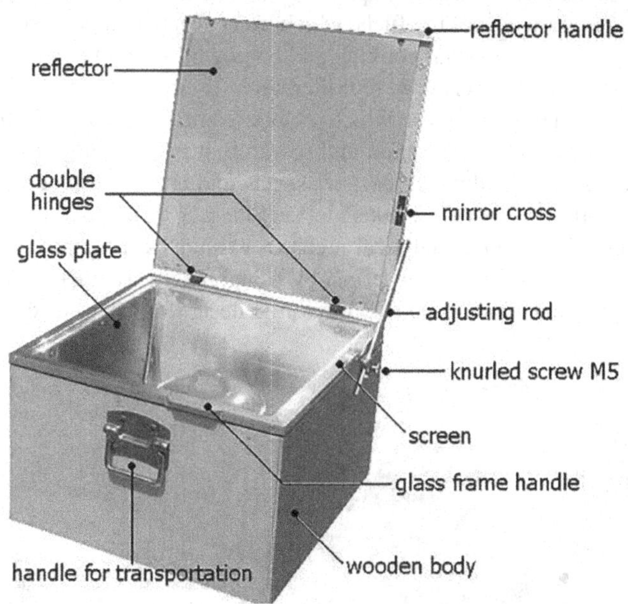

FIGURE 8.6 (a) Image of the solar furnace at Odeillo in France (Mbison & Phillthorpe, 2020). (b) Solar cooker with FTSRR (adapted from Getnet et al., 2023 with permission from Elsevier).

by the integration of STA into the bakery system (Lentswe et al., 2021; Mishra et al., 2019). The same way solar energy can be used for baking of food with the use of a solar oven which works on the same principle of a conventional oven instead of using solar energy than conventional energy sources. Solar ovens, which are specifically designed for baking, provide a controlled and enclosed environment to achieve consistent baking temperatures. The US-based company, Sun Oven International®, has built "*The Villager Sun Oven*," which works on the same principle of a solar oven (Welch et al., 2015). Figure 8.7 shows the actual image of the Villager Sun Oven, which is designed and built for durability and ease of use. It is highly adaptable to changing weather, as it can be rotated to face the sun, and comes with a propane backup system to be used when it rains. From the literature, it is found that it can bake up to 500 bread loaves each day and allows users to save over 150 tonnes of firewood per year, resulting in a reduction of roughly 277 tonnes of greenhouse gas emissions per year. The bread baked in the oven is 30% less expensive than normal bread purchased elsewhere (Welch et al., 2015).

Finally, the thermal use of solar energy is getting a lot of interest in sustainable cooking because of the inefficiencies of traditional cooking and the ever-increasing expense of fossil fuels (Abu-malouh et al., 2023). Apart from that, cooking energy accounts for 36% of global primary energy use. Stone, charcoal, agricultural waste, and dung cakes are prevalent cooking fuels in rural regions, with 86.1% of rural families utilizing fuel wood and cow dung (Pohekar et al., 2005). Every year, around 61,771 square miles of woods are burned to provide this cooking fuel. Household air pollution is connected to such cooking fuels, which has a negative impact on rural inhabitants. The advancements in solar cooking and baking have significantly increased the utilization of renewable energy. A few literatures used a flat plate reflector/mirror with the conventional solar cooker, which are briefed in Table 8.3. To further enhance these processes and foster innovation and research, it is essential to have a comprehensive understanding of detailed design aspects and analytical models (Aragaw & Adem, 2022; Hilda & Daniel, 2019; Patel & Singh, 2015). This involves utilizing various numerical tools and analytical methods to study and improve solar cooking and baking techniques. In simpler terms, by using analytical models and numerical tools, researchers can analyse and optimize the design of solar cookers and ovens,

FIGURE 8.7 Installation of the Villager Sun Oven at the new bakery site (Welch et al., 2015).

TABLE 8.3
Brief Summary of Flat Plate Reflector Used for Different STAs

Authors & References	Objective of Study	Observation
Wassie et al. (2022)	To predict the effect of flat mirror on the performance of solar cooker.	The cooker with three flat mirrors increased the thermal efficiency by about 35% more than the conventional cooker.
Zhu & Huang (2014)	To perform numerical ray tracing simulation to predict the performance of LFR	There is an improvement of 9.3% in the output temperature in the case of LFR used with two plane reflectors.
Zhu et al. (2017)	To design and analyse a scalable LFR with use of a multiple flat reflector for home cooking applications.	By use of scalable LFR, the maximum thermal efficiency is achieved by 64% than simple LFR system.
Bellos & Tzivanidis (2018)	To study the performance of LFR with nanoparticles.	The heat transfer coefficient is found to be enhanced close to 30%–35% due to the combination of nanofluids and LFR
Lin et al. (2013)	To predict the performance of a flat plate reflector-based solar collector prototype with a modified V-shaped cavity receiver	This hybrid system increased efficiency by 18% than a simple system with the highest temperature reach of 150°C.
Tawfik et al. (2021)	To examine the performance of a solar cooker by incorporating flat plate internal reflectors.	The thermal efficiency of a cooker with reflector is increased to 12% than the conventional system.
Apaolaza-Pagoaga et al. (2023)	To investigate the effect of partial load on the performance of a funnel solar cooker made out of flat reflector.	Overall 11% decrement in the thermal performance of the funnel cooker is observed which can be overcome with the use of glass dome-type cover on the cooker.
Harmim et al. (2013)	To analyse the effect of plane and curved reflector on the performance of solar cookers.	The solar cooker performed 7% better when it is augmented with the curved reflector than the flat plate reflector.
Guidara et al. (2017)	To investigate numerically and experimentally the performance of solar furnace with a flat reflector.	The performance of solar system is increased by 24% with the use of four-sided flat reflector over the system.

leading to improved efficiency and better utilization of renewable energy in cooking and baking. The following discussion explains the different analytical methods and numerical tools used in solar cooking and baking (Bauer, 2016; Suharta et al., 1996; Zafar et al., 2019).

8.3 MATHEMATICAL METHODS AND TECHNIQUES

Mathematical and analytical studies are crucial for scientific study as they provide a structured approach to problem-solving, enabling precise analysis and accurate predictions. They offer valuable tools for understanding complex phenomena, guiding research direction, and facilitating informed decision-making. The analysis of the optical behaviour of the solar system provides valuable insights into the characteristics of solar rays. It helps us understand how solar rays behave, including their reflection patterns and the energy they carry as they travel from the sun to the receiver's surface. In the subsequent discussion, these aspects are explained in detail, shedding light on the intricate workings of solar radiation within the solar system.

8.3.1 Optical Analysis of STA

Optical analysis of solar beam ray gives information about the path followed by the ray into the STS. It will give the information about the number of times the reflection is done to reach the ray to its final destination/receiver or it leaves to the system. There are mainly two approaches which have been used to analyse the behaviour of ray, (a) Ray tracing simulation (RTS) and (b) Monte Carlo optimization method (MCOM), in which the RTS is a widely used method to forecast the behaviour of sun beam rays to solar system while MCOM is used for examining the exact number that is important to study critically. For many years, large-scale solar furnaces have been used with various optical structure designs consideration to allow for experiments in various fields in which the flat plate reflector worked as the most important fragment of the system (Cui et al., 2019). In the next section, the discussion on the RTS technique and MCOM is done in detail to understand and explain the characteristics and importance of solar rays in a simpler way.

8.3.1.1 Solar Ray and Its Behaviour

Sun rays reach to earth's surface in two different formats at earth (a) beam rays and (b) diffuse rays, where beam rays play a major role in solar radiation heat transfer due to their straight line traveling approach (Bellos & Tzivanidis, 2019b). The method and equation used to calculate the intensity of solar beam ray on the Earth's horizontal surface is discussed in the following discussion (Duffi & Beckman, 1989; Rehman, 2019). The pictorial representation of the angle made by the solar beam ray to the horizontal surface is shown in Figure 8.8. The amount of radiation received in the outer atmosphere of the earth is calculated with the help of Eq. (8.2), where G_{on} is represented as the extra-terrestrial radiation incident on the plane (W/m^2).

The declination angle (δ) and the angle of inclination (θ) can be calculated with the use of Eqs. (8.3) and (8.4) simultaneously. In actual practice, there may be horizontal ($\beta = 0°$) or vertical ($\beta = 90°$) solar radiation receiving surfaces, in that case, Eqs. (8.5) and (8.6) help to find the value of θ.

$$G_{on} = \begin{cases} G_{sc}\left(1 + 0.033\cos\dfrac{360n}{365}\right) \\ G_{sc}\begin{pmatrix} 1.000110 + 0.034221\cos\delta + 0.001280\sin\delta + \\ 0.000179\cos 2\delta + 0.000077\sin 2\delta \end{pmatrix} \end{cases} \tag{8.2}$$

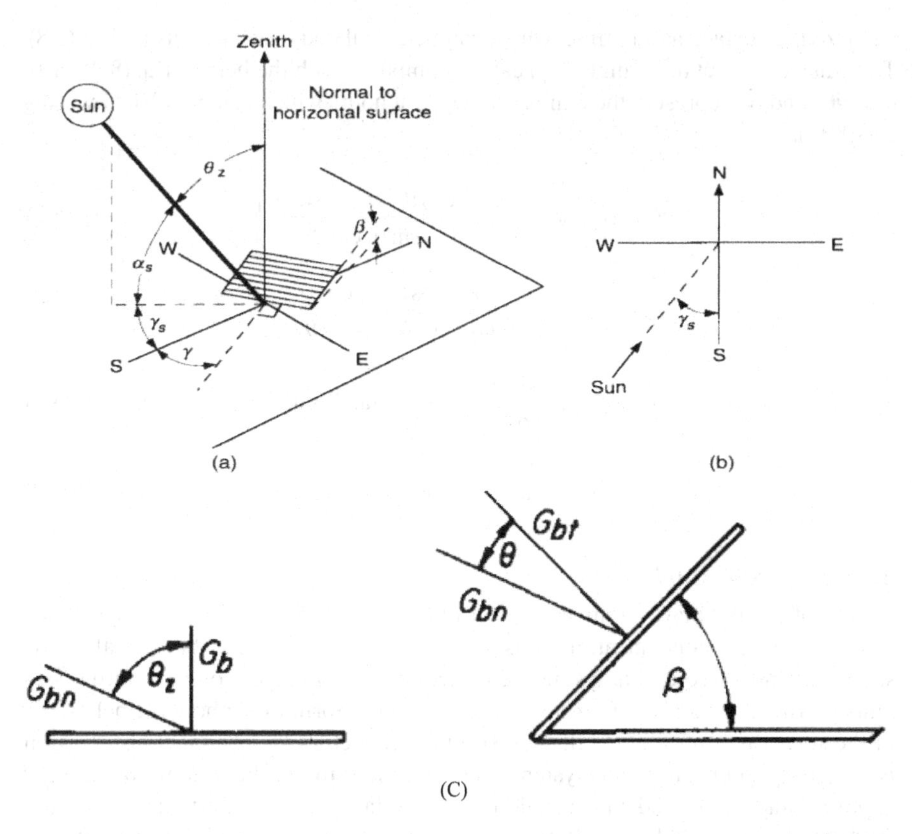

(C)

FIGURE 8.8 (a) Zenith angle, slope, surface azimuth angle, and solar azimuth angle for a tilted surface, (b) plan view showing solar azimuth angle, and (c) beam radiation on horizontal and tilted surfaces (Duffi & Beckman, 1989).

$$\delta = 23.45 \sin\left(360 \frac{284+n}{365}\right) \tag{8.3}$$

$$\begin{aligned}\cos\theta = {} & \sin\delta \sin\phi \cos\beta - \sin\delta \cos\phi \sin\beta \cos\gamma + \cos\delta \cos\phi \cos\beta \cos\omega \\ & + \cos\delta \sin\phi \sin\beta \cos\gamma \cos\omega + \cos\delta \sin\beta \sin\gamma \sin\omega\end{aligned} \tag{8.4}$$

$$\cos\theta = -\sin\delta \cos\phi \cos\gamma + \cos\delta \sin\phi \cos\gamma \cos\omega + \cos\delta \sin\gamma \sin\omega \tag{8.5}$$

$$\cos\theta_z = \cos\phi \cos\delta \cos\omega + \sin\phi \sin\delta \tag{8.6}$$

For purposes of solar process design and performance calculations, it is often necessary to calculate the hourly radiation on a tilted surface of a collector from measurements or estimates of solar radiation on a horizontal surface. For better calculation of the study and to get insight data, the following quantities will help, and the value of solar azimuth angle (γ_s) will be calculated by using Eq. (8.7). The geometric factor (R_b) can be defined as the ratio of beam radiation on the tilted surface to that on a horizontal surface at any time, which can be calculated exactly by using Eq. (8.8). The value of sunset hour angle (ω_s) can be computed with the help of Eq. (8.9) when $\theta_z = 90°$ and N_d represent the number of daylight hours which can be solved by using Eq. (8.10).

$$\gamma_s = \text{sign}(\omega)\left|\cos^{-1}\left(\frac{\cos\theta_z \sin\phi - \sin\delta}{\sin\theta_z \cos\phi}\right)\right| \tag{8.7}$$

$$R_b = \frac{G_{b,T}}{G_b} = \frac{G_{b,n}\cos\theta}{G_{b,n}\cos\theta_z} = \frac{\cos\theta}{\cos\theta_z} \tag{8.8}$$

$$\cos\omega_s = -\frac{\sin\varnothing \sin\delta}{\cos\varnothing \cos\delta} = -\tan\varnothing \tan\delta \tag{8.9}$$

$$N_d = \frac{2}{15}\cos^{-1}\left(-\tan\varnothing \tan\delta\right) \tag{8.10}$$

8.3.1.2 RTS Method

In case of the flat plate augmented solar reflector system, it reflects all rays which fall on it without concentration, thus to understand that process the optical analysis plays a major role. Analysing the optical performance of solar reflecting systems is crucial for several reasons. It will give the information about the behaviour of the solar ray after it gets reflected from the reflectors (Georgiou et al., 2013). It is very easy to optimize the system's design to maximize the amount of sunlight captured and converted into usable energy. Solar reflecting systems are mostly deployed in large-scale installations, such as solar power plants or concentrated solar power facilities. Understanding how sunlight is managed and redirected by these systems allows us to minimize light pollution and unwanted glare, ensuring that the surrounding ecosystem and human activities are not negatively affected (Bellos & Tzivanidis, 2019a). Researchers have proposed several theories in which RTS methods are easy to implement and accurate. RTS method is widely utilized in STA to accurately model and analyse the behaviour of light rays within complex optical systems. By tracing the paths of individual rays, these simulations provide valuable insights into the performance and efficiency of STS (de Bruin & van Sark, 2023).

In the following section, the equation and steps used to perform RTS method are explained with the use of solar still model as shown in Figure 8.9. As shown in the figure, the solar still has four opaque sides and the top surface is covered with transparent glass. However, Figure 8.10 helps in the understanding angle associated with the single beam ray when it gets interacted with any kind of titled surface. In RTS, the line is considered to be a ray and the plane is considered to be a reflector. However, the intersection of line and plane is called as intersection point and it can be obtained by solving line, plane, and its intersection equation simultaneously (Fossa et al., 2021). The line equation can be represented by Eqs. (8.11)–(8.13), where x, y, z represent the coordinate of point on the line, x_0, y_0, z_0 represent the origin point of the line, t shows the non-dimensional number which is used to represent the

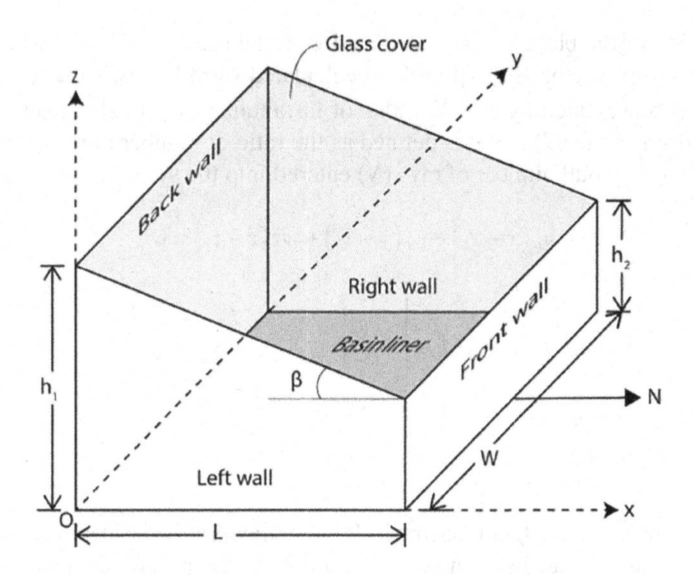

FIGURE 8.9 Schematic of a single slope solar still (adapted from Rehman, 2019 with permission from Elsevier).

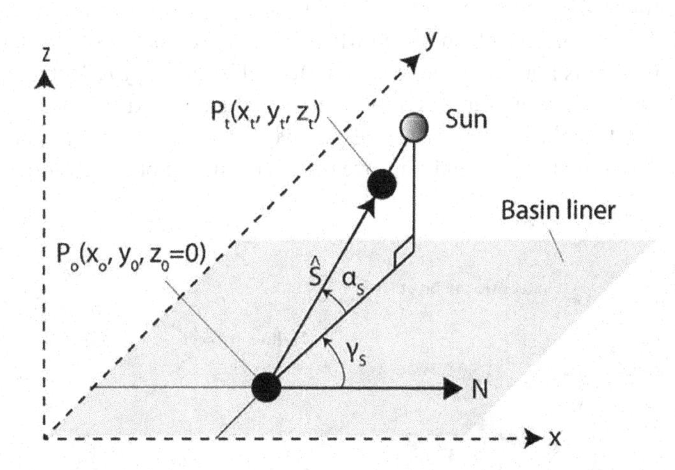

FIGURE 8.10 Basic understanding of solar beam ray follows the vector direction of sun ray (adapted from Rehman, 2019 with permission from Elsevier).

location of point (x, y, z) on the line, and u_s, v_s, w_s show the unit vector of the line and its value can be obtained by using Eq. (8.14).

$$x = x_o + u_s t \tag{8.11}$$

$$y = y_o + v_s t \tag{8.12}$$

$$z = z_o + w_s t \tag{8.13}$$

$$\hat{S} = u_s\hat{i} + v_s\hat{j} + w_s\hat{k} = (\cos\gamma_s \cos\alpha_s)\hat{i} + (\sin\gamma_s \cos\alpha_s)\hat{j} + (\sin\alpha_s)\hat{k} \tag{8.14}$$

The equation of the plane can be calculated with the use of Eq. (8.15), where u_p, v_p, w_p show the unit vector of the plane and calculated from Eq. (8.16). The normal of the plane is represented by $\hat{N}$. The value of instantaneous optical efficiency (η_o) is obtained from Eq. (8.17) and it is defined as the ratio of number of rays received (n) at the base to the total number of rays (N) entered into the system.

$$u_p(x-x_1)+v_p(y-y_1)+w_p(z-z_1)=0 \tag{8.15}$$

$$\hat{N}=u_p\hat{i}+v_p\hat{j}+w_p\hat{k}=\begin{bmatrix} \hat{i} & \hat{j} & \hat{k} \\ x_2-x_1 & y_2-y_1 & z_2-z_1 \\ x_3-x_1 & y_3-y_1 & z_3-z_1 \end{bmatrix} \tag{8.16}$$

$$\eta_o=n/N \tag{8.17}$$

However, Cui et al. (2019) demonstrated RTS method with consideration of a heliostat reflector and a receiver as shown in Figure 8.11. The proposed equation to solve the RTS model by considering diffuse radiation can be found in the following discussion. The solar incidence optic cone centre vector $(\overrightarrow{S_s})$ represents the relation between the altitude angle of the receiver's global axis (α_s) and the receiver rotating axis (γ_s). It is calculated with the use of Eq. (8.18) where R_s is represented as the solar rotation matrix and [0, 0, 1] is known as the axis of system. The primary reflected optic cone center $(\overrightarrow{t_s})$ and the value of unit vector of the heliostat global axis $(\overrightarrow{\eta_H})$ are calculated with the use of Eqs. (8.19) and (8.22), respectively. Besides, a few literatures have used the RTS method to study and improve the performance of solar furnace (SF-60)

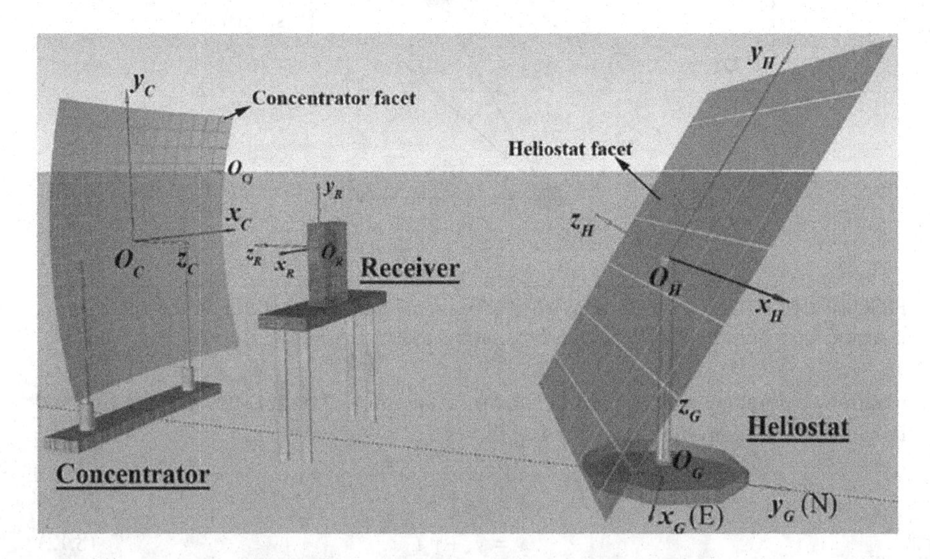

FIGURE 8.11 Schematic diagram of the solar furnace and coordinate systems (adapted from Cui et al., 2019 with permission under the Creative Commons Attribution 4.0 International (CC BY 4.0) license).

at Almera, Spain (Jafrancesco et al., 2014; Pereira et al., 2019). A specific C program has been developed based on the RTS analytical method and from the results they observed that there is a 12% improvement in the thermal efficiency of numerically constructed models when suitable assumptions are taken into account. However, several RTS technique limitations, such as the need for statistical analysis and the modelling of complicated geometry, have been solved by employing MCOM, which is discussed in the discussion that follows.

$$\overrightarrow{S_s} = [0,0,1] R_s = [\cos\alpha_s \cos\gamma_s, \cos\alpha_s \sin\gamma_s, \sin\alpha_s] \tag{8.18}$$

$$\overrightarrow{t_s} = [\cos\alpha_{h2t} \cos\gamma_{h2t}, \cos\alpha_{h2t} \sin\gamma_{h2t}, \sin\alpha_{h2t}] \tag{8.19}$$

$$\overrightarrow{S_s} + \overrightarrow{t_s} = 2 \cdot \overrightarrow{\eta_H} \cdot \cos\theta_s \tag{8.20}$$

$$\cos 2\theta_s = \overrightarrow{S_s} \cdot \overrightarrow{t_s} \tag{8.21}$$

$$\overrightarrow{\eta_H} = [\cos\alpha_H \cos\gamma_H, \cos\alpha_H \sin\gamma_H, \sin\alpha_H] \tag{8.22}$$

8.3.1.3 Monte Carlo Optimization Method

MCOM is more commonly applied to studies where it is needed for statistical analysis of data (simulation and validation) and to estimate probabilities associated with different scenarios. It is a large family of computational algorithms that use repeated random sampling to obtain approximation numerical results. The basic idea is to employ randomness to solve problems that are, in principle, deterministic (Kocharov et al., 1998; Makhmutov et al., 2007; Schüler et al., 2007; Wang et al., 2015). The three main applications of Monte Carlo methods are optimization, numerical integration, and generating draws from a probability distribution. In the following discussion, the steps and equations are discussed. The idea of Lattice Monte Carlo optimization method for determining Normalized Effective Thermal Conductivity was discussed by Li & Gariboldi (2019) in late 20s. They have suggested that the probable expectation ($E(x)$) is calculated from the approximation of random variable (x) as shown in Eq. (8.23). The effective-D (D_{eff}) is known as the dimension material based on random walks of energy particles calculated from Eq. (8.24). However, the effective thermal conductivity of particular d dimension material (λ_{eff}) and the maximum possible conductivity of all phase ($\lambda_{\max}$) are obtained from Eq. (8.26). The total specific heat (I_{eff}) in W/m^2 is obtained from Eq. (8.25).

$$E(x) \approx \frac{1}{N} \sum_{n=1}^{N} x_n \tag{8.23}$$

$$D_{\text{eff}} = \frac{[R^2]}{N} \cdot \frac{\lambda_{\max}}{I_{\text{eff}}} \tag{8.24}$$

$$I_{\text{eff}} = (\rho C_p)_{\text{eff}} = \sum_{j=1}^{n} \rho_j C p_j f_j \tag{8.25}$$

$$\lambda_{\text{eff}} = D_{\text{eff}} * I_{\text{eff}} \tag{8.26}$$

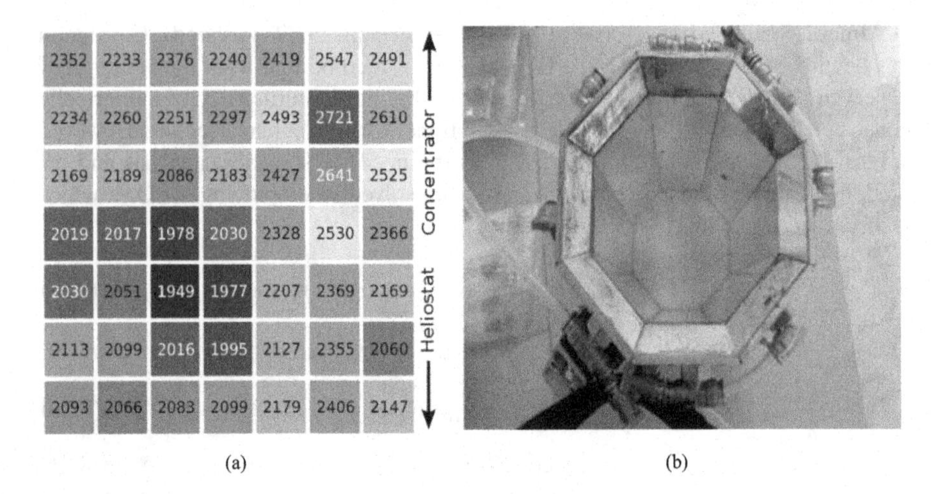

FIGURE 8.12 (a) Energy flux received at base while using a homogenizer, (b) radiation homogenizers with eight sides, seen from the top (adapted from Pereira et al., 2020 with permission under the Creative Commons Attribution 4.0 International (CC BY 4.0) license).

A handful of papers have used MCOM for STA, such as Pereira et al. (2020), who did an experiment with MCOM and the RTS technique to demonstrate the influence of homogeneous flux distribution on the receiver surface. The details are shown in Figure 8.12a and b. According to the findings, a multi-mirror homogenizer (four sides and eight sides) was successfully used to avoid radiation distribution at the solar furnace's focal point. Jafrancesco et al. (2014) conducted a numerical analysis of the impact of mirror reflectors with spherical and parabolic aperture areas on the thermal performance of the STS. They came to the conclusion that spherical mirrors may provide greater temperatures whereas parabolic reflectors can produce a uniform dispersion of sun radiation. Additionally, a few articles (Cui et al., 2019; Riveros-Rosas et al., 2010; Guidara et al., 2017) asserted that optical factors such as focal length, curvature angle, and reflectivity of reflector had a significant impact on the overall performance of the STS. Table 8.4 gives the information about the several STS optical simulation techniques. The numerical method used to solve the STA models is explained in the section that follows.

8.4 NUMERICAL MODELLING TOOL USED FOR STS AND ITS LIMITATION

Numerical methods provide a way to solve problems quickly and easily compared to analytic solutions. There are several numerical techniques and tools available to reduce the time required to solve the complex problem. An overview of the many numerical tools available for STS is provided in this section (Amit R et al., 2014). Numerical model of STS helps in deciding the appropriate size of solar system components by considering the impact of different variables like weather data, solar

TABLE 8.4
Summary of RTS Method and MCOM Used for Numerical Analysis of STS

Authors & Reference	Method	Objective	Observation
Cui et al. (2019)	RTS	To numerically investigate the effect of tilt error on the heliostat reflector used for solar furnace.	It is found that with increasing the tilt error of the heliostat, the concentrated heat flux distribution and spot are diverging.
Nandwani (1988)	RTS	To analyse the performance of a box-type solar oven with use of four booster reflectors.	It is observed that the effective temperature of the receiving surface can be reached up to 150°C with use of reflector.
Chen et al. (2010)	RTS	To examine the different geometry to obtain a higher value of *C* with use of heliostat reflectors.	The concentrator with the ellipsoidal mirror is performed better by 12% than hyperboloid mirror in system augmented with heliostat reflectors.
Chong & Tan (2012)	RTS	To analyse the effect of azimuthal alignment system with the fully tracked solar flat plate reflecting solar system.	There is a 12% decrement in the solar thermal performance of the solar azimuthal alignment system, which is 23% cheaper than the fully tracked system.
Terrón-Hernández et al. (2018)	RTS	To experiment and validate different position setups for CPC with flat plat reflectors to achieve the temperature in the range of 40°C–60°C.	CPC works satisfactory when it makes an angle of 21°C with the horizontal axis.
Yurcheko et al. (2015)	RTS	To perform numerical analysis of non-tracking type CPC with a heliostat reflector to observe its thermal performance.	Performance-to-cost ratio is greater than 1.2 and during solar peak hours the system is achieving 146°C of temperature.
Cheng et al. (2014)	MCOM	To numerically analyse the performance of the PTC system with a flat plate reflector under different solar altitudes.	The reflecting system worked efficiently in low solar altitude angle when the value of focal length is 0.47 m.
Tanaka (2011)	MCOM	To investigate the effect of the value of *C* on the performance of a funnel-type solar cooker.	There is a 21% increment in the thermal performance of the solar system and observed the value of *C* is from 1.2 to 2.3.
Petrasch (2014)	MCOM	To develop an open source numerical model for testing cylindrical, spherical, and parabolic concentration solar system with a flat plate reflector.	The code has been thoroughly validated and tested with pre-existing code, which shows good agreement. It has been successfully used to design and optimize a series of novel STS.

(Continued)

TABLE 8.4 (*Continued*)
Summary of RTS Method and MCOM Used for Numerical Analysis of STS

Authors & Reference	Method	Objective	Observation
Duan et al. (2020)	MCOM	To develop a numerical model based on MCOM for simulating the radiative flux distribution of the receiver, which is reflected by a large heliostat field.	A comparative investigation of the various configurations of the heliostat mirror on the field produced the best results for the tool's ultimate performance.
Zhao et al. (2018)	MCOM	To analyse the performance of a solar cooker with a FRL reflector.	The inside max. Temperature reaches up to 220°C which is quite good for normal cooking application.
Nydal (2014)	MCOM	To investigate the effect of two flat reflectors on the performance of PTC augmented with sensible heat storage material.	Thermal performance of the novel system is 21% more compared to the conventional system.

irradiation data, geometric input data, and the location of the territory. The following are some of the most significant advantages of STS computer modelling: (a) Save money by forgoing the need to create prototypes. (b) Logical organization is present in complex systems. (c) Exhibit a thorough knowledge of the system's operation, the interactions between its numerous parts, and how to best use each part. (d) Determine the anticipated energy production of the system. (e) Provide a range of temperatures for the system. (f) Using the same weather circumstances, determine how a design-dependent variable will affect the system's performance (Kalogirou & Papamarcou, 2000; Patel et al., 2016).

The fundamental steps of numerical modelling are presented in this subsection. The first is to develop a structure that will be used to represent the system that has to be analysed. While developing a model of system to analyse, the following things should keep in mind. Like, the structure will always be a skewed representation of reality. Both the process of constructing a system structure and the structure itself will aid in people's comprehension of the actual system (Patel et al., 2020; Tiwari et al., 2015). Computing speed, low cost, quick turnaround—which is especially important during iterative design phases—and non-technical users' ease of use are key benefits of simplified analysis methodologies. The lack of control over assumptions and a limited number of systems that can be analysed are all disadvantages; hence, an extensive computer simulation may be necessary to produce accurate results if the system application, setup, or load characteristics under consideration are notably non-standard. The following sections provide a brief overview of software tools such as TRNSYS, Polysun, F-Chart, MATLAB, and Mathematica used in the modelling and prediction of the overall performance of the STS.

8.4.1 TRNSYS Simulation Program

The acronym TRNSYS stands for "transient simulation," which is also referred to as a quasi-steady simulation model. This programme was created by the team at the University of Wisconsin's Solar Energy Laboratory. The mathematical models of the subsystem components are represented in terms of algebraic or conventional differential equations. The entire challenge of system simulation comes down to figuring out all the parts that make up the given system and coming up with a general mathematical description for each one. After all of the system's components have been identified and a mathematical description of each component has been established, it is critical to create an information flow diagram for the system. However, it has a number of drawbacks, such as the fact that it is not an application that is user-friendly (Levy, 2018).

8.4.2 PloySun Simulation Program

PloySun is a tool that offers dynamical annual simulations to aid in the optimization of STS systems. It uses dynamic time steps that range from 1 s to 1 h, which increases simulation stability and accuracy. The program's graphical user interface makes it easy to enter all system parameters and is user-friendly. Physical models, on which the simulation is based, do not in any way make use of empirical correlation terms. The application also maintains ecological balance and analyses economic viability (Ploysun & Velasolaris, 2015).

8.4.3 F-Chart Method

Peña-Cruz et al. (2014) had developed this method, which allows us to estimate the percentage of total heating and cooling that will be provided by the solar energy utilization system. The primary design variable is the collector area; other variables include collector type, storage capacity, fluid flow rates, and the diameters of the load and collector heat exchangers. The technique allows for the simulation of typical water and air system setups.

8.4.4 MATLAB Simulation Program

The multi-paradigm programming language and environment for numerical computing used by MathWorks is called MATLAB. With MATLAB, it is able to perform matrix operations, visualize functions and data, develop algorithms, create user interfaces, and communicate with other programming languages. Thus, ray tracing can be accomplished by using equations for calculating solar rays and concentrating reflectors. This is a highly laborious operation that requires a solid foundation in both science and mathematics (Mathworks et al., 1984).

8.4.5 Mathematica Simulation Program

Mathematica is also considered the world's most powerful global computing environment. This enables you to create technical computing web services, such as numerical, symbolic, and graphical apps, that quickly and easily address your daily technical computing concerns. Mathematica is unique among technical computing platforms in that it comprises a massive library of carefully curated data of all kinds, which is updated and increased on a regular basis. By implementing solar basic radiation equations and few geometric calculative equations, the calculation of the STS can be possible (Torrence & Torrence, 2009; Walfram, n.d.).

8.4.6 SolTrace Software

SolTrace Software is commonly used to perform mathematical modelling and simulation of solar incident radiation on a CTSRR using RT simulation method, especially it helps in displaying the pattern of flux concentration on the receiver. SolTrace results were used to establish the position of this maximum and minimum flux distribution on the receiver because the flux dispersion is typically not uniform. The impact of modifying the flux distribution by changing the aperture width was further investigated by building numerous models in the SolTrace software. These findings will aid in the development of a heat receiver based on maximal heat extraction from the receiver area (More et al., 2018).

8.4.7 Tonatiuh Software

The Tonatiuh software seeks to provide an open-source simulation tool, which uses the MCRT method that is precise and simple to use for the optical modelling of CTSRR. The aim of this work is to enhance the current state of simulation tools for the design and analysis of solar concentrating systems and to make such tools freely available. It is advantageous to have a clear and adaptable software design that makes it simple for users to modify, expand, and increase its features. However, it is primarily employed to analyse only curved-type reflectors (Blanco, 2017).

8.4.8 OptiCAD Software

OptiCAD is a 100% non-sequential, illumination, optical analysis, and visualization programme. OptiCAD can carry out the analysis on arbitrarily placed optical components, with the capability to do diffuse ray tracing with consideration of reflection, scattering luminance, and polarization. Light sources can be modelled as equations, surfaces, volumes, lines, points, and more. Multiple sources may be positioned everywhere and sources may be divergent, convergent, or collimated. Additionally, users may define sources using tables (Osório et al., 2016; Overview, n.d.).

8.4.9 OTSun Software

OTSun is a forward ray tracing numerical tool based on the MC method. With the use of this programme, the trajectory of solar beam radiation can be measured. This tool can evaluate four various surfaces, including opaque, absorptive, specular, and reflective surfaces (Pujol-Nadal et al., 2017).

8.4.10 Raytrace3D Software

Raytrace3D is an in-house software suite that is used for the assessment and optimization of concentrating collectors. It is modified and continually improved through continuous development to meet the requirements of project work. A fast yet adaptable combination of C/C++, Bash/Python, and other languages is used to create both the simulation engine itself and the pre- and post-processing tools. The software package has been used with great success for many different and quite diverse applications. It is mostly utilized for sun-reflecting surfaces that are curved (Qian & Science, 2006).

8.4.11 STRAL Software

STRAL is a ray tracing software with a graphical user interface written in C++ that has been developed at DLR (German Aerospace Centre). The software's primary use is to simulate the flux density of heliostat fields (i.e., a large number of heliostats). The software is principally made to analyse parameter data sets for the whole reflecting surface in the field and real highly resolved data sets of sun shaped distributions. The performance of a single heliostat with precise sun reflection cannot be predicted by using this numerical tool (Osório et al., 2016)

8.4.12 SPRAY Software

SPRAY is a command-line-based RT programme created internally at DLR and written in FORTRAN. It was initially designed by Sandia National Laboratories for testing the optical performance of the solar tower system. The fundamental idea behind the ray tracing code is to produce a large number of solar rays and then track them as they interact with every element of the system that is being evaluated. There are two fundamental ways to generate rays: generating rays on an "insolation box" for forward ray tracing and generating rays on concentrator elements for backward ray tracing. It is mostly utilized for curved reflectors (Osório et al., 2016).

8.4.13 SPEOES Software (ANSYS)

SPEOS is a numerical computing tool used to analyse the optical behaviour of reflectors of both the flat and curved types. It was introduced by ANSYS Inc. in 2002 and is continuously being developed. Its primary application is to analyse the reflected light from the surface. Consequently, it is not entirely able to address flat surface reflection (Mill, 2020). Additionally, the details of availability of tool (commercially/open source) and the limitations of the present tool are shown in Table 8.5. However, Table 8.6 shows the real application of the mentioned numerical tool used for different STS.

The need for precise numerical simulations to predict the system's overall performance arises when the geometry is constructed using several flat plate-type reflectors. The results of such a numerical analysis are extremely helpful in creating

TABLE 8.5
Summary of Different Numerical Tool Used for STS and Its Availability in the Market

Name of Tool & Reference	Availability	Simulation Methods	Limitation
F-Chart (Klein, n.d.)	Commercially available	RTS	Only used for CTSRR surface-based system. Not designed for flat plate-based solar reflecting systems.
OptiCAD (Program, 1999)	Commercially available	RTS	Used for the simulation of Scheffler disc. Not used for the flat plate reflector.
OTSun (Pujol-Nadal et al., 2019)	Open source	RTS	Mainly used for curved type of reflecting surfaces.
Raytrace3D (Filatov, 2019)	Open source	RTS	Only used for curved surface not designed to be used for flat surface.
SPEOES (Mill, 2020)	Commercially available	RTS	Used to analyse the simple optical system available in standard size only.
Tonatiuh (Blanco, 2017)	Open source	RTS	Difficult to perform simulation for a single heliostat reflector.
TRNSYS (Levy, 2018)	Commercially available	RTS	Used for the simulation of heliostat field and difficult to give a precise result of single heliostat reflection
PloySun (Ploysun & Velasolaris, 2015)	Commercially available	MCOM	Used for simulation of large heliostat field but inefficient to give a precise result for a single flat plate reflector.
SolTrace (Wendelin, 2003)	Open source	MCOM	Difficult to perform simulation for a single heliostat reflector.
SPRAY (Richter et al., 2019)	Commercially available	MCOM	Difficult to perform simulation for a single heliostat reflector.
STRAL (Belhomme et al., 2009)	Commercially available	MCOM	Not designed for analysing the single heliostat reflector.

TABLE 8.6
List of the Several Numerical Tools Used for STS Analysis

Authors and Reference	Numerical Tool Used	Objective Behind Study	Observation
Georgiou et al. (2013)	SolTrace	To analyse the performance of a large number of heliostats by computing the solar image size and flux density on a stationary receiver.	The results show that the peak concentration is achieved at the received when the solar altitude angle is from 65°C to 88°C.

(Continued)

TABLE 8.6 (*Continued*)
List of the Several Numerical Tools Used for STS Analysis

Authors and Reference	Numerical Tool Used	Objective Behind Study	Observation
More et al. (2018)	SolTrace	To develop a mathematical model of the solar CPC in order to locate the best suitable receiver's locations.	The flux distribution pattern seen on the absorber was mathematically correlated, according to the findings. The same conclusions are helpful for creating a powerful heat receiver.
Kincaid et al. (2018)	SolTrace	To investigate the optical performance of three representative concentrating solar power collector designs including linear Fresnel, parabolic trough, and central-receiver technologies.	The parabolic trough has the highest optical performance among all.
Belhomme et al. (2009)	STRAL	To build a numerical model to analyse the FTSRR system inaccuracy when the inclination angle is slightly skewed.	It was determined that when there is a change of 2% in the titled angle of flat plate, there is an average 23% decrease in the effectiveness of the system.
Fossa et al. (2021)	Tonatiuh	To analyse the numerical model of solar Fresnel collector for home application.	When the sun was at its brightest, a peak efficiency of 74% was attained. The developed model helps to solve the same kind of reflecting system.
Bode et al. (2014)	Tonatiuh	To develop an openly available algorithm for the STS to decide the precise number of rays for the simulation.	The developed model gave the number of rays required to simulate the RTS methods for specific STS.
Peña-Cruz et al. (2014)	Tonatiuh	To develop a mathematical model for a parabolic trough mirror using the fringe reflection theory.	The suggested findings aid in choosing the number of mirrors and the slope of the mirror for the mentioned solar system.
Bonanos et al. (2019)	Tonatiuh	To project the precise reflection of the heliostat on the receiver with the use of two techniques, photogrammetry and laser scanning.	With the usage of the two proposed procedures, the final output is 20% better.
Giovinazzo et al. (2015)	Tonatiuh	To design a mathematical model of a flat plate collector with flat plate reflectors on both sides based on MCOM.	The performance of the collector is substantially impacted by the use of plane reflectors. In comparison to the conventional model, the modified model is 21% higher in the overall performance.

(*Continued*)

TABLE 8.6 (*Continued*)
List of the Several Numerical Tools Used for STS Analysis

Authors and Reference	Numerical Tool Used	Objective Behind Study	Observation
Hertel et al. (2020)	OTSun	To develop a numerical model of single heliostat reflecting CPC with the use of RTS method.	The proposed numerical model is 21% accurate and there is good agreement between experimental and numerical results.
Tsvetkov et al. (2020)	PolySun	To analyse the mathematical model developed to test the thermal performance of solar hot water systems.	The developed model is validated with experimental results and it will be used to conduct preliminary evaluations of the efficiency of solar hot water supply systems.

accurate experimental setups that can efficiently assess and optimize the operation of the STS. The existing shortcomings of numerical tools for flat plate-type reflectors point to the necessity for more study and advancement in this area. Future studies should focus on creating more sophisticated and robust numerical models that can accurately simulate the behaviour of these FTSRR. Additionally, these numerical tools must be capable of taking into consideration a number of factors, including reflector size, shape, and material characteristics, as well as interactions with incident light or radiation. The design and optimization of such systems can develop significantly if researchers can fill this gap and improve the numerical tools available for flat plate-type reflectors. As a result, flat plate-type reflector applications in a variety of industries, such as solar energy, wireless communication, and radar systems, will perform better, be more efficient, and be more reliable.

8.5 CONCLUSION

The current study offers comprehensive information on the numerous numerical tools used for STS design and analysis that are accessible in the public and private domains. Based on open literature, it is found that these tools, developed by different commercial and institutional publishers using various computer languages, primarily focus on forecasting the optical performance of CTSRRs using the RTM and MCOM. However, there is a lacuna in the availability of specialized software or equipment created especially for monitoring the behaviour of individual solar beam rays. Furthermore, the existing tools primarily provide total sun irradiation information through the shadow reflection method, rather than offering comprehensive insights into the performance of a single flat plate reflector or a complex/typical STS constructed using FTSRR.

Several methods and numerical tools exist for analysing the performance of large heliostat reflector fields. However, the absence of specifically tailored numerical tools for individual reflectors or complex STS configurations is a significant research gap that needs to be addressed. In order to advance utilization of FTSRR-based STS systems in India, it is imperative to develop new and distinctive computing codes that can cater to the specialized numerical analysis requirements of these systems. By creating dedicated tools that focus on individual reflectors and complex STS configurations, researchers and practitioners will be better equipped to optimize their designs, evaluate their performance accurately, and drive advancements in solar energy utilization. This work emphasizes the demand for specialized numerical tools capable of precisely assessing the performance of a single flat plate reflector or a sophisticated solar thermal system utilizing FTSRR.

REFERENCES

Abdullah, A. S., Omara, Z. M., Essa, F. A., Alarjani, A., Mansir, I. B., & Amro, M. I. (2021). Enhancing the solar still performance using reflectors and sliding-wick belt. *Solar Energy*, *214*, 268–279. https://doi.org/10.1016/j.solener.2020.11.016.

Abu-malouh, R., Abdallah, I., & Abdallah, S. (2023). Food preparation using solar oven with automated two axes sun following. *Materials Today: Proceedings*. https://doi.org/10.1016/j.matpr.2023.04.307.

Akorede, M. F., Hizam, H., & Pouresmaeil, E. (2010). Distributed energy resources and benefits to the environment. *Renewable and Sustainable Energy Reviews*, *14*(2), 724–734. https://doi.org/10.1016/j.rser.2009.10.025.

Aktar, M. A., Alam, M. M., & Al-Amin, A. Q. (2021). Global economic crisis, energy use, CO2 emissions, and policy roadmap amid COVID-19. *Sustainable Production and Consumption*, *26*, 770–781. https://doi.org/10.1016/j.spc.2020.12.029.

Alam, A., Malik, I. A., Abdullah, A. Bin, Hassan, A., Faridullah, Awan, U., Ali, G., Zaman, K., & Naseem, I. (2015). Does financial development contribute to SAARC'S energy demand? from energy crisis to energy reforms. *Renewable and Sustainable Energy Reviews*, *41*, 818–829. https://doi.org/10.1016/j.rser.2014.08.071.

Ali, M. Y., Hassan, M., Rahman, M. A., Kafy, A. A., Ara, I., Javed, A., & Rahman, M. R. (2019). Life cycle energy and cost analysis of small scale biogas plant and solar PV system in rural areas of Bangladesh. *Energy Procedia*, *160*, 277–284. https://doi.org/10.1016/j.egypro.2019.02.147.

Alternative Energy Tutorials. (2020). https://www.alternative-energy-tutorials.com/solar-hot-water/flat-plate-collector.html.

Amit, P. (2021). A solar system for alignment with the sun, Indian Patent Act -1970, Indian Patent.

Amit, P., Sarkar, P., Singh, H., & Tyagi, H. (2018). A Method for Determining a Single Emission Discounting Rate for process involving environmentl emissions.

Amit, R. P., Prabir, S., Harpreet, S., Himanshu, T., & Sagi, S. (2014). Life cycle assessment of intermediate pyrolysis of wheat straw for sustainable energy alternate and emission mitigation. *International Review of Applied Engineering Research*, *4*(4), 325–330.

Apaolaza-Pagoaga, X., Carrillo-Andrés, A., Ruivo, C. R., & Fernández-Hernández, F. (2023). The effect of partial loads on the performance of a funnel solar cooker. *Applied Thermal Engineering*, *219*. https://doi.org/10.1016/j.applthermaleng.2022.119643.

Apostolopoulos, N., Chalvatzis, K. J., Liargovas, P. G., Newbery, R., & Rokou, E. (2020). The role of the expert knowledge broker in rural development: Renewable energy funding decisions in Greece. *Journal of Rural Studies*, *78*, 96–106. https://doi.org/10.1016/j.jrurstud.2020.06.015.

Aragaw, Y. T., & Adem, K. D. (2022). Development and performance evaluation of tube-type direct solar oven for baking bread. *Heliyon*, *8*(11), e11502. https://doi.org/10.1016/j.heliyon.2022.e11502.

Bauer, G. (2016). Evaluation of usage and fuel savings of solar ovens in Nicaragua. *Energy Policy*, *97*, 250–257. https://doi.org/10.1016/j.enpol.2016.07.041.

Belhomme, B., Pitz-Paal, R., Schwarzbözl, P., & Ulmer, S. (2009). A new fast ray tracing tool for high-precision simulation of heliostat fields. *Journal of Solar Energy Engineering, Transactions of the ASME*, *131*(3), 0310021–0310028. https://doi.org/10.1115/1.3139139.

Bellos, E. (2019). Progress in the design and the applications of linear Fresnel reflectors - A critical review. *Thermal Science and Engineering Progress*, *10*, 112–137. https://doi.org/10.1016/j.tsep.2019.01.014.

Bellos, E., & Tzivanidis, C. (2018). Multi-criteria evaluation of a nanofluid-based linear Fresnel solar collector. *Solar Energy*, *163*, 200–214. https://doi.org/10.1016/j.solener.2018.02.007.

Bellos, E., & Tzivanidis, C. (2019a). A review of concentrating solar thermal collectors with and without nanofluids. *Journal of Thermal Analysis and Calorimetry*, *135*(1), 763–786. https://doi.org/10.1007/s10973-018-7183-1.

Bellos, E., & Tzivanidis, C. (2019b). A review of concentrating solar thermal collectors with and without nanofluids. *Journal of Thermal Analysis and Calorimetry*, *135*(1), 763–786. https://doi.org/10.1007/s10973-018-7183-1.

Bhandari, B., Poudel, S. R., Lee, K., & Ahn, S. (2014). Mathematical modeling of hybrid renewable energy system: A review on small hydro-solar-wind power generation. *International Journal of Precision Engineering and Manufacturing-Green Technology*, *1*(2), 157–173. https://doi.org/10.1007/s40684-014-0021-4.

Blanco, M. J. (2017). Tonatiuh. In *International Energy Agency's SolarPACES Symposium*, Celebrated in Berlin, Germany. https://iat-cener.github.io/tonatiuh/.

Blanco, M.J., & Miller, S. (2017). Introduction to Concentrating Solar Thermal (CST) technologies. In *Advances in Concentrating Solar Thermal Research and Technology* (pp. 3–25). Cambridge: Woodhead Publishing.

Bode, S. J., Gauché, P., & Griffith, D. (2014). A novel approach to reduce ray tracing simulation times by predicting number or rays. *Energy Procedia*, *49*, 2454–2461. https://doi.org/10.1016/j.egypro.2014.03.260.

Bonanos, A. M., Faka, M., Abate, D., Hermon, S., & Blanco, M. J. (2019). Heliostat surface shape characterization for accurate flux prediction. *Renewable Energy*, *142*, 30–40. https://doi.org/10.1016/j.renene.2019.04.051.

Borah, R. R., Palit, D., & Mahapatra, S. (2014). Comparative analysis of solar photovoltaic lighting systems in India. *Energy Procedia*, *54*, 680–689. https://doi.org/10.1016/j.egypro.2014.07.309.

Carmona, M., Palacio Bastos, A., & García, J. D. (2021). Experimental evaluation of a hybrid photovoltaic and thermal solar energy collector with integrated phase change material (PVT-PCM) in comparison with a traditional photovoltaic (PV) module. *Renewable Energy*, *172*, 680–696. https://doi.org/10.1016/j.renene.2021.03.022.

Chai, S., Yao, J., Liang, J. De, Chiang, Y. C., Zhao, Y., Chen, S. L., & Dai, Y. (2021). Heat transfer analysis and thermal performance investigation on an evacuated tube solar collector with inner concentrating by reflective coating. *Solar Energy*, *220*, 175–186. https://doi.org/10.1016/j.solener.2021.03.048.

Chen, C. F., Lin, C. H., & Jan, H. T. (2010). A solar concentrator with two reflection mirrors designed by using a ray tracing method. *Optik*, *121*(11), 1042–1051. https://doi.org/10.1016/j.ijleo.2008.12.010.

Cheng, Z. D., He, Y. L., Cui, F. Q., Du, B. C., Zheng, Z. J., & Xu, Y. (2014). Comparative and sensitive analysis for parabolic trough solar collectors with a detailed Monte Carlo ray-tracing optical model. *Applied Energy*, *115*, 559–572. https://doi.org/10.1016/j.apenergy.2013.11.001.

Chong, K. K., & Tan, M. H. (2012). Comparison study of two different sun-tracking methods in optical efficiency of heliostat field. *International Journal of Photoenergy*, *2012*. https://doi.org/10.1155/2012/908364.

Christine Stork, D. A. (1965). Carbon monoxide. *American Industrial Hygiene Association Journal*, *26*(4), 431–434.

Costa, B. A., Lemos, J. M., & Guillot, E. (2018). Solar furnace temperature control with active cooling. *Solar Energy*, *159*, 66–77. https://doi.org/10.1016/j.solener.2017.10.017.

Coventry, J. S. (2005). Performance of a concentrating photovoltaic / thermal solar collector. *Solar Energy*, *78*, 211–222. https://doi.org/10.1016/j.solener.2004.03.014.

Cui, Z., Bai, F., Wang, Z., & Wang, F. (2019). Influences of optical factors on the performance of the solar furnace. *Energies*, *12*(20). https://doi.org/10.3390/en12203933.

Damm, A., Greuell, W., Landgren, O., & Prettenthaler, F. (2017). Impacts of +2°C global warming on winter tourism demand in Europe. *Climate Services*, *7*, 31–46. https://doi.org/10.1016/J.CLISER.2016.07.003.

de Bruin, T. A., & van Sark, W. G. J. H. M. (2023). Investigation of quantum dot luminescent solar concentrator single, double and triple structures: A ray tracing simulation study. *Ceramics International*. https://doi.org/10.1016/j.ceramint.2022.12.084.

Dincer, I. (1998). Energy and environmental impacts: Present and future perspectives. *Energy Sources*, *20*(4–5), 427–453. https://doi.org/10.1080/00908319808970070.

Dincer, I. (1999). Environmental impacts of energy. *Energy Policy*, *27*(14), 845–854. https://doi.org/10.1016/S0301-4215(99)00068-3.

Dincer, I. (2000). Renewable energy and sustainable development: A crucial review. *Renewable & Sustainable Energy Reviews*, *4*(2), 157–175. https://doi.org/10.1016/S1364-0321(99)00011-8.

Dincer, I., & Rosen, M. A. (1998). A worldwide perspective on energy, environment and sustainable development. *International Journal of Energy Research*, *22*(15), 1305–1321. https://doi.org/10.1002/(SICI)1099-114X(199812)22:15<1305::AID-ER417>3.0.CO;2-H.

Dincer, I., & Rosen, M. A. (2011). *Thermal Energy Storage Systems and Applications*, 2nd ed., Wiley, West Sussex, United Kingdom.

Donev, J. (n.d.). *Energy Education*. Universiy of Calgary. https://energyeducation.ca/encyclopedia/Solar_collector.

Duan, X., He, C., Lin, X., Zhao, Y., & Feng, J. (2020). Quasi-Monte Carlo ray tracing algorithm for radiative flux distribution simulation. *Solar Energy*, *211*, 167–182. https://doi.org/10.1016/j.solener.2020.09.061.

Duffi, J.A., & Beckman, W. A. (1989). Solar engineering of thermal processes. *Clinics in Laboratory Medicine*, 9(2), 255–267. https://doi.org/10.1016/s0272-2712(18)30627-9.

Filatov, A. A. (2019). Complete ray tracing simulation model of the solar simulator as a launch pad for prosperous Virtual lab on thermal vacuum testing simulation. *Journal of Physics: Conference Series*, *1313*(1). https://doi.org/10.1088/1742-6596/1313/1/012019.

Fossa, M., Boccalatte, A., & Memme, S. (2021). Solar Fresnel modelling, geometry enhancement and 3D ray tracing analysis devoted to different energy efficiency definitions and applied to a real facility. *Solar Energy*, *216*, 75–89. https://doi.org/10.1016/j.solener.2020.12.047.

Gandhi, T., Bhavsar, S., Shah, J., Solanki, T., & Patel, J. (2017). Pneumatic gear system. *International Journal of Advance Engineering and Research Development*, *4*, 630–636.

Georgiou, M. D., Bonanos, A. M., & Georgiadis, J. G. (2013). Caustics as an alternate of ray tracing to evaluate heliostat mirrors. In *Conference Papers in Energy*, 2013, 1–7. https://doi.org/10.1155/2013/395659.

Getnet, M. Y., Gunjo, D. G., & Sinha, D. K. (2023). Experimental investigation of thermal storage integrated indirect solar cooker with and without reflectors. *Results in Engineering*, *18*, 101022. https://doi.org/10.1016/j.rineng.2023.101022.

Giaconia, A., & Grena, R. (2021). A model of integration between PV and thermal CSP technologies. *Solar Energy*, *224*, 149–159. https://doi.org/10.1016/j.solener.2021.05.043.

Giovinazzo, C., Bonfiglio, L., Gomes, J., & Karlsson, B. (2015). Ray tracing modelling of an asymmetric concentrating PVT. In *Eurosun 2014*, September 2014, 16–19. https://doi.org/10.18086/eurosun.2014.21.01.

Yanqin, M. (2011). An empirical analysis on energy consumption and economic growth in Henan Province. In *2011 International Conference on Management Science and Industrial Engineering, MSIE 2011*, pp. 303–306. https://doi.org/10.1109/MSIE.2011.5707722.

Grodsky, S. M., & Hernandez, R. R. (2020). Reduced ecosystem services of desert plants from ground-mounted solar energy development. *Nature Sustainability*, *3*(12), 1036–1043. https://doi.org/10.1038/s41893-020-0574-x.

Guidara, Z., Souissi, M., Morgenstern, A., & Maalej, A. (2017). Thermal performance of a solar box cooker with outer reflectors: Numerical study and experimental investigation. *Solar Energy*, *158*, 347–359. https://doi.org/10.1016/j.solener.2017.09.054.

Harmim, A., Merzouk, M., Boukar, M., & Amar, M. (2013). Design and experimental testing of an innovative building-integrated box type solar cooker. *Solar Energy*, *98*, 422–433. https://doi.org/10.1016/j.solener.2013.09.019.

Herrando, M., Wang, K., Huang, G., Otanicar, T., Mousa, O. B., Agathokleous, R. A., Ding, Y., Kalogirou, S., Ekins-Daukes, N., Taylor, R. A., & Markides, C. N. (2023). A review of solar hybrid photovoltaic-thermal (PV-T) collectors and systems. *Progress in Energy and Combustion Science*, *97*, 101072. https://doi.org/10.1016/j.pecs.2023.101072.

Hertel, J. D., Canals, V., & Pujol-Nadal, R. (2020). On-site optical characterization of large-scale solar collectors through ray-tracing optimization. *Applied Energy*, *262*, 114546. https://doi.org/10.1016/j.apenergy.2020.114546.

Hilda, D. M. F., & Daniel, M. M. (2019). Experimental performance assessment of a solar powered baking oven. *Procedia Manufacturing*, *35*, 535–540. https://doi.org/10.1016/j.promfg.2019.05.076.

Hong, S., Bradshaw, C. J. A., & Brook, B. W. (2013). Evaluating options for the future energy mix of Japan after the Fukushima nuclear crisis. *Energy Policy*, *56*, 418–424. https://doi.org/10.1016/j.enpol.2013.01.002.

Hunt, J. D., Stilpen, D., & de Freitas, M. A. V. (2018). A review of the causes, impacts and solutions for electricity supply crises in Brazil. *Renewable and Sustainable Energy Reviews*, *88*, 208–222. https://doi.org/10.1016/j.rser.2018.02.030.

International Energy Agency. (2021). Oil 2021. International Energy Agency, 167. www.iea.org.

Jafrancesco, D., Sansoni, P., Francini, F., Contento, G., Cancro, C., Privato, C., Graditi, G., Ferruzzi, D., Mercatelli, L., Sani, E., & Fontani, D. (2014). Mirrors array for a solar furnace: Optical analysis and simulation results. *Renewable Energy*, *63*, 263–271. https://doi.org/10.1016/j.renene.2013.09.006.

Jebasingh, V. K., & Herbert, G. M. J. (2016). A review of solar parabolic trough collector. *Renewable and Sustainable Energy Reviews*, *54*, 1085–1091. https://doi.org/10.1016/j.rser.2015.10.043.

Jignesh, P., Pujak, S., Aejaj, S., Jha, V., & Patel, J. (2018). Investigation on thermal performance of solar air heater by using artificial roughness - A review. *International Journal of Innovative Research in Science, Engineering and Technology*, *3*(5), 12574–12579.

Kalogirou, S. A. (2004). Solar thermal collectors and applications. *Progress in Energy and Combustion Science*, *30*(3). https://doi.org/10.1016/j.pecs.2004.02.001.

Kalogirou, S. A., & Papamarcou, C. (2000). Modelling of a thermosyphon solar water heating system and simple model validation. *Renewable Energy*, 21, 471–493.

Katircioglu, S., Köksal, C., & Katircioglu, S. (2021). The role of financial systems in energy demand: A comparison of developed and developing countries. *Heliyon*, *7*(6). https://doi.org/10.1016/j.heliyon.2021.e07323.

Khatri, R., Goyal, R., & Sharma, R. K. (2021). Advances in the developments of solar cooker for sustainable development: A comprehensive review. *Renewable and Sustainable Energy Reviews*, *145*, 111166. https://doi.org/10.1016/j.rser.2021.111166.

Kincaid, N., Mungas, G., Kramer, N., Wagner, M., & Zhu, G. (2018). An optical performance comparison of three concentrating solar power collector designs in linear Fresnel, parabolic trough, and central receiver. *Applied Energy*, *231*, 1109–1121. https://doi.org/10.1016/j.apenergy.2018.09.153.

Klein, S. A. (n.d.). *F-Chart Software : Engineering Software*. Retrieved August 6, 2022, from https://www.fchartsoftware.com/ees/.

Kocharov, L., Vainio, R., Kovaltsov, G. A., & Torsti, J. (1998). Adiabatic deceleration of solar energetic particles as deduced from Monte Carlo simulation of interplanetry transport. *Solar Physics*, 182, 195–215.

Kumar, Y., & Arora, V. K. (2020). Energy and exergy analysis and its application of different drying techniques: A review. *ISME Journal of Thermofluid*, *5*(2), 10–29.

Kusch, W., Schmidla, T., & Stadler, I. (2012). Consequences for district heating and natural gas grids when aiming towards 100% electricity supply with renewables. *Energy*, *48*(1), 153–159. https://doi.org/10.1016/j.energy.2012.06.054.

Lentswe, K., Mawire, A., Owusu, P., & Shobo, A. (2021). A review of parabolic solar cookers with thermal energy storage. *Heliyon*, *7*(10), e08226. https://doi.org/10.1016/j.heliyon.2021.e08226.

Levy, G. (2018). TRaNsient SYstem Simulation Program (TRNSYS 17). *Energy Power Risk*, 4. https://doi.org/10.1108/978-1-78743-527-820181015.

Li, J., & Just, R. E. (2018). Modeling household energy consumption and adoption of energy efficient technology. *Energy Economics*, *72*, 404–415. https://doi.org/10.1016/j.eneco.2018.04.019.

Li, Z., & Gariboldi, E. (2019). Reliable estimation of effective thermal properties of a 2-phase material by its optimized modelling in view of Lattice Monte-Carlo simulation. *Computational Materials Science*, *169*, 109125. https://doi.org/10.1016/j.commatsci.2019.109125.

Lin, M., Sumathy, K., Dai, Y. J., Wang, R. Z., & Chen, Y. (2013). Experimental and theoretical analysis on a linear Fresnel reflector solar collector prototype with V-shaped cavity receiver. *Applied Thermal Engineering*, *51*(1–2), 963–972. https://doi.org/10.1016/j.applthermaleng.2012.10.050.

Liu, J., Möller, M., & Schuttelaars, H. M. (2021). Balancing truncation and round-off errors in FEM: One-dimensional analysis. *Journal of Computational and Applied Mathematics*, *386*, 113219. https://doi.org/10.1016/j.cam.2020.113219.

Liu, X., Huang, J., & Mao, Q. (2015). Sensitive analysis for the efficiency of a parabolic trough solar collector based on orthogonal experiment. *International Journal of Photoenergy*, *2015*. https://doi.org/10.1155/2015/151874.

Makhmutov, V. S., Desorgher, L., Bazilevskaya, G. A., & Flückiger, E. (2007). Evaluation of the energy spectrum of solar protons according to the balloon measurement data and Monte Carlo simulation. *Bulletin of the Russian Academy of Sciences: Physics*, *71*(7), 950–952. https://doi.org/10.3103/s1062873807070179.

Manieniyan, V., Thambidurai, M., & Selvakumar, R. (2009). Study on energy crisis and the future of fossil. *Proceedings of SHEE*, October, 7–12. https://doi.org/10.13140/2.1.2234.3689.

Mannhardt, J., Gabrielli, P., & Sansavini, G. (2023). Collaborative and selfish mitigation strategies to tackle energy scarcity: The case of the European gas crisis. *Iscience*, *26*(5), 106750. https://doi.org/10.1016/j.isci.2023.106750.

Masud, M. H. 2021. "Solar Pyrolysis Parabolic Dish", Hybrid Energy System Models.

Mathworks, MATLAB, Anonymous, MATLAB, Sataloff, R. T., Johns, M. M., Kost, K. M., Gorrostieta, E., Matlab, U., Carlos, W., Leite, F., Costa, G. A., Da, J. V., Neto, F., Moura, J. P. De, Franklin, E., Ferreira, M., Castro, I. L. De, Yu, Y., ... Mahfouz, A. A. (1984). Computation visualization programming getting started with MATLAB the language of technical computing. *Manual*, 7(1). https://dx.doi.org/10.1016/j.mechatronics.2012.10.002%0Awww.mathworks.com%0Ahttps://www.mathworks.de/products/matlab/%0Awww.mathworks.com.

Mbison, Phillthorpe, P. (2020). *World's Largest Solar Furnace - Font-Romeu-Odeillo-Via, France - Atlas Obscura*. Atlas Obscura. https://www.atlasobscura.com/places/worlds-largest-solar-furnace.

McDaniel, S. (2019). *"Archimedes's Death Ray" Debunked*. Tales of Times Forgotten. https://talesoftimesforgotten.com/2019/04/11/archimedess-death-ray-debunked/

Mill, J. (2020). *Ansys SPEOS : Illuminating the Possibilities*. www.ansys.com.

Mishra, A., Powar, S., & Dhar, A. (2019). Solar thermal powered bakery oven. *Energy, Environment, and Sustainability*, 577–592. https://doi.org/10.1007/978-981-13-3302-6_19.

Mlatho, J. S. P., McPherson, M., Mawire, A., & Van den Heetkamp, R. J. J. (2010). Determination of the spatial extent of the focal point of a parabolic dish reflector using a red laser diode. *Renewable Energy*, *35*(9), 1982–1990. https://doi.org/10.1016/j.renene.2010.01.030.

Mohajan, H. K. (2018). Acid rain is a local environment pollution but global concern human rights view project acid rain is a local environment pollution but global concern. *Open Science Journal of Analytical Chemistry*, *3*(5), 47–55. https://www.openscienceonline.com/journal/osjac.

Mohanakrishna, G. (2016). Bioprocesses for Waste and Wastewater Remediation for Sustainable Energy. In *Bioremediation and Bioeconomy* (pp. 537–565). Elsevier. https://doi.org/10.1016/B978-0-12-802830-8.00021-6. https://www.sciencedirect.com/science/article/abs/pii/B9780128028308000216

Monreal, A., Riveros-Rosas, D., & Sanchez, M. (2015). Analysis of the influence of the site in the final energy cost of solar furnaces for its use in industrial applications. *Solar Energy*, *118*, 286–294. https://doi.org/10.1016/j.solener.2015.05.032.

More, S. S., Ravindranath, G., More, S. E., & Thipase, S. S. (2018). Mathematical modeling and analysis of compound parabolic concentrator using soltrace. *International Journal of Mechanical Engineering and Technology*, *9*(6), 113–121.

Nandwani, S. S. (1988). Experimental and theoretical analysis of a simple solar oven in the climate of Costa Rica - I. *Solar and Wind Technology*, *5*(2), 159–170. https://doi.org/10.1016/0741-983X(88)90075-6.

Nydal, O. J. (2014). Ray tracing for optimization of a double reflector system for direct illumination of a heat storage. *Energy Procedia*, *57*, 2211–2220. https://doi.org/10.1016/j.egypro.2014.10.188.

Oluoch, S., Lal, P., Susaeta, A., & Vedwan, N. (2020). Assessment of public awareness, acceptance and attitudes towards renewable energy in Kenya. *Scientific African*, *9*, e00512. https://doi.org/10.1016/j.sciaf.2020.e00512.

Osório, T., Horta, P., Larcher, M., Pujol-nadal, R., Hertel, J., Rooyen, D. W. Van, Schneider, S., Benitez, D., Frein, A., Denarie, A., Osório, T., Horta, P., Larcher, M., & Pujol-nadal, R. (2016). Ray-tracing software comparison for linear focusing solar collectors. *AIP Conference Proceedings*. https://doi.org/10.1063/1.4949041.

Overview, F. (n.d.). *FRED ™ Overview*. https://photonengr.com/wp-content/uploads/2022/05/FREDOverview.pdf

Panchal, A., Dave, N., Parmar, B., Patel, J. H., & Patel, J. (2017). An assessment of phase change materials for domestic applicatons. *International Journal of Advance Engineering and Research Development*, *2*(2), 61–67.

Paramati, S. R., Shahzad, U., & Doğan, B. (2022). The role of environmental technology for energy demand and energy efficiency: Evidence from OECD countries. *Renewable and Sustainable Energy Reviews*, *153*. https://doi.org/10.1016/j.rser.2021.111735.

Patel, A., Sarkar, P., Tyagi, H., & Singh, H. (2016). Time value of emission and technology discounting rate for off-grid electricity generation in India using intermediate pyrolysis. *Environmental Impact Assessment Review*, *59*, 10–26. https://doi.org/10.1016/j.eiar.2016.03.001.

Patel, A. R., Sarkar, P., Tyagi, H., & Singh, H. (2020). Triple bottom line analysis, methodology, and its implementation. In *Green Energy and Infrastructure* (Issue August). https://doi.org/10.1201/9781003095811-13.

Patel, J., & Patel, A. R. (2023). Effect of reflector height to base ratio on the optical performance of square shaped aperture area based flat plat solar reflecting system. In *Proceeding of International Conference on Recent Advances in (Applied) Sciences & Engineering (Raise)*, Vol. IV, 9–16. ISBN: 978-81-962938-1-9.

Patel, J., & Singh, V. (2015). PCM augumented composite wall for base transceiver station (BTS) for telecom towers. In *Proceedings of the 17th ISME Conference ISME17*, October 3–4, 2015, IIT Delhi, New Delhi, 3–8.

Peña-Cruz, M. I., Arancibia-Bulnes, C. A., Vidal, A. M., & González, M. S. (2014). Improving optical qualification of solar concentrator by FOCuS tool. *Energy Procedia*, *57*, 427–436. https://doi.org/10.1016/j.egypro.2014.10.196.

Pereira, J. C. G., Fernandes, J. C., & Guerra Rosa, L. (2019). Mathematical models for simulation and optimization of high-flux solar furnaces. *Mathematical and Computational Applications*, *24*(2), 65. https://doi.org/10.3390/mca24020065.

Pereira, J. C. G., Rodríguez, J., Fernandes, J. C., & Rosa, L. G. (2020). Homogeneous flux distribution in high-flux solar furnaces. *Energies*, *13*(2). https://doi.org/10.3390/en13020433.

Petrasch, J. (2014). A Free and open source Monte Carlo ray tracing program for concentrating solar enrgy research. In *ASME 2010 4th International Conference on Energy Sustainability*, 1–8. https://asmedigitalcollection.asme.org/ES/proceedings/ES2010/43956/125/348077

Ploysun, & Velasolaris. (2015). Polysun Simulation Software (User manual). In *Vela Solaris*. https://www.velasolaris.com/wp-content/uploads/2022/01/Tutorial_EN.pdf

Pohekar, S. D., Kumar, D., & Ramachandran, M. (2005). Dissemination of cooking energy alternatives in India - A review. *Renewable and Sustainable Energy Reviews*, *9*(4), 379–393. https://doi.org/10.1016/j.rser.2004.05.001.

Priyanshu, S., Abhishek, S., Akshay, T., Vijay, T., & Patel, J. (2018). Development of multipurpose commercial solar furnace - A review. *2nd International Conference on Current Research Trends in Engineering and Technology*, *4*(5), 579–585. https://doi.org/10.32628/IJSRSET.

Program, O. A. (1999). OptiCAD,"Optical Analysis Program", User's Guide, Vrsion 9,USA, www.opticad.com.

Pujol-Nadal, R., Bonnín-Ripoll, F., Hertel, J. D., Martínez-Moll, V., & Cardona, G. (2019). *OTSun: An Open Source Code for Optical Analysis of Solar Thermal Collectors and PV Cells*. 1–7. https://doi.org/10.18086/eurosun2018.12.05.

Pujol-Nadal, R., Martínez-Moll, V., Moià-Pol, A., Cardona, G., Hertel, D. J., & Bonnin, F. (2017). *OTSun Project: Development of a Computational Tool for High-Resolution Optical Analysis of Solar Collectors*. 1–11. https://doi.org/10.18086/eurosun.2016.06.04.

Qian, F., & Science, C. (2006). Tutorial project of computer graphics 3D renderer (ray tracing engine). *Computer*, *2*, 1–21.

Ramalingam, K., & Indulkar, C. (2017). Solar energy and photovoltaic technology. *Distributed Generation Systems*, 69–147. doi: 10.1016/B978-0-12-804208-3.00003-0.

ur Rehman, N. (2019). Optical-irradiance ray-tracing model for the performance analysis and optimization of a single slope solar still. *Desalination*, *457*, 22–31. https://doi.org/10.1016/j.desal.2019.01.026.

Reji, A. K., Kumaresan, G., Mukundhan, A., Menon, A. S., Harikrishna, A. P., & Parappadi, J. J. (2021). Experimental analysis of a four slope solar still augmented with a parabolic reflector array with V-shaped mirror arrangement for pre-heating brackish water. *Materials Today: Proceedings*, *45*, 1834–1838. https://doi.org/10.1016/j.matpr.2020.09.036.

Richter, P., Kepp, F., Büsing, C., & Kuhnke, S. (2019). Optimization of robust aiming strategies in solar tower power plants. *AIP Conference Proceedings*, *2126*. https://doi.org/10.1063/1.5117557.

Riveros-Rosas, D., Herrera-Vázquez, J., Pérez-Rábago, C. A., Arancibia-Bulnes, C. A., Vázquez-Montiel, S., Sánchez-González, M., Granados-Agustín, F., Jaramillo, O. A., & Estrada, C. A. (2010). Optical design of a high radiative flux solar furnace for Mexico. *Solar Energy*, *84*(5), 792–800. https://doi.org/10.1016/j.solener.2010.02.002.

Ruivo, C. R., Apaolaza-Pagoaga, X., Carrillo-Andrés, A., & Coccia, G. (2022). Influence of the aperture area on the performance of a solar funnel cooker operating at high sun elevations using glycerine as load. *Sustainable Energy Technologies and Assessments*, *53*. https://doi.org/10.1016/j.seta.2022.102600.

Sánchez-González, A. (2022). Analytic function for heliostat flux mapping with astigmatism and defocus. *Solar Energy*, *241*, 24–38. https://doi.org/10.1016/j.solener.2022.05.045.

Sayigh, A. A. M. (2000). Renewable will meet the challenges in the year 2000. *3*(4), 297–304.

Schüler, A., Kostro, A., Galande, C., Valle Del Olmo, M., De Chambrier, E., & Huriet, B. (2007). Principles of Monte-Carlo ray-tracing simulations of quantum dot solar concentrators. In *ISES Solar World Congress 2007, ISES 2007*, Vol. 2, 1033–1037. https://doi.org/10.1007/978-3-540-75997-3_200.

Selvakumar, N., & Barshilia, H. C. (2012). Review of physical vapor deposited (PVD) spectrally selective coatings for mid- and high-temperature solar thermal applications. *Solar Energy Materials and Solar Cells*, *98*, 1–23. https://doi.org/10.1016/j.solmat.2011.10.028.

Sharma, N. K., & Savithri, A. K. (2021). *Energy Statistucs India 2021*. www.mospi.gov.in.

Shen, N., Deng, R., Liao, H., & Shevchuk, O. (2020). Mapping renewable energy subsidy policy research published from 1997 to 2018: A scientometric review. *Utilities Policy*, *64*, 101055. https://doi.org/10.1016/j.jup.2020.101055.

Shukla, J. B., Verma, M., & Misra, A. K. (2017). Effect of global warming on sea level rise : A modeling study. *Ecological Complexity*, *32*, 99–110. https://doi.org/10.1016/j.ecocom.2017.10.007.

Singh, G., & Chandra, L. (2018). On the flow stability in a circular cylinder based open volumetric air receiver for solar convective furnace. *Energy Procedia*, *144*, 88–94. https://doi.org/10.1016/j.egypro.2018.06.012.

Singh, H. R., Sharma, D., & Soni, S. L. (2021). Dissemination of sustainable cooking: A detailed review on solar cooking system. *IOP Conference Series: Materials Science and Engineering*, *1127*(1), 012011. https://doi.org/10.1088/1757-899x/1127/1/012011.

Singer, L. E., & Peterson, D. (2011). International energy outlook 2010. In *International Energy Outlook and Projections* (Vol. 0484, Issue May).

Suharta, H., Parangtopo, & Sayigh, A. M. (1996). Solar Oven, design, and its field testing in West Lombok regency, Indonesia. *Renewable Energy*, *9*(1–4 SPEC. ISS.), 749–753. https://doi.org/10.1016/0960-1481(96)88392-3.

Tanaka, H. (2011). Solar thermal collector augmented by flat plate booster reflector: Optimum inclination of collector and reflector. *Applied Energy*, *88*(4), 1395–1404. https://doi.org/10.1016/j.apenergy.2010.10.032.

Tawfik, M. A., Sagade, A. A., Palma-Behnke, R., El-Shal, H. M., & Abd Allah, W. E. (2021). Solar cooker with tracking-type bottom reflector: An experimental thermal performance evaluation of a new design. *Solar Energy*, *220*, 295–315. https://doi.org/10.1016/j.solener.2021.03.063.

Terrón-Hernández, M., Peña-Cruz, M. I., Carrillo, J. G., Diego-Ayala, U., & Flores, V. (2018). Solar ray tracing analysis to determine energy availability in a CPC designed for use as a residentialwater heater. *Energies*, *11*(2). https://doi.org/10.3390/en11020291.

Tiwari, A. K., Patel, A. R., & Kumar, N. (2015). Investigation of strain rate on residual stress distribution. *Materials and Design*, *65*, 1041–1047. https://doi.org/10.1016/j.matdes.2014.09.051.

Torrence, B. F, & Torrence, E. A. (2009). *The Student's Introduction to MATHEMATICA®: A Handbook for Precalculus, Calculus, and Linear Algebra.* Cambridge University Press. https://www.researchgate.net/publication/269107473.

Tsvetkov, N. A., Krivoshein, U. O., Tolstykh, A. V. Y., Khutornoi, A. N., & Boldyryev, S. (2020). The calculation of solar energy used by hot water systems in permafrost region: An experimental case study for Yakutia. *Energy*, *210*, 118577. https://doi.org/10.1016/j.energy.2020.118577.

Verma, R., Vinoda, K. S., Papireddy, M., & Gowda, A. N. S. (2016). Toxic pollutants from plastic waste - A review. *Procedia Environmental Sciences*, *35*, 701–708. https://doi.org/10.1016/j.proenv.2016.07.069.

Viviescas, C., Lima, L., Diuana, F. A., Vasquez, E., Ludovique, C., Silva, G. N., Huback, V., Magalar, L., Szklo, A., Lucena, A. F. P., Schaeffer, R., & Paredes, J. R. (2019). Contribution of variable renewable energy to increase energy security in Latin America: Complementarity and climate change impacts on wind and solar resources. *Renewable and Sustainable Energy Reviews*, *113*. https://doi.org/10.1016/j.rser.2019.06.039.

Voronkin, A. F., Lisochkina, T. V., Malinina, T. V., Taratin, V. A. and Rozova, V. I. (1995). Economic renewable efficiency of power stations using energy sources. *Hydrotechnical Construction*, *29*(6), 347–352.

Walfram. (n.d.). *Wolfram webMathematica 3*. WALFRAM. https://www.wolfram.com/products/webmathematica/.

Wang, C.-H., Padmanabhan, P., & Huang, C.-H. (2022). The impacts of the 1997 Asian financial crisis and the 2008 global financial crisis on renewable energy consumption and carbon dioxide emissions for developed and developing countries. *Heliyon*, *8*(2), e08931. https://doi.org/10.1016/j.heliyon.2022.e08931.

Wang, G., Ge, Z., & Lin, J. (2022). Design and performance analysis of a novel solar photovoltaic/thermal system using compact linear Fresnel reflector and nanofluids beam splitting device. *Case Studies in Thermal Engineering*, *35*, 102167. https://doi.org/10.1016/j.csite.2022.102167.

Wang, G., Wang, F., Shen, F., Chen, Z., & Hu, P. (2019). Novel design and thermodynamic analysis of a solar concentration PV and thermal combined system based on compact linear Fresnel reflector. *Energy*, *180*, 133–148. https://doi.org/10.1016/j.energy.2019.05.082.

Wang, H., Zhu, N., & Bai, X. (2015). Reliability model assessment of grid-connected solar photovoltaic system based on Monte-Carlo. *Applied Solar Energy (English Translation of Geliotekhnika)*, *51*(4), 262–266. https://doi.org/10.3103/S0003701X15040192.

Wassie, H. M., Getie, M. Z., Alem, M. S., Kotu, T. B., & Salehdress, Z. M. (2022). Experimental investigation of the effect of reflectors on thermal performance of box type solar cooker. *Heliyon*, *8*(12), e12324. https://doi.org/10.1016/j.heliyon.2022.e12324.

Welch, D., Sebakunzi, J., Carla O'Donnell, G. R. (2015). The Gitega Solar Bakery Project. *Regions of Climate Action*. https://regions20.org/wp-content/uploads/2016/10/r20-Solar-Bakery-Case-Study-GNL.pdf.

Weldu, A., Zhao, L., Deng, S., Mulugeta, N., Zhang, Y., Nie, X., & Xu, W. (2019). Performance evaluation on solar box cooker with reflector tracking at optimal angle under Bahir Dar climate. *Solar Energy*, *180*, 664–677. https://doi.org/10.1016/j.solener.2019.01.071.

Wendelin, T. (2003). Soltrace: A new optical modeling tool for concentrating solar optics. In *International Solar Energy Conference*, March 2003, 253–260. https://doi.org/10.1115/ISEC2003-44090.

Winzer, C. (2012). Conceptualizing energy security. *Energy Policy*, *46*, 36–48. https://doi.org/10.1016/j.enpol.2012.02.067.

Wrighton, M. S., & Society, A. C. (1978). BOOK REVIEWS, Solar energy "*Fundamentals in building designt Solid state chemistry of energy conversion and storaget*", *21*, 161–162.

Yao, S., Zhang, S., & Zhang, X. (2019). Renewable energy, carbon emission and economic growth: A revised environmental Kuznets Curve perspective. *Journal of Cleaner Production*, *235*, 1338–1352. https://doi.org/10.1016/j.jclepro.2019.07.069.

Yurcheko, V., Yurcheko, E., Çiydem, M., & Totuk, O. (2015). Ray tracing for optimization of compound parabolic concentrators for solar collectors of enclosed design. *Turkish Journal of Electrical Engineering & Computer Sciences*, *23*, 1761–1768. https://doi.org/10.3906/elk-1404-267.

Zafar, H. A., Badar, A. W., Butt, F. S., Khan, M. Y., & Siddiqui, M. S. (2019). Numerical modeling and parametric study of an innovative solar oven. *Solar Energy*, *187*, 411–426. https://doi.org/10.1016/j.solener.2019.05.064.

Zahedi, A. (2011). A review of drivers, benefits, and challenges in integrating renewable energy sources into electricity grid. *Renewable and Sustainable Energy Reviews*, *15*(9), 4775–4779. https://doi.org/10.1016/j.rser.2011.07.074.

Zhao, L., Yu, R., Wang, Z., Yang, W., Qu, L., & Chen, W. (2020). Development modes analysis of renewable energy power generation in North Africa. *Global Energy Interconnection*, *3*(3), 237–246. https://doi.org/10.1016/j.gloei.2020.07.005.

Zhao, Y., Zheng, H., Sun, B., Li, C., & Wu, Y. (2018). Development and performance studies of a novel portable solar cooker using a curved Fresnel lens concentrator. *Solar Energy*, *174*, 263–272. https://doi.org/10.1016/j.solener.2018.09.007.

Zhu, J., & Huang, H. (2014). Design and thermal performances of Semi-Parabolic Linear Fresnel Reflector solar concentration collector. *Energy Conversion and Management*, *77*, 733–737. https://doi.org/10.1016/j.enconman.2013.10.015.

Zhu, W., Xu, Y., Du, H., & Li, J. (2019). Thermal performance of high-altitude solar powered scientific balloon. *Renewable Energy*, *135*, 1078–1096. https://doi.org/10.1016/j.renene.2018.12.083.

Zhu, Y., Shi, J., Li, Y., Wang, L., Huang, Q., & Xu, G. (2017). Design and thermal performances of a scalable linear Fresnel reflector solar system. *Energy Conversion and Management*, *146*, 174–181. https://doi.org/10.1016/j.enconman.2017.05.031.

9 Energy Poverty and the Sustainable Development of Renewable Energy Systems

Ishan Suvarna
Birla Institute of Technology and Science
University of Windsor

David S-K. Ting
University of Windsor

Shibu Clement
Birla Institute of Technology and Science

9.1 INTRODUCTION

Over the past decades, the rapid increase in human population, the expansion of industries and the rising effects of global warming and climate change have resulted in a rapid increase in global energy consumption. Energy consumption has continued to rise at an average rate of 2% since the 1950s, slowing down only twice: during the oil crisis in the 1980s and the COVID-19 pandemic [1,2]. Much of the corresponding rise in energy production has been through non-renewable sources. This has, in turn, resulted in greater global carbon emissions and further damage to the environment.

Today, access to modern energy remains low in developing countries across the world, from Asia to South America. Over 700 million people do not have access to electricity, and nearly 2 billion still depend on traditional forms of energy, such as biomass and wood, for their daily energy needs [9]. Even in developed nations, there are a significant number of people spending a large portion of their income in fulfilling their energy needs, increasing financial burdens and contributing to economic inequality.

To combat energy poverty, many countries have set up programs for improving energy access and creating modern electricity infrastructure. A lot of these programs also focus on harnessing energy from renewable sources, reducing the environmental impact caused by increased energy production. However, many programs end up unsuccessful due to lack of funding, failure of installed renewable energy systems

DOI: 10.1201/9781003403456-9

or insufficient energy production. For this reason, there is an increased number of studies into the sustainability of renewable energy systems in order to improve their design and create procedures to ensure that projects are more successful in the future at providing better energy access in the long run.

Highlighting some of the studies that were carried out in developing sustainability assessment, Terrapon-Pfaff et al. in 2014 [3] studied the sustainability of small-scale projects in developing countries. The study clearly laid out the benefits of small-scale projects, including the ease of increasing energy access and the lower costs, while also discussing the problems faced by small-scale projects. Important factors were identified that determined the success of projects, and the impact of the projects was studied using the Millennium Development Goals as the basis. While not explicitly setting sustainability as a goal, the study nevertheless unearthed key factors that were used in later studies of sustainability models.

In 2015, with the adoption of the UN Sustainable Development Goals (SDGs), sustainability had a much clearer definition. With the goal of improving the performance of small-scale renewable energy systems, several studies were conducted on sustainability analysis. In 2018, Bhandari et al. [4] developed a novel sustainability analysis and used it to study a micro-hydropower plant in Nepal. The study used some of the broad goals of sustainability set out under the UN SDGs, breaking sustainability into four broad themes: technical, environmental, economic and social. Indicators were created to assess the impact of the project on each of the themes.

Later studies were focused on improving the models to better suit the projects being studied. Dauenhauer et al. [5] created a new model that scored all projects between 0 and 1 without giving any specific weightage to indicators. The new model was more project-centric rather than goal-centric, with the indicators and the scores designed to reflect the performance of the project rather than studying the impact of the project in advancing a particular goal. This method was argued to be better for studying sustainability. Meanwhile, Lee and Shepley [6] used a more subjective, survey-based study to study the sustainability of photovoltaic (PV) projects in Seoul. Their study was highly focused on studying the economic and social effects of the project, using interviews with groups of stakeholders to find specific issues that the project faced.

Despite the improvements made in sustainability models, there is a lack of broad equations that could be applied to diverse kinds of projects across the board. Such models could allow comparisons to be made between projects and allow us to study how specific factors influence sustainability. A uniform analysis could also help in designing projects and deciding what kind of renewable source would be most sustainable in a particular application.

This study aims at explaining the global impact of energy poverty and how renewable energy systems are well-suited to reducing energy poverty and improving energy access globally. This study also aims to develop a universal model to assess the sustainability of small-scale projects, and then use the model to assess the sustainability of various projects. The model could then be further applied to a large number of projects and subsequently be refined to improve its accuracy.

9.2 ENERGY POVERTY

Energy poverty is a lack of access to modern energy services. This includes modern power plants, electricity infrastructure, and clean sources of energy. It is one of the most significant issues we face today [7]. Along with climate change, food security and overpopulation, energy access is increasingly being recognised as one of the defining global problems of the 21st century.

Affordable and Clean Energy is one of the 17 SDGs adopted by the United Nations in 2015. [9] One of the targets under this goal is to ensure universal access to affordable and reliable energy services. According to the 2022 Sustainable Development Goal Report, nearly 700 million people do not have access to electricity, a number that is far above the targets set up by the UN. Over three billion people depend on traditional energy sources, such as firewood and crop waste, to fulfil their energy requirements. Figure 9.1 illustrates the extent to which inequality in energy consumption exists between developing countries and the developed world. The COVID-19 pandemic has only exacerbated the effects of energy poverty in the developing world.

9.2.1 Geographical Distribution

Energy access varies widely across the world. Countries in North America and Europe see some of the highest levels of electricity access in the world. At the same time, regions in Asia and Africa have some of the lowest [2]. Due to the large size of the population in countries like India and China, Asia has more people without access to electricity and is dependent on traditional sources. However, the countries with the lowest levels of access in the world are in Africa, particularly in sub-Saharan Africa. This includes countries like Malawi, the Democratic Republic of Congo, and Mozambique. Figure 9.2 represents the geographic distribution of per capita energy consumption, which serves as a metric to demonstrate the lack of energy access. It is important to note that there are also wide variations within countries, especially between rural and urban regions. Most of the people without access to

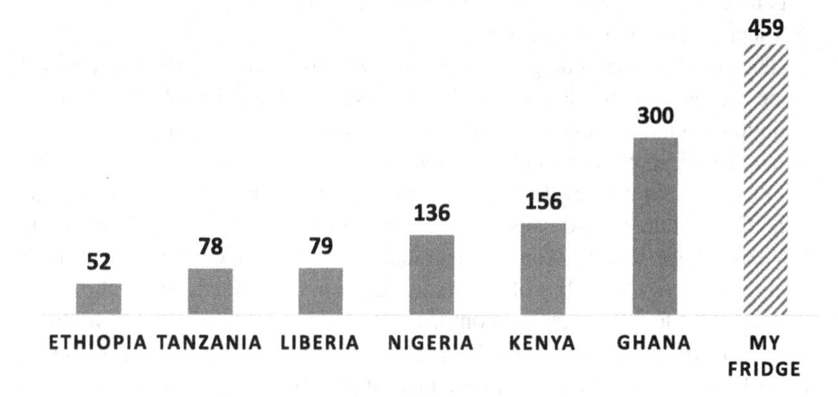

FIGURE 9.1 Per capita annual energy consumption in kWh for an average citizen in six sub-Saharan African countries, compared to the annual energy consumption in kWh of a refrigerator in a developed nation. Created based on data from Centre for Global Development [8].

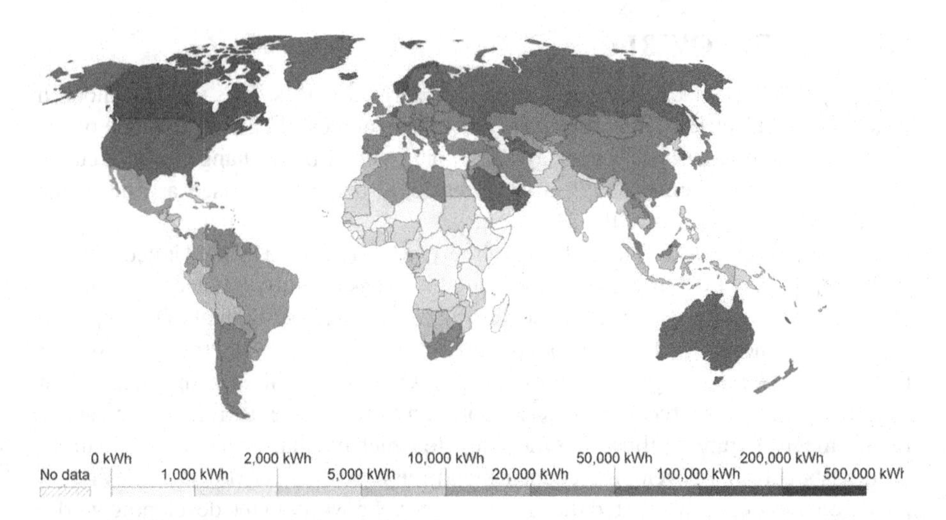

FIGURE 9.2 Per capita energy consumption, 2021 [2].

electricity live in rural regions. In many countries, energy grids only serve urban regions, leaving large parts of the population in rural areas with insufficient electricity for their needs.

9.2.2 Causes of Energy Poverty

1. **Energy Sources:**

   The lack of viable energy sources is a major cause of energy poverty. Generally, a larger variety of energy sources are associated with higher levels of energy access. Developed countries produce electricity from a wide variety of sources, generally with a large share of energy generation from renewable and clean sources. Meanwhile, in developing countries, people depend on primary energy sources such as wood and coal for basic necessities such as cooking and heating.

   The use of primary energy sources has several detrimental effects. Burning wood causes indoor pollution, which can cause severe health problems. Indoor pollution is a major cause of death in several low-income nations. Indoor pollution also disproportionately affects younger children who spend more time indoors, causing the development of cardiovascular and respiratory diseases [10]. These sources also cause environmental damage. The use of firewood has led to mass deforestation and degradation of the land. Cutting down trees for fuel reduces the ability of forests to maintain carbon balance in the ecosystem and affects the nutrient composition of the soil, leaving it unfit for any other use. In cases where wood supply is insufficient, people tend to burn crops for fuel, which can cause food insecurity [11].

   Another aspect of primary energy sources is the quality of energy produced. The highest quality within context is electrical energy [12], typically used for powering all kinds of electronic devices. Mechanical energy is of lower quality, and thus, converting it to electricity involves significant losses. The low-quality energy is thermal energy, i.e., heat. Low-quality (temperature) thermal energy is best used directly for heating purposes.

2. **Insufficient Electricity Infrastructure:**

   Providing widespread access to energy requires the development of modern power plants, energy grids and distribution centres. This requires the investment of significant resources and funding. Many low-income countries lack the ability to build this infrastructure, leaving them dependent on international loans and external sources of funding. Figure 9.3 clearly illustrates the distribution of people without access to electricity in the world, which is mostly in countries in Asia and sub-Saharan Africa, which are developing regions. The construction and development of modern energy systems also take time, so there needs to be more immediate solutions to control the effects of energy poverty.

   In some countries, geography limits the construction of large infrastructure projects. For example, in Nepal, the mountainous terrain prevents the extension of the energy grid further into rural regions [4]. Cases like this necessitate the development of off-grid solutions.

3. **Resource Management:**

   Energy resources need to be efficiently managed to ensure long-term use. Electricity grids and power plants require regular maintenance to prevent unexpected failures in energy supply. In Venezuela, decades of inadequate regulatory frameworks and mismanagement of resources have resulted in an acute energy crisis [13]. Figure 9.4 shows the extent to which recent energy policies have drastically reduced per capita energy consumption in Venezuela. Despite having the largest oil reserves in the world, as well as massive potential for renewable sources such as hydropower and wind

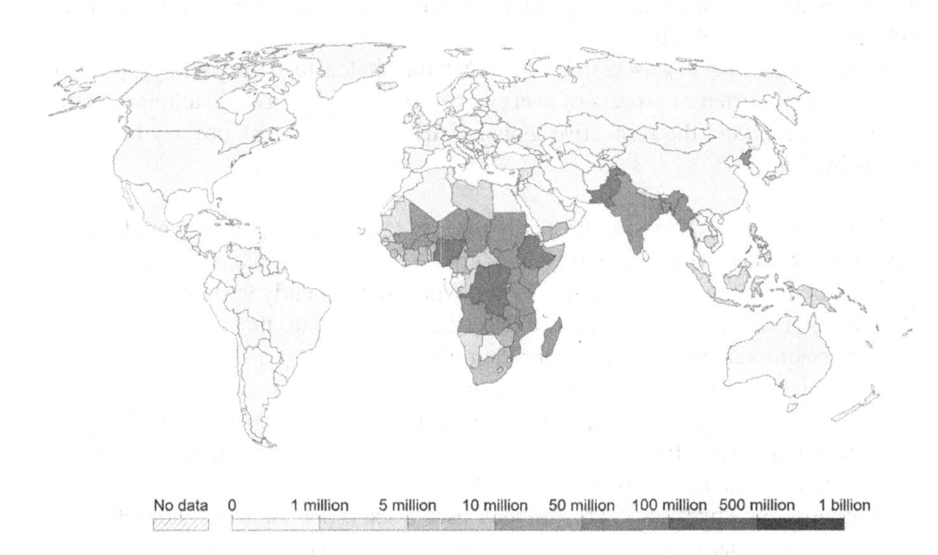

FIGURE 9.3 Number of people without access to electricity, 2020 [2].

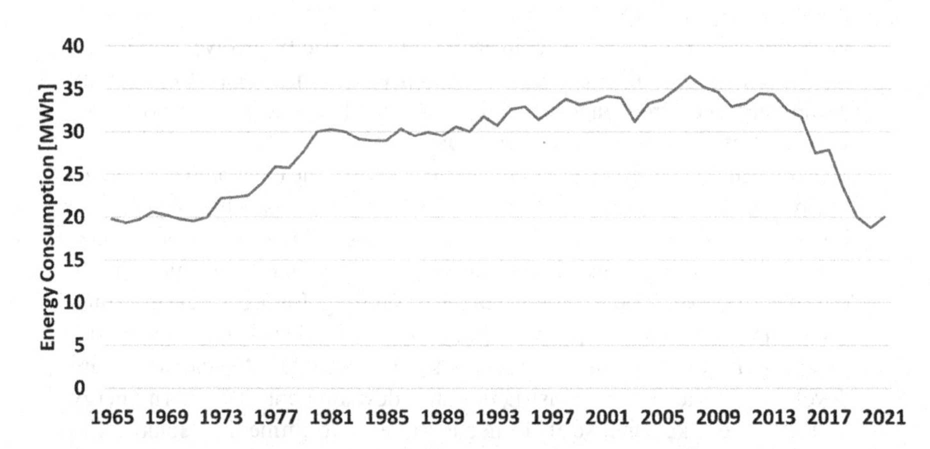

FIGURE 9.4 Per capita energy consumption in Venezuela from 1965 to 2020. The decrease in recent years is an indicator of the growing energy crisis. (Data taken from "Our World in Data" [2].)

energy, delays in the construction of electricity infrastructure, flawed implementation of energy policies and lack of proper maintenance have resulted in inefficient usage of these resources. Venezuelans suffer from rolling blackouts and unexpected grid failures, coupled with rising socio-economic issues that have exacerbated the impact of the energy crisis.

9.2.3 Measurement of Energy Poverty

To study the extent of energy poverty and provide practical solutions, it is essential to quantify and measure it. Several metrics are used to quantify energy poverty, including energy use per capita and the percentage of energy access. However, there needs to be a more basic definition of energy poverty.

González-Eguino [7] used three different thresholds to define energy poverty based on three different aspects of energy poverty. Table 9.1 briefly summarises the three thresholds and the respective global distribution of energy poverty based on that definition.

1. **Technological Threshold:** Energy poverty is a lack of access to "modern" energy sources. This includes modern electric power plants and energy grids. Based on data published by the World Bank, nearly 900 million people globally do not have access to electricity. This means they are dependent on traditional energy sources for their energy needs.
2. **Physical Threshold:** A minimum energy need is calculated based on the energy requirements to supply the basic necessities of households. These necessities vary from region to region since a single definition of "basic necessity" cannot be used everywhere.
3. **Economic Threshold:** This method uses a fixed percentage of household income to define energy poverty. Generally, if a household spends more than 10% of its income on energy spending, it is defined as energy poverty.

TABLE 9.1
Thresholds for Defining Energy Poverty

Threshold	Definition	Global Distribution
Economic	More than 10% of household income spent on energy	• Mostly in low-income nations, but large percentages even in high-income regions
Physical	Energy supply insufficient for basic necessities	• Around 1.6–2 billion people globally • Basic necessities vary between regions
Technological	Lack of access to modern energy sources	• Nearly 800 million people with no access to electricity, and 2 billion people dependent on biomass for energy needs

Source: Based on Ref. [7].

Most studies use the economic threshold to define energy poverty since it is a more objective metric than the physical and technological threshold, which is more subjective and varies widely. However, it is difficult to compare energy poverty across different countries due to the difference in economic conditions.

9.2.4 Solutions to Energy Poverty

Most practical solutions to energy poverty focus on providing energy access. There are several ways to do this. Since most regions experiencing energy poverty are not connected to the central energy grid, one proposed solution is to expand the current energy grids. This would be in conjunction with increasing the capacity of existing power plants or constructing new power plants. However, such a solution requires the investment of significant resources and funding and also needs several years of development. Most energy-poor nations are developing low-income nations, which do not have the financial and technological capacity to undertake projects on such a large scale. This leaves them dependent on loans and technical support from other nations. Additionally, we cannot expand conventional sources of electricity in a world increasingly impacted by the effects of global warming and climate change caused by the burning of fossil fuels.

Several alternate solutions are proposed that can be better implemented for energy access [14]. Renewable energy systems provide a more sustainable way to increase energy production without impacting the environment. To solve the problem of infrastructure, small-scale off-grid systems have been developed, which do not need an energy grid for functioning.

9.2.5 International Role in Reducing Energy Poverty

As seen in Figure 9.2, the countries with the highest levels of energy consumption in the world tend to be highly developed nations with advanced economies. These countries are able to afford the construction of modern power plants and extensive power grids, resulting in virtually 100 per cent energy access in many of these nations.

Another advantage of these countries is that some of the most advanced research in sustainable and efficient energy production happens here. Therefore, they are likely to continue improving their energy systems. Meanwhile, less developed nations remain unable to increase energy production in a sustainable manner, as already constrained government budgets, they tend to focus on providing the more basic necessities of life for their citizens.

A major focus of foreign aid programs led by developed nations tends to be the development of renewable energy projects in order to improve energy access. Developing nations often do not have the resources and excess energy capacity to develop renewable energy systems by themselves. The cheapest way to increase energy production tends to be through the use of non-renewable sources such as fossil fuels, that is, governments have little choice but to develop their non-renewable energy sector. However, increased energy production from conventional sources is harmful to our environment. One of the solutions to this issue is through foreign investment in the development of renewable energy industries. This would ensure that energy access is improving through sustainable means, reducing the strain on the developing economy. Studies have shown that increased foreign investment into energy sectors of developing nations has often led to an increase in the production of electricity through renewable sources [15]. Foreign aid also allows developing nations to provide incentives for the development of local renewable energy projects, which serves the dual purpose of providing employment and reducing energy poverty.

International programs have an important role in improving energy access across the world. Global organisations such as the United Nations are key in organising such international efforts towards energy poverty reduction efforts. UN states in the seventh SDG: *Ensure access to affordable, reliable, sustainable and modern energy for all* [16]. In 2021, during a historic High-Level Dialogue on Energy in New York, in front of a delegation of over 130 heads of state and representatives from member nations, the UN Secretary-General laid out an aggressive timeline to ensure the achievement of sustainable energy access to everyone [17]. The timeline aims to provide energy access to 500 million more people by 2025 by aggressively scaling up investments in developing renewable energy. This includes creating 30 million new jobs in the sector, re-directing investment from fossil fuel subsidies towards green energy solutions and an immediate halt to the development of new coal power plants. Roadmaps like these and other efforts by the United Nations help in creating an organised international effort towards solving the problem of energy poverty and serve as a guideline that other nations can follow for their own programs.

9.3 RENEWABLE ENERGY

Renewable energy is produced from sources that are replenished on a timescale compared to the human lifetime. Conventional energy sources like fossil fuels and natural gas reserves accumulate over millions of years and once utilised cannot be restored easily. Meanwhile, renewable sources like hydro power, wind and solar energy provide a continuously replenishing source of energy.

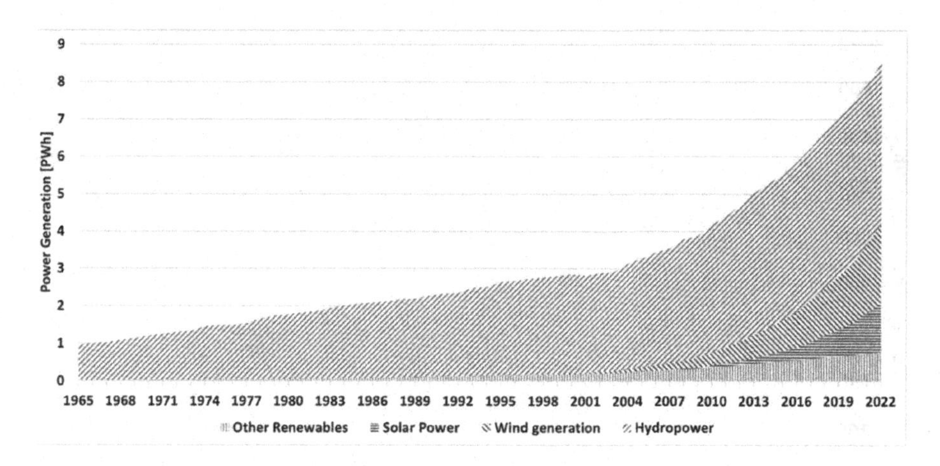

FIGURE 9.5 Distribution of power generated from various renewable energy sources. (Data taken from "Our World in Data" [2].)

Renewable energy has gained significant importance in recent times, with increased awareness of the environmental and economic effects of fossil fuels. Figure 9.5 illustrates the increasing amount of power generated from renewable sources and the increasing diversity in the types of renewable sources as well. This rise has led to innovation and development of several new systems that use renewable sources of energy. The reduced impact of renewable systems on the environment makes them one of the best solutions for increasing energy access.

The two largest sources of renewable energy outside of hydropower are solar and wind energy. They are also rising rapidly in popularity, especially due to their low environmental impact and cost of installation. Hydroelectric projects in most regions require large-scale deployment of resources, and the construction of dams has a widespread environmental impact. Solar and wind energy, by comparison, cause little impact on the environment, either in terms of emissions or damage to the ecosystem. In the last few decades, a combination of government subsidies and technical innovation has made these sources of energy highly affordable, resulting in a rapid increase in power generation from these sources, as seen in Figure 9.6. A large part of this increase is seen in countries like China and India, where governments have been heavily encouraging the growth of renewable energy sources.

9.3.1 Small-Scale vs Large-Scale Projects

There is a distinction between the two broad types of renewable energy systems being developed today. Large-scale renewable energy projects are seen in the form of large solar farms, dams for hydropower and offshore wind farms. They are high-capacity projects, generating enough electricity to power large regions. On the other hand, small-scale renewable projects produce lesser electricity, primarily for one or two

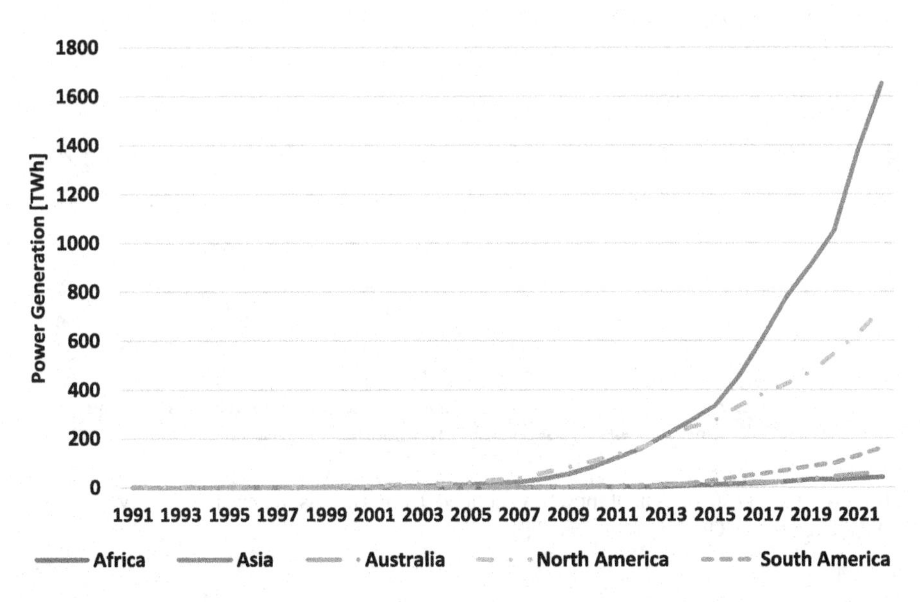

FIGURE 9.6 Increase in power generation solar and wind energy. (Data taken from "Our World in Data" [2].)

households or a small town. These projects are smaller in size and are usually implemented to provide energy access independent of an energy grid.

Small-scale renewable energy projects offer a better solution in terms of improving energy access compared to large projects [3]. Many of the larger projects face the same structural problems as conventional sources of energy. Developing countries, which suffer most from energy poverty, have limited energy grids. Rolling blackouts and grid failures are common problems. With large energy projects, these issues would still persist, as the electricity generated would still need to be transported to the regions that are needed. This reduces the impact of these projects. A significant amount of capital and massive development in infrastructure is required to build large-scale projects. Since energy poverty often correlates with lower GDP [18], this means that countries do not have the capacity to build projects at this scale, leaving them dependent on foreign aid and loans from external sources of funding, like the World Bank.

Small-scale projects do not need large investments or the development of infrastructure. They are decentralised systems that can function without being connected to a central energy grid. This is particularly useful in remote and inaccessible regions, where it is difficult to build electricity grids. Small projects are also simple systems with low maintenance and operational costs. This reduces the amount of investment required to provide energy access. Large-scale projects also require a lot of years, or in some cases decades, to be built up and become operational at full capacity. Meanwhile, small projects can be set up much faster, providing a quicker solution to the growing energy crisis. Small-scale projects are also more adaptable and can be customised best to suit the energy consumption characteristics of the community.

9.3.2 Challenges Faced by Small-Scale Projects

1. **Technical challenges:** Small-scale projects often provide energy access to regions that have never had electricity access before. This means there is not sufficient data to project the growth in energy demand accurately. In some cases, the development of the region resulted in an increase in electricity consumption to the point that the small-scale systems powering the region could no longer match the demand. Continued usage under these circumstances would result in overloading, causing damage to critical components of the system and eventual failure of the system itself.
2. **Operational problems:** Like any other technical system, small-scale renewable energy systems require regular maintenance to keep them operational. However, since these systems are often set up in remote rural regions, it can be difficult to create proper maintenance strategies. Local technicians need to be trained in maintenance procedures, and maintenance schedules need to be followed.

9.4 SUSTAINABILITY

Sustainability refers to designing systems for long-term use and development. Over the last few decades, a large number of small-scale renewable energy projects have been implemented around the world. We now have a lot of data on the performance of these systems, the causes of failures and the effects of these projects on the communities. Many studies have been done on the sustainability of small-scale projects, especially on the causes of failures.

A large number of projects end up being unsustainable in the long term [3], either failing entirely or not providing sufficient energy in the long term. Small-scale projects are built using funding from various organisations or government loans. These projects are funded for their installation and covering operational costs for one or two years. After this, the project is handed off to a local administration. Often, these local organisations lack the technical knowledge and personnel capacity for proper maintenance of these systems. In some cases, inefficient administrative practices are also responsible for failures, such as using low-quality parts or not maintaining correct schedules. Under sizing of systems is also a major problem.

The breakdown of these systems can rapidly undo any progress made in improving energy access and is also a waste of money and resources. Therefore, in recent years, there has been a push to design and develop projects in a way that accounts for these problems and ensures long-term utilisation of the benefits of renewable energy. Measures such as community involvement in the design stage [5] and commercial applications to increase income generation [19] are often found to be beneficial in improving the sustainability of small-scale renewable energy projects.

9.4.1 Measurement of Sustainability

In order to study the sustainability of projects and create effective solutions for better use, we need to be able to quantify and measure sustainability. This is a difficult endeavour, primarily because "sustainability" as a term is highly subjective. According to Kuhlman and Farrington [20], the term "sustainability" began to be used primarily in public policy as a means to ensure that current development does not affect the ability of future generations to meet their needs. This term has since then gained a wider reputation and is now widely used across industries. However, it still remains subjective in its meaning and application, making its quantification difficult. There are many different aspects of sustainability, as illustrated in Figure 9.7.

One of the most widely used methods to study sustainability is through the use of indicators. Indicators are measures of the various aspects of the performance of the system. This includes both technical measures like efficiency and lifespan and non-technical factors such as accessibility and entrepreneurial potential. Generally, indicators are defined using broader goals of sustainability, such as the UN SDGs. Multiple indicators show how a system is contributing to the achievement of that goal and how it compares to other systems.

Scores are assigned to indicators using a variety of different methods. Sometimes, a series of questions are designed, and responses are collected using surveys and interviews. These interviews are conducted with people from the communities where these projects are installed and deal with understanding the level of satisfaction with the performance and other social impacts of the project. In other cases, especially for technical indicators, physical metrics such as efficiency can be used [6].

In order to account for differences between the importance of various kinds of indicators, each of the scores is also assigned a weightage. The weightage represents the significance of those particular indicators to the sustainability assessment of the entire project. For instance, projects in low-income nations tend to have high weightage for economic sustainability indicators since the financial success of the project is important given the low amount of capital that can be invested into such projects.

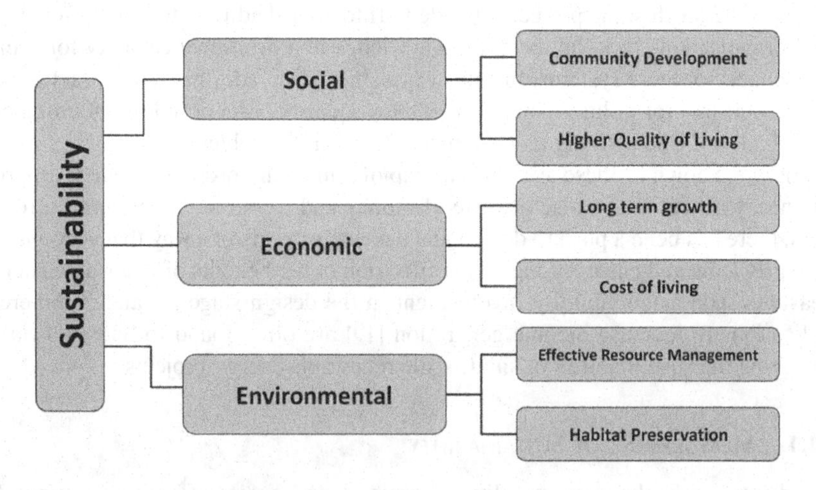

FIGURE 9.7 Various themes of sustainability.

The weightage and the scores are then summed to create a final score, which is used for comparison. Sustainability,

$$S = w_1 I_1 + w_2 I_2 + w_3 I_3 + w_4 I_4 \ldots \tag{9.1}$$

Here, I_1, I_2, I_3, I_4, … refer to the scores assigned to each of the indicators, and w_1, w_2, w_3, w_4, … are the corresponding weightages of the indicators, based on the relative importance of each indicator.

9.4.2 Sustainability Assessment Model

There are four broad themes of indicators that are used to thoroughly assess the sustainability of any project or system: Environmental, social, technical and economic. Environmental indicators show the effect on the environment in terms of emissions or by-products generated. Social indicators measure how sustainable the project is for the community that uses the energy produced. Technical indicators quantify the lifespan of the project, including factors such as reliability and efficiency. Finally, economic indicators assess if a project is financially sustainable over the long term.

Reviewing sustainability assessment models used in previous studies, there is a wide variation in assigning weightage to each indicator. Some projects average scores of all indicators, while others assign higher weightage to some indicators. Geographic, economic and social factors often determine the weightage given for each indicator in projects. In some cases, like Dauenhauer et al. [5], environmental sustainability was not considered because the environmental impact of PV panels was considered minimal enough to be omitted from a sustainability assessment. Therefore, it is difficult to assign a common weightage to all projects.

However, a common fact observed is that the primary cause for the failure of most projects is not technical factors. Small-scale renewable projects are built to be simple in design to save costs and make installation and maintenance easy. For many projects, technicians and mechanics in the community can perform regular maintenance locally. Also, in modern times, the reliability of renewable energy systems, especially solar PV, has increased considerably, lowering the chances of equipment failure in the short term. Another issue with technical factors is that a project with a very efficient design without considering external factors would get a high sustainability score. This would make comparisons between projects complex and biased. There are also varying levels of technical complexity involved with projects. Wind turbines offer some of the highest efficiency in energy generation, but they cannot be used in regions without sufficient continuous wind currents. Therefore, it is logical that technical factors should be assigned a low weightage.

One major technical factor that should be considered is under sizing. Projects are installed with some future rise in capacity built in. However, for most regions where these projects are installed, there needs to be more previous data regarding energy consumption characteristics, which makes it difficult to predict what the increase in future consumption will be. In the future, as access to electricity spurs development in the region, the energy demand might exceed the maximum capacity of the renewable systems providing electricity. This would lead to overloading of critical components, causing damage and lowering the lifespan of projects. Therefore, proper models for predicting growth in energy demand are essential.

In general, the environmental impact of renewable energy projects is very low. Conventional sources produce vast amounts of carbon emissions, increasing atmospheric pollution and contributing to increasingly severe global warming. They also produce toxic by-products, which, if not correctly disposed of, cause a severe impact on the environment. In comparison, renewable sources have no by-products of energy generation and do not contribute to carbon emissions. For large-scale renewable projects, such as wind farms and dams for hydroelectric power generation, the environmental impact is primarily seen in how it affects the natural cycles of the ecosystem. Dams prevent the free movement of aquatic animals along the rivers.

Similarly, wind farms affect the flow of wind currents, which affects the vegetation and animal habitats in the wake of the flow [21]. However, this study is focused on small-scale projects. The impact of small-scale projects on the ecosystem is very low. Projects like micro-hydropower plants and urban wind turbines have some effects, but it is still very low compared to large-scale projects. Also, since small-scale renewable projects do not necessitate the construction of electric grids, the environmental impact is further lowered. Therefore, we assign a lower weightage to environmental indicators.

Economic indicators include the investment needed to set up the project, the regular maintenance costs and the project's economic benefits. Overall, it assesses whether the project can financially sustain itself in the long run, which is an important factor in the long-term sustainability of the project. Most small-scale renewable energy systems, particularly in developing countries, are built using funding from various organisations or loans from governments. Typically, the funding is for the installation of the project and one or two years of operation. Beyond this point, it is expected that the project should be able to fund itself. The project's economic benefits are seen in the development of local businesses and the expansion of existing businesses due to the availability of low-cost electricity. A part of this extra revenue generated is then utilised for keeping the system operational, either directly by the businesses or in the form of additional taxes collected by the local administration overseeing the project. Ensuring that the project is financially sustainable is very important since, without sufficient money, it is not possible to pay for maintenance, spare parts and the technicians needed to keep the systems operational. Apart from maintenance, the economic sustainability of the project is also seen in the long-term reduction in energy spending in households powered by the project. Energy poverty also has an economic component, so lower energy spending is crucial to any program that intends to alleviate energy poverty. Therefore, economic indicators should have a high weightage in any sustainability assessment model.

The social sustainability of a project is another critical factor in the long term. This is represented in the degree of community involvement and ownership in the project's design and operation. One of the most significant examples of this is seen in Dauenhauer et al. [5]. While comparing 65 PV projects across Malawi, it was seen that the projects with higher sustainability scores had higher levels of community involvement, which made it easier to train local technicians to keep the systems operational and ensure that the local administration can maintain the projects well in the long term. Community priorities need to be considered in the project's design. It also helps in assessing the rise in energy demand in the future, which would reduce the problem of under sizing. Community support for a project also ensures financial support for the project because the users understand the necessity of keeping the project operational. A higher awareness of the benefits of renewable energy also results in broader acceptance of these projects, which greatly improves energy access. Therefore, social sustainability would have a higher weightage in the model.

To summarise, in analysing the importance of each factor for the long-term functioning of the project, it is seen that social and economic factors should have a higher weightage over technical and environmental factors. However, the difference cannot be large; for there needs to be a sufficient representation of all factors to provide a better overall picture of a project's sustainability. For the model of assessment in this study, we will assign a weightage of 30% to social and economic factors, 25% to technical and 15% to environmental factors.

Therefore, using the decided weightages in Eq. (9.1), we get a final Sustainability Assessment Equation. Sustainability,

$$S = 0.25\, I_{\text{tech}} + 0.15\, I_{\text{env}} + 0.3\, I_{\text{eco}} + 0.3\, I_{\text{soc}} \tag{9.2}$$

Here, I_{tech} is the final score for technical indicators, I_{env} is the final score for environmental indicators, I_{eco} is the final score for economic indicators and I_{soc} is the score for social indicators. For easy comparison, we will score the indicators between 0 and 1.

9.4.3 Analysis of Small-Scale Renewable Energy Projects

In order to study the applications of the equation, we will now analyse three different small-scale renewable projects and create sustainability scores for the projects.

9.4.3.1 Solar PV in Malawi

Malawi has some of the lowest levels of energy access in the world. The central grid is limited to urban areas, leaving more than 80% of the population without access to electricity. In order to improve energy access, solar PV projects are being built across Malawi. These projects are particularly suited to Malawi due to the large levels of solar energy and the relatively low cost of installation of PV panels.

Dauenhauer et al. [5] conducted a sustainability analysis of 65 PV projects across Malawi. The projects were grouped under three main categories based on the implementer of the project. First were the two main groups under the Malawi Renewable Energy Acceleration Programme: the Community Energy Development Program (CEDP) and the Strategic Energy Project (SEP) led by the University of Malawi Polytechnic. Projects implemented under these programs differed from other projects, grouped together as "Other." The differences resulted in significant differences in the project's sustainability, which makes studying these factors very important.

PV systems have nearly no environmental impact. Large solar farms might affect vegetation growth over large parts of the land, but small-scale PV are some of the lowest-impact renewable systems. For this reason, we will assign a score of 1 for environmental sustainability.

Solar PV panels are simple systems that require minimal maintenance. There are no moving parts and no high-energy reactions occurring, which means there is low damage over the long term. The major technical sustainability factors are the battery and panel quality, under-sizing and ability to meet user demands. Using data collected by Dauenhauer et al. [5], we see that on nearly every metric, CEDP and SEP scores are higher than other projects. Projects built by SEP and CEDP had better panel and battery health and matched the actual user demands more closely. The most significant difference was seen in the quality of the panels, with both SEP and CEDP projects having better quality panels, and their panels were also much better sized than other projects, so their predicted user demand was closer to actual consumption.

The economic sustainability of the projects was found to be very low. This was primarily due to the need for more commercial usage of the electricity generated by the project. Very few projects generated enough income to sustain themselves over the long-term sufficiently.

The best way to measure economic sustainability is by using a ratio of the income generated to the expenses incurred in keeping the projects functioning. Of the three major groups, only the projects under CEDP were designed to include activities which could produce an income for the project. This meant that CEDP projects had the highest scores among all three projects. SEP projects often had a negative net revenue, while projects under "Other" had low revenue but also had low operating expenses due to the relatively lower cost of components.

Social sustainability mainly measures the degree of community involvement in the projects. This included the variety of stakeholders involved in the project, the contribution of the community in keeping the project operational and the frequency of meetings of the stakeholders. Once again, projects under CEDP and SEP score higher under this metric because of the higher degree of community involvement. The CEDP and SEP projects were relatively newer, so the higher scores showed the increasing prioritisation of community involvement in renewable energy projects. Projects under SEP, which often had a net negative income, had a high degree of community involvement, suggesting that social sustainability is independent of the project's economic viability, so they should be considered separately.

Table 9.2 summarises the scores assigned to each of the sustainability themes, among the three major divisions of projects. Using the equation with the weightages for each of the different themes, we get a final sustainability score of 0.512. This score is due to the high social sustainability scores of the projects. An important observation here is the relative scores of the three major implementers. The results match the conclusions of the study, which have indicated a high degree of sustainability for CEDP and SEP projects.

9.4.3.2 Urban PV Systems in Seoul

Most of the research on the sustainability of small-scale renewable energy systems is focused on remote, rural regions. There is a dearth of studies on small-scale systems installed in urban residential areas, where their primary function is not to provide

TABLE 9.2
Summary of Sustainability Scores

Sustainability Scores	CEDP	SEP	Other	Average
Technical	0.748	0.734	0.451	0.644
Economic	0.374	0.019	0.132	0.175
Environmental	1	1	1	1
Social	0.611	0.542	0.328	0.494
Final	0.632	0.502	0.4	0.512

Source: Data taken from Ref. [5].

energy access but to supplement the energy supply from the main grid and reduce energy costs for households. Low-income households, even in urban areas, often spend more than 10% of their income for fulfilling their energy needs. This affects their ability to spend money on other basic needs; thus, this is a serious issue.

Lee and Shepley studied PV panels installed in low-income housing in Seoul, South Korea. The government subsidised the installation of these projects, and the aim was to provide cost savings for the households. The study collected data in the form of responses from voluntary surveys, mainly concerning satisfaction with the projects and their expectations.

The projects performed effectively on a technical level. Users were generally happy with the general functioning of the system. However, they desired an increase in the capacity of the systems. The cost savings provided by the systems were not seen as sufficient for the households. Apart from this, the panels were installed at an angle of 25° to reduce shadows, whereas, for maximum efficiency, the panels should be installed perpendicular. There was also a lack of proper maintenance of the systems after installation, so there was no way to know if the system was functioning properly. Considering all these factors, the project is assigned a technical sustainability score of 0.7.

Solar PV panels have a minimal environmental impact. In addition, more than half the electricity generated in South Korea is from non-renewable sources such as coal, oil and natural gas. Therefore, by reducing the electricity consumption from the electricity grid, PV systems reduce the carbon emissions caused by the energy consumption of households. This means that the systems have a net positive impact on the environment. Therefore, environmental sustainability is given a score of 1.

Unlike off-grid projects, the PV systems installed here supplement the energy from the central grid. Therefore, to measure economic sustainability, the project's impact is seen in the reduction of households' energy costs. Figure 9.8 shows the reduction in energy bills for households and its variation across the year. Note the rise in cost savings from February to April, despite constant solar generation, due to lower consumption of electricity.

A large number of households desired a higher reduction in cost savings. The reduction in energy bills was not entirely sufficient and varied widely between

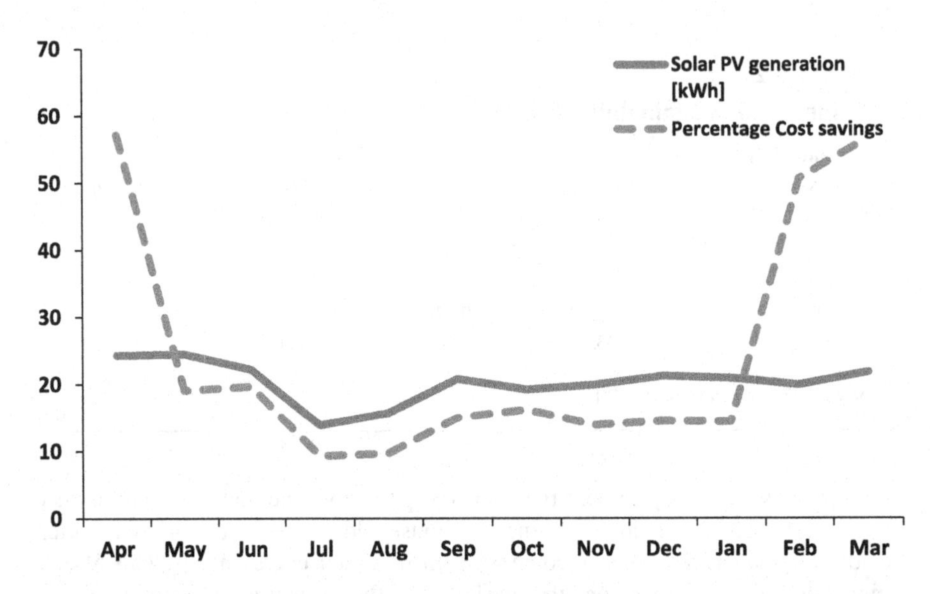

FIGURE 9.8 Comparison of solar PV generation and cost savings produced by the photovoltaic systems, from April 2018 to March 2019. (Created using data from [6].)

months. Since the primary reason for most households to install such a system was to reduce energy bills, the economic factor plays a vital role in determining the long-term effectiveness of the project. Further improvements in the design or positioning of the solar PV systems or further increasing the systems' capacity would help further reduce energy poverty. For this project, we will consider the ratio of cost savings to the net cost as the metric for economic sustainability. This results in a score of 0.249.

One of the biggest benefits seen for the project was the resulting rise in awareness of the benefits of renewable energy systems. Users were satisfied with the reduction in carbon emissions that their use of renewable systems would bring about. There was also an increased awareness of the environmental benefits of reducing energy consumption. The advantage of this awareness is greater support in the future for any such renewable energy programs and conscious support for reducing wasteful energy consumption. The results of the study also show a high degree of satisfaction with their involvement in the program and with the administration of the project itself. However, they did desire a more systematic maintenance plan for the project, as well as better customer support. Overall, social sustainability gets a score of 0.7.

Table 9.3 summarises the final scores for the sustainability themes. Using the sustainability assessment equation results in a final sustainability score for the project of 0.609. This score is higher than the average scores from the PV projects in Malawi, and the primary reason is the higher economic sustainability of these projects. There is a significant increase in cost savings, although slightly below the initial user expectations.

TABLE 9.3
Overall Sustainability Scores

Theme	Score
Social	0.7
Technical	0.7
Environmental	1
Economic	0.249
Final	0.609

Source: Using data from Ref. [6].

9.4.3.3 Micro-Hydropower in Nepal

Nepal is a mountainous country, with the Himalayas passing through the nation with an average altitude of 3,265 m. Building widespread energy grids is challenging, especially in regions with rugged terrain and high elevation. However, Nepal is rich in inland water resources, with over 6,000 rivers. Taking advantage of this, the government has incentivised the development of several micro-hydropower plants. These plants are small in scale, with minimal environmental disruption, and they are off-grid systems, which eliminates the need for an energy grid. However, over the years of operation of these plants, there have been failures, and several studies have been undertaken to improve the design.

Bhandari et al. conducted a sustainability assessment of a micro-hydropower plant [4]. Their study used a new assessment model, developing indicators specifically in the context of micro-hydropower plants. Scores were assigned for indicators based on interviews with several community stakeholders, as well as quantitative data collected on the plant since 2006.

Micro-hydropower plants usually require a single operator. However, if the plant breaks down when the operator is on leave or otherwise unavailable, it results in unexpected disruptions in the power supply. There is also a lack of proper maintenance schedules for the plant, which causes even more disruptions. Since villagers depend on continuous power supply from the plant, such unexpected disruptions seriously affect the plant's performance. The plant, in general, is well-designed, and there are no fluctuations in the power generated. One of the major improvements that could be made to the plant is installing a more efficient turbine, which could further increase the plant's capacity. Overall, the technical sustainability of the plant, based on data collected by the study, is 0.612.

The environmental sustainability of the project is very high. Traditional hydroelectric projects tend to have large effects on the ecology of the surrounding regions, especially due to the large-scale change in the flow of water. However, for micro-hydropower plants, there is minimal disruption to the movement of fish along the river, and the effects of sedimentation and erosion of soil are low. There is some negative impact on the region surrounding the river due to the construction of infrastructure to supply power from the plant. There is also a drastic reduction in fossil fuel consumption in the area, an added benefit of power supply from renewable sources. Therefore, the environmental sustainability score is 0.8.

The plant has resulted in a significant increase in business activities in the village and expanded operations of existing businesses due to the availability of low-cost electricity. There was also time saved from collecting firewood which could now be used for businesses and the development of local enterprises. The costs of keeping the plan operational were also low, combined with the funding for the project itself being provided by the government, which meant that the project could sustain itself economically for the long term. The economic sustainability score for the plant is 0.75.

There was a strong social impact of the project. There is a high sense of ownership over the project, with most of the project management being made by the local community. There is also extensive involvement of the community in the maintenance and operation of the plant itself. The cost of electricity supply is fixed based on a community consensus, and residents are satisfied with the quality of the power supply. However, the disruption caused by insufficient maintenance is a problem. There was limited application of the project in improving access to electronic devices; however, the electricity access allowed the local school to open a computer lab. Of all the factors, social sustainability scores the highest, at 0.834.

Table 9.4 summarises the scores under each of the sustainability themes. Using the sustainability assessment equation, we get an aggregate sustainability score of 0.748, considering all the different indicators with their scores. This is a considerably high score, compared to the other two projects. Despite the lower environmental impact, the economic and social sustainability score is significantly higher than other projects, which pushes the score higher.

9.4.4 Comparison of Scores

Comparing the results of sustainability scores for the three projects from Table 9.5, we observe that the micro-hydropower plants have the highest sustainability, followed by the PV projects in Seoul and Malawi. Despite solar energy being generally

TABLE 9.4
Overall Sustainability Scores

Theme	Score
Social	0.834
Technical	0.612
Environmental	0.8
Economic	0.75
Final	0.748

Source: Using data from Ref. [4].

TABLE 9.5
Overall Sustainability Scores for All Three Projects

Project	Score
Off-grid PV project in Malawi	0.512
Urban PV systems in South Korea	0.609
Micro-hydropower plant in Nepal	0.748

considered the most sustainable form of renewable energy due to its very low environmental impact, we find that in actual applications, project sustainability is governed by a host of factors, which vary widely.

The results given by the model generally agree with the conclusions of the studies themselves. The sustainability of PV projects in Malawi is very low, with many systems failing and a host of issues. There are several aspects of improvement, and there are several flaws in the implementation of the projects highlighted in the study. Meanwhile, micro-hydropower plants are successful off-grid renewable energy projects, with a large part of the hydro-power generation in Nepal coming from these low-capacity plants. It had a significant impact for more than a decade after its installation without any major problems or failures. The sustainability score reflects the better overall implementation of the project.

An interesting observation in the results is that the final sustainability scores for CEDP projects in Malawi are higher than the aggregate sustainability scores for the PV systems in Seoul. This is primarily due to the higher economic sustainability of the CEDP projects. CEDP required the development of income-generating businesses as part of the project's implementation, while the South Korean project did not satisfy the savings expectations that the users had. Using the sustainability scores, it is possible to identify specific factors which are causing a project to have a higher or lower success rate.

Using the sustainability assessment model, we can quantitatively assess the long-term success of the project. The model also reflects the importance of factors that should be considered during the development of small-scale renewable energy projects. The economic and social sustainability of projects must be taken into account and are relatively more important than environmental and technical sustainability. The primary reason for this is the inherently low impact of renewable energy systems, as well as the simpler technical design and low maintenance cost of small-scale projects. With an increasing focus on sustainability in the modern world, the model helps clarify and prioritise different aspects of a project in order to ensure operation in the long term.

There is always a certain degree of uncertainty associated with the scoring of sustainability, but there is a lack of sufficient information on the variation of the different factors affecting sustainability. Therefore, the uncertainty of these scores is not quantified here.

9.5 CONCLUSION

Energy poverty is a major global issue, affecting nearly two billion people around the world, especially in low-income developing countries. There needs to be a sustained effort to solve this problem, as it affects the development of the world as a whole and hampers the progress made in other issues. However, we must ensure that the solution does not end up being a problem itself. Therefore, we must work on creating using renewable sources to provide energy access. This would decrease the environmental impact of increased energy consumption and also spur further improvement in renewable energy technology.

Small-scale renewable energy projects are often found to be better at improving energy access since they work independently of an energy grid and require less funding, two factors that are crucial in regions most severely affected by energy poverty. Large-scale implementation of these systems can significantly increase the energy supply.

The sustainability of these systems is an important factor in their success. If projects fail after one or two years of installation, then they are simply a waste of money and resources that could have been better utilised. Therefore, the causes of failure must be studied in order to improve the systems and ensure they act as an effective long-term solution.

The sustainability equation developed in this study shows reliable results when compared with other studies. However, future work needs to focus on testing this model on a diverse variety of small-scale renewable energy projects and adapt the weightages and indicators in order to more accurately reflect the actual data. Once the model is sufficiently accurate, it can then be used as a general equation for any kind of renewable energy system.

ACKNOWLEDGEMENTS

This work was made possible in part by Mitacs Globalink and Natural Sciences and Engineering Research Council of Canada.

REFERENCES

1. Karpov, K.A. Development trends of global energy consumption. *Studies on Russian Economic Development* 30(1): 38–43, 2019.
2. Ritchie, H., Roser, M., Rosado, P. Energy. Published online at OurWorldInData.org, 2022. Retrieved from https://ourworldindata.org/energy on June 28, 2023.
3. Terrapon-Pfaff, J., Dienst, C., König, J., Ortiz, W. A cross-sectional review: Impacts and sustainability of small-scale renewable energy projects in developing countries. *Renewable and Sustainable Energy Reviews*, 40: 1–10, 2014.
4. Bhandari, R., Saptalena, L.G., Kusch, W. Sustainability assessment of a micro hydropower plant in Nepal. *Energy, Sustainability and Society*, 8(3): 1–15, 2018.
5. Dauenhauer, P.M., Frame, D., Eales, A., Strachan, S., Galloway, S., Buckland, H. Sustainability evaluation of community-based, solar photovoltaic projects in Malawi. *Energy Sustainability and Society*, 10: 12, 2020.
6. Lee, J., Shepley, M.M. Benefits of solar photovoltaic systems for low-income families in social housing of Korea: Renewable energy applications as solutions to energy poverty. *Journal of Building Engineering*, 28: 101016, 2020.

7. González-Eguino, M. Energy poverty: An overview. *Renewable and Sustainable Energy Reviews*, 47: 377–385, 2015.
8. Moss, T. My Fridge Versus Power Africa, Centre for Global Development, September 09, 2013. https://www.cgdev.org/blog/my-fridge-versus-power-africa. Accessed on July 12, 2023.
9. United Nations. The Sustainable Development Goals Report 2022, pp. 40–41, available from https://unstats.un.org/sdgs/report/2022/The-Sustainable-Development-Goals-Report-2022.pdf. Accessed on July 19, 2023.
10. Tran, V.V., Park, D., Lee, Y-C. Indoor air pollution, related human diseases, and recent trends in the control and improvement of indoor air quality. *International Journal of Environmental Research and Public Health*, 17(8): 2927, 2020.
11. Sovacool, B.K. The political economy of energy poverty: A review of key challenges. *Energy for Sustainable Development*, 16(3): 272–282, 2012.
12. Ting, D.S-K. *Thermofluids: From Nature to Engineering*, pp 24–27. Academic Press, London, United Kingdom, 2022.
13. Pietrosemoli, L., Rodríguez-Monroy, C. The Venezuelan energy crisis: Renewable energies in the transition towards sustainability. *Renewable and Sustainable Energy Reviews*, 105: 415–426, 2019.
14. Ruiz-Rivas, U., Tahri, Y., Arjona, M.M., Chinchilla, M., Castaño-Rosa, R., Martínez-Crespo, J. Energy poverty in developing regions: Strategies, indicators, needs, and technological solutions. In: Rubio-Bellido, C., Solis-Guzman, J. (eds.) *Energy Poverty Alleviation*. Springer, Cham, 2022.
15. Villanthenkodath, M.A., Mahalik, M.K., Patel, G. Effects of foreign aid and energy aid inflows on renewable and non-renewable electricity production in BRICS countries. *Environmental Science and Pollution Research*, 30: 7236–7255, 2023.
16. United Nations. Transforming our world: the 2030 Agenda for Sustainable Development, pp. 19, available from https://undocs.org/en/A/RES/70/1. Accessed on August 26, 2023.
17. United Nations. Global Roadmap for Accelerated SDG7 Action in Support of the 2030 Agenda for Sustainable Development and the Paris Agreement on Climate Change, available from https://www.un.org/sites/un2.un.org/files/2021/11/hlde_outcome_-_sdg7_global_roadmap.pdf. Accessed on August 27, 2023.
18. Igawa, M., Managi, S. Energy poverty and income inequality: An economic analysis of 37 countries. *Applied Energy*, 306: 118076, 2022.
19. Hiremath, R.B., Kumar, B., Balachandra, P., Ravindranath, N.H., Raghunandan, B.N. Decentralised renewable energy: Scope, relevance and applications in the Indian context. *Energy for Sustainable Development*, 13(1): 4–10, 2009.
20. Kuhlman, T., Farrington, J. What is sustainability? *Sustainability*, 2(11): 3436–3448, 2010.
21. Diffendorfer, J.E., Vanderhoof, M.K., Ancona, Z.H. Wind turbine wakes can impact down-wind vegetation greenness. *Environmental Research Letters*, 17(10): 104025, 2022.

10 The Challenges of Stakeholder Engagement in Climate Change Adaptation in Nigeria

E. D. Oruonye
Taraba State University

H. K. Ayuba
Nasarawa State University

10.1 INTRODUCTION

Climate change is one of the development challenges confronting Nigeria and the global community in recent times. The problem of climate change is visible in our day-to-day life. Nigeria responded by creating appropriate policies, strategies, and action plans through the National Adaptation Strategy and Plan of Action for Climate Change in Nigeria (NASPA CCN) (2015) and the NAP Framework (2020) to fulfill its adaptation priorities. These policy documents contained in-depth descriptions of the adaptation strategies, policies, and action plans for each sector and for all the stakeholders in the country. These numerous policy instruments have been produced over the years to combat climate change impact on the country.

Nigeria continues to struggle to get the intended results despite its modest efforts to integrate climate change adaptation into its socioeconomic agenda and policies. Despite all these endeavors, Nigeria has not been able to make significant progress in terms of coordinated climate change adaptation efforts. These issues include a lack of finance, the need to expand capacity, and inadequate technical capabilities. There is a lack of cooperation, coordination, and synergy among parties. Additionally, the absence of target setting, monitoring, and evaluation has resulted in overlaps, effort duplication, and increased costs. Another issue affecting the efficiency of the nation's adaptation efforts is poor communication. A significant obstacle to the country's implementation of the NAP is the exclusion of the sub-national governments (states and local governments), Civil Society Organizations, indigenous people, women, youth, and people with disabilities (FGN, 2021). In short, lack of adequate engagement of critical stakeholders.

Climate change adaptation can be defined as the process whereby certain groups or individuals employ various strategies to cope with or/and adapt to the consequences

 DOI: 10.1201/9781003403456-10

of climatic impacts (Adger et al., 2009 and IPCC, 2014). The process of adaptation involves improving, developing, and putting into practice techniques to lessen, manage, and benefit from the effects of climatic occurrences. Governmental agencies or non-governmental organizations (NGOs) and their networks make choices and formulate public policies that affect persons, groups, and organizations. To determine the most suitable kinds of adaptation, relevant parties must be brought together. Characterizing existing and potential future vulnerability requires a thorough examination of stakeholders' resilience to and ability to respond to climatic disasters. Implementing adaptation policies will be made easier if stakeholders' roles in the decision-making process are understood and incorporated into any action plan.

Adaptation is a process by which strategies to moderate, cope with, and take advantage of the consequences of climatic events are enhanced, developed, and implemented. Adaptation occurs through public policymaking and decisions made by stakeholders, i.e., individuals, groups, organizations (governmental agencies or NGOs), and their networks. Relevant stakeholders need to be brought together to identify the most appropriate forms of adaptation. Analyzing the capacity of stakeholders to cope with and adapt to climatic events is fundamental to characterizing current and possible future vulnerability.

The major importance of the adaptation process is that it helps in increasing resilience among individuals and groups in the community (Rao et al., 2011; Mkonda et al., 2018; Vaidya et al., 2021). Scholars have observed that sustainable climate change adaptation can only take place through the thorough involvement of critical stakeholders such as individuals, groups, scientists, governmental agencies, private sectors, international agencies, and their networks in the process of decision and policymaking (Wamsler, 2017).

Rainey (2006) defined a stakeholder as "any individual or group that is directly or indirectly influenced by the products, programs, processes, and/or systems, but does not immediately benefit as an economic participant, such as a customer or supplier." Stakeholders are additionally described by Brown et al. (1998) as people or organizations with an interest in a problem or specific natural resource. Therefore, the stakeholder includes both individuals who have the ability to influence how resources are used and others whose livelihoods are affected by climate change and changing resource use but who have no such authority (Mkonda, 2022).

The term "stakeholder" in climate change studies refers to policymakers, scientists, administrators, communities, and managers in the economic sectors most at risk (Conde & Lonsdale, 2004). In this context, the stakeholder can be brought together from both public and private enterprises to develop a joint understanding of the issues and to create adaptations. The definition of stakeholder used here is "those who have interests in a particular decision, either as individuals or as representatives of a group of people. This includes people who influence a decision, or can influence it, as well as those affected by it" (Hemmati, 2002).

Stakeholder engagement is a central aspect of climate change adaptation planning and related activities. Stakeholder engagement is a process that an individual, group, or organization can follow in order to listen to, collaborate with, or inform their existing stakeholders (Brown et al., 2001). This process involves identifying, mapping, and prioritizing stakeholders in order to determine the best tactics for effective communication while making the best use of available resources. Consequently, stakeholder engagement helps organizations to proactively consider the needs and desires of anyone who has a stake in their organization, which can foster connections, trust, confidence, and buy-in for your organization's key initiatives (Brown et al., 1998; Adger et al., 1999).

Stakeholder engagement, when done well, can reduce possible risks and conflicts with stakeholder groups, including doubt, discontent, misalignment, disengagement, and resistance to change. However, because stakeholders frequently fit into more than one category, stakeholder identification can be complicated. Citizens or groups may believe that a plan does not represent the interests and welfare of the community as a whole without the involvement of stakeholders. Stakeholder engagement can be very challenging. They play a crucial role in the process of adaptation and policy creation and as such, it is essential to actively involve them in addressing the implications of climate change from the beginning. However, there is only weak empirical evidence to suggest that the stakeholder is actively participating in creating different adaptation plans.

All of these require stakeholder analysis, a system for gathering data on groups or people who are impacted by decisions, classifying that data, and describing potential conflicts between key groups and places where trade-offs may be feasible. It might be done just to identify the parties involved or to look into potential collaboration opportunities (Brown et al., 2001). It is crucial to involve stakeholders in research projects, especially those focusing on climate change and the environment, as numerous studies clearly demonstrate. This is very important because the research community and stakeholders may enhance the current adaptive methods by combining professional scientific knowledge with indigenous knowledge about various environmental changes (Akompab et al., 2012; Filho et al., 2018).

Depending on the degree of interaction where the stakeholders can offer information, different stakeholder engagement strategies may be used. It entails determining one's vulnerability to climate adaptation, involving stakeholders in the process, assessing present and future climate threats, assessing socioeconomic conditions as they are now and in the future, assessing and enhancing one's ability to adapt, formulating adaptation policy, and maintaining one's current adaptation process (Tsufac et al., 2021).

Understanding the role of stakeholders in the decision-making process will assist in the implementation of adaptation policies. This study is important as it involves and streamlines important stakeholders in the adaptation process to climate change. It can be useful to the country in its effort to adopt sustainable adaptation to climate change impacts (IPCC, 2014).

10.2 SIGNIFICANCE OF THE STUDY

Recent development has shown that climate change is impacting the livelihood of most people in Nigeria, particularly the rural dwellers, women, and marginalized groups. Although the NAP Framework for Nigeria (2020) has been developed, there has yet to be a stakeholder engagement strategy and road map that will guarantee successful implementation of the framework. A stakeholder engagement strategy is therefore crucial to the development of any climate change adaptation plan because an "adaptation community" is needed to sustain any adaptation process.

For adaptation to be successful and sustainable, it has to sufficiently involve the relevant critical stakeholders in all the processes (Filho et al., 2018; Nicholson, 2018). In addition, assessing stakeholders' resilience to climatic stress and ability to adapt is crucial for identifying current and prospective future vulnerabilities (Paavola, 2008; Vogel & Henstra, 2015; Mkonda, 2021). This has to do with the role of stakeholder in the process of decision-making, which, in turn, determines the success of the implementation of the laid-out adaptation strategies. The aim of a stakeholder engagement strategy therefore is to identify and engage every critical stakeholder at the subnational (local or regional), national, and international level in the adaptation process. Many countries including Nigeria have developed various policy documents to address the challenges of climate change, exposure, vulnerability, and adaptation (IPCC, 2014; Mkonda et al., 2018). Many have also developed the national adaptation plan and nationally determined contributions to greenhouse gas (GHG) emission reduction policy documents to show their commitments to mitigating and adapting to climate challenges. However, the development of such a vital climate change adaptation plan over the years has not sufficiently involved stakeholder engagement. It is widely accepted that engaging stakeholders in assessing vulnerability and in the implementation of climate change adaptation plans is an important factor for enhancing adaptation implementation and success. It is against this background that this study examines the challenges of stakeholder engagement in climate change adaptation in Nigeria.

10.3 METHODOLOGY

The study employed the use of primary and secondary sources of data. The secondary sources included an in-depth review of relevant literature which was designed to gather and synthesize the existing literature related to stakeholder engagement strategies, processes, and outcomes from different countries and identify good lessons around adaptation engagement at the global level. The review brought to the fore a number of issues relating to stakeholder engagement on climate change adaptation that are beneficial to Nigeria.

Primary sources of data were generated from key informant (KI) interviews, which were carried out physically and virtually to generate more detailed information on issues that were not addressed through a literature review. Individuals who have knowledge and expertise on engaging a broad range of stakeholders were interviewed. The KIs that could not be reached physically either because of distance or the nature of their work were interviewed online. Through online engagement information on the level of knowledge, attitudes and practices related to climate change and adaptation were received. The interviews helped to provide good perspectives on how best to engage with stakeholders, on challenges and practices that are unique to stakeholders' engagement on climate change. Additional information was generated through questionnaires administered to sixty (60) stakeholders from Ministries, Departments and Agencies (MDAs) civil society organizations, the media, legislators, academia, and local communities during their interactive session at the meeting organized by the Department of Climate Change and the NAP Secretariat on 4th April 2023 at Sandralia Hotel, Jabi, Abuja. The interactive meeting provided an avenue for generating additional information used in this study. An analysis of strengths, weaknesses, opportunities, and threats (SWOT) was undertaken to help appreciate the challenges of stakeholder engagement in climate change adaptation in Nigeria. The SWOT analysis was based on the results of an in-depth review of literature, interactive meetings with stakeholders, and interviews with KI. Descriptive statistics and content analysis were used to analyze the data collected.

10.4 RESULT OF THE FINDINGS

10.4.1 Potential Climate Change Stakeholders and Organizations in Nigeria

Although climate change is an all-encompassing issue that affects everyone in the society, both the low and mighty, the respondents listed the following as stakeholders and organizations that must be involved in the development of stakeholder's engagement strategy in Nigeria.

i. Federal Ministry of Information and Culture and its Agencies (NTA, NAN, NOA, NCM&M, etc.);
ii. Private Broadcasting organizations
iii. Print Media Organizations
iv. State Broadcasting Corporations and
v. Information Offices at the Local Council levels
vi. National Emergency Management Agency (NEMA)
vii. Nigerian Meteorological Agency (NiMet)
viii. Ministry of Environment (Department of Climate Change)
ix. Ministry of Health
x. Ministry of Trade and Industry
xi. Ministry of Women affairs
xii. Federal Ministry of Agriculture and Natural Resources
xiii. Ministry of Water Resources
xiv. Ministry of Power and Energy
xv. Educational institutions
xvi. Network operators
xvii. National Inland Water Ways
xviii. Coastal/littoral States
xix. Farmers Associations
xx. Communities in flood-prone areas
xxi. NESREA

10.4.2 Nigeria's Strengths on Stakeholders' Engagement

The respondents view on Nigeria's strength on stakeholder engagement is presented in Table 10.1.

TABLE 10.1
Nigeria's Strengths on Stakeholders' Engagement

S/No	Nigeria's Strengths	Frequency	Percentage (100%)
1	Large youth population	24	40.0
2	Large number of higher educational institutions	14	23.3
3	The community settings	08	13.4
4	Available human resources	14	23.3
	Total	**60**	**100**

From Table 10.1, 40% of the respondents believed that the strength of Nigeria's stakeholder engagement in climate change adaptation and mitigation efforts is its large youthful population, 23.3% large number of higher educational institutions (Universities, Colleges of Education and Polytechnics), 13.4% its diverse community settings (vast and multi-cultural settings), and 23.3% available human resources. Nigeria has an estimated population of over 220 million people, more than 103 million of whom are children (UNPD, 2019). Nigeria is the most populous country in Africa. The country has the largest population of youths in the world, with 70% of the population under 30, and 42% are under the age of 15 and a median age of 18.1 years. The country has a large youthful population that can be easily harnessed in climate change adaptation and mitigation efforts. The size and youthfulness of the population offer great potential for Nigeria to leverage in climate actions if properly harnessed. The country also has a vast diverse and multi-cultural ethnic groups and communities with a wide range of indigenous knowledge on the effect of climate change that can be harnessed in adaptation and mitigation efforts. Nigeria also has 170 Universities, 159 polytechnics, 47 accredited monotechnic, and 82 colleges of education as of June 2022. This large number of higher education if properly harnessed can strengthen its climate change adaptation efforts.

10.4.3 Nigeria's Weaknesses on Stakeholders Engagement

The respondents view on the weaknesses of stakeholder engagement in Nigeria is presented in Table 10.2.

The findings of the study revealed that despite Nigeria's strength in stakeholder engagement in climate change adaptation and mitigation, there are lots of weaknesses that constrain the country's effort. From Table 10.2, it can be seen that 15% of the respondents believed that the weaknesses in stakeholder engagement in Nigeria are the lack of specific enabling policies, laws, regulations, and structures on effective stakeholder engagement in climate change adaptation and mitigation, 15% non-implementation of legal mandates on stakeholders' engagement, and 11.7% argued that the facilitation of effective stakeholder engagement requires an investment of time and resources. It is doubtful if Nigeria has the resources required for effective stakeholders' engagement in climate change adaptation. 20% of the respondents are of the view that inadequate capacity among sector stakeholders is the weakness, 18.3% suggest a high level of corruption, 8.3% lack of implementation of

TABLE 10.2
Nigeria's Weaknesses on Stakeholder Engagement

S/No	Nigeria's Weaknesses	Frequency	Percentage (100%)
1	Lack of specific enabling policies, laws, regulations, and structures	09	15.0
2	Non-implementation of legal mandates on stakeholder engagement	09	15.0
3	Facilitation of effective stakeholder engagement requires an investment of time and resources	07	11.7
4	Inadequate capacity among sector stakeholders	12	20.0
5	High level of corruption	11	18.3
6	Lack of implementation of good project	05	8.3
7	Non-release of funds	07	11.7
	Total	**60**	**100**

good projects, and 11.7% non-release of funds. Nigeria lacks the capacity (in terms of infrastructure and resources) in many of its sub-national states to adequately respond to the impacts of climate change. There is inadequate capacity among sector stakeholders that inhibits the development of strategic partnerships between private sector, civil society, academia, and media organizations that would enable them to increase productivity. This includes the lack of capacity (in terms of infrastructure and resources) in many states to adequately respond to the impacts of climate change issues around them. Stakeholder engagements are weakened by corrupt leadership, especially at community levels. The country suffers from high-level corruption in all fabrics of Nigerian society.

10.4.4 Opportunities of Stakeholder Engagement in Nigeria

The respondents view on opportunities of stakeholder engagement in Nigeria is presented in Table 10.3.

The findings of the study in Table 10.3 revealed that the opportunities for stakeholder engagement according to 26.6% of the respondents is that it provides for an all-inclusive approach in project preparation, planning, implementation, and monitoring processes in climate actions in Nigeria. 11.7% of the respondents are of the view that stakeholder engagement is geared toward ensuring meaningful and wide consultative process according to best practices in climate actions, while 21.7% are of the opinion that stakeholder engagement provides an opportunity for building buy-in and support for adapting to climate change actions, 11.7% opportunity for gathering technical and local knowledge from diverse stakeholders in addressing climate change issues in the country, and 5% opportunity for access to a wide range of human capacity with skills, knowledge, and experiences to enhance climate actions in the country. Furthermore, 10% of the respondents are of the view that stakeholder engagement provides an opportunity to leverage partnerships and networks for climate actions and 13.3% contribute to empowerment of individuals, communities, and organizations.

TABLE 10.3
Opportunities of Stakeholder Engagement in Nigeria

S/No	Nigeria's Opportunities	Frequency	Percentage (100%)
1	Provides for all-inclusive approach	16	26.6
2	Geared toward ensuring meaningful and a wide consultative process	07	11.7
3	Building buy-in and support for adapting to climate change	13	21.7
4	Gathering technical and local knowledge from diverse stakeholders	07	11.7
5	Access to a wide range of human capacity	03	5.0
6	Leverage partnerships and networks for climate actions	06	10.0
7	Contribution to empowerment of individuals, communities, and organizations	08	13.3
	Total	**60**	**100**

TABLE 10.4
Threats to Stakeholders' Engagement in Climate Actions in Nigeria

S/No	Threats to Stakeholders' Engagement	Frequency	Percentage (100%)
1	Lack of education (large illiterate population)	18	30.0
2	Inappropriate methodology of disseminating information at the lowest levels	11	18.3
3	Inadequate tools	06	10.0
4	Lack of good communication	11	18.3
5	Working in silos	14	23.4
	Total	**60**	**100**

10.4.5 Threats to Effective Stakeholder Engagement in Climate Actions in Nigeria

The respondents view on the threats to effective stakeholder engagement in Nigeria is presented in Table 10.4.

From Table 10.4, it can be seen that 30% of the respondents believed that the threats to effective stakeholder engagement in climate actions in Nigeria are lack of education (large illiterate population), inappropriate methodology of disseminating information at the lowest levels, 10% inadequate tools, 18.3% lack of good communication, and 23.4% working in silos by most of the potential stakeholders to be engaged in climate actions in the country.

10.4.6 Ways Nigeria Can Effectively Engage Stakeholders in Climate Change Actions

The findings of the study revealed that some of the ways Nigeria can effectively engage stakeholders in climate change actions include involving all stakeholders from the Federal level down to the community level with appropriate and sustainable funding arrangements. Timely release of project funds and involvement of stakeholders in the planning process can go a long way in effectively engaging stakeholders in climate change actions in Nigeria. Other ways include bringing all stakeholders together to work as a team, enforcing participation from all relevant stakeholders, regular meetings to enhance understanding, and workshop and allocation of specific assignments to stakeholders that will be engaged.

10.4.7 Ways of Monitoring and Evaluating Stakeholders' Engagement in Nigeria

The findings of the study revealed that some of the ways of monitoring and evaluating stakeholder engagements in Nigeria include the following;

i. Use of independent M&E framework dedicated for Climate Change stakeholder engagement;
ii. Appropriate evaluation teams should be constituted which is independent of the Steering or Technical Committees;
iii. Having desk officers connecting with coordinating ministries and states;
iv. By interfacing with various MDAs;
v. M&E to be carried out by policy designers and planners;
vi. Track progress, outcome/output;
vii. Regular engagement with stakeholder;
viii. Proper communication channels instituted among stakeholder;
ix. Proper linkages from top to bottom, Federal to LGA levels;
x. Track progress, outcome, and output;
xi. Working on feedback mechanism through each sector involved.

10.5 CONCLUSIONS

This study has examined the challenges of stakeholders' engagement in climate change adaptation in Nigeria. The study employed SWOT analysis to appreciate the challenges of stakeholder engagement in climate change adaptation in Nigeria based on the results of an in-depth review of literature, meetings with stakeholder, and interviews with KI. The findings of the study revealed that Nigeria's over 220 million people with large youthful population, 170 Universities, 159 polytechnics, 47 accredited monotechnic, and 82 colleges of education present a strong strength for the country's climate actions. The findings of the study revealed that the states (sub-nationals) in Nigeria lack the capacity to adequately respond to the impacts of climate change. There is inadequate capacity among sector stakeholder, which inhibits the development of strategic partnerships between private sector, civil society, academia, and media organizations that would enable them to increase productivity. Stakeholder engagements are weakened by corrupt leadership at all levels, especially at the community level.

10.6 RECOMMENDATIONS

Based on the findings of the study, the following recommendations are made;

i. Involving all stakeholders from the Federal level down to the community level with appropriate and sustainable funding arrangements.
ii. Providing adequate and timely release of project funds and involvement of stakeholders in the planning process can go a long way in effectively engaging stakeholders in climate change actions in Nigeria.
iii. Bringing all stakeholders together to work as a team and enforcing participation from all relevant stakeholders.
iv. Holding regular meetings to enhance understanding and allocation of specific assignments to stakeholders that will be engaged.
v. Use of independent M&E framework dedicated for Climate Change stakeholder engagement. Appropriate evaluation teams should be constituted which is independent of the Steering or Technical Committees.

REFERENCES

Adger, W.N., Brown, K., Tompkins, E. and Young, K. (1999), "Report of the Consensus Building Stakeholder Workshop for Buccoo Reef Marine Park, Tobago", ODG Research Working Paper, April 1999, Overseas Development Group (ODG), University of East Anglia, Norwich.

Adger, W., Lorenzoni, I. and O'Brien, M. (2009), *Adapting to Climate Change: Thresholds, Values and Governance*, Cambridge University Press, Cambridge.

Akompab, D., Peng, B., Williams, S., Saniotis, A., Walker, I. and Augoustinos, M. (2012), "Engaging stakeholders in an adaptation process: governance and institutional arrangements in heat health policy development in Adelaide, Australia", *Mitigation and Adaptation Strategies for Global Change*, Vol. 18, No. 7, pp. 1001–1018, Springer, doi: 10.1007/s11027-012-9404-4.

Brown, K.W., Adger, E., Tompkins, P., Bacon, D. and Shim, K. (1998), "A framework for incorporating stakeholder participation in marine resource management: a case study in Tobago", CSERGE Working Paper GEC 98-23, Centre for Social and Economic Research on the Global Environment, University of East Anglia, Norwich.

Brown, K., Tompkins, E. and Adger, W.N. (2001), Trade-off Analysis for Participatory Coastal Zone Decision-Making, Overseas Development Group University of East Anglia, Norwich.

Conde, C. and Lonsdale, K. (2004). Engaging Stakeholders in the Adaptation Process. Technical Paper 2: Engaging Stakeholders in the Adaptation Process. pp. 49–66.

Federal Republic of Nigeria (FGN) (2021). Initial Adaptation Communication to the United Nations Framework Convention on Climate Change. Nigeria's Federal Ministry of Environment, Department of Climate Change.

Filho, W.F., Azeiteiro, U., Alves, F., Pace, P., Mifsud, M., Brandli, L., Caeiro, S. and Disterheft, A. (2018), "Reinvigorating the sustainable development research agenda: the role of the sustainable development goals (SDG)", *International Journal of Sustainable Development and World Ecology*, Vol. 25, No. 2, pp. 131–142, doi: 10.1080/13504509.2017.1342103.

Hemmati, M. (2002). *Multi-stakeholder Processes for Governance and Sustainability*, London: Earthscan.

IPCC (2014), "Climate change impacts. Adaptation, and vulnerability. Part A: Global and sectoral aspects", in Field, C. B., Barros, V. R., Estrada, R. C., Genova, B., Girma, E. S., Kissel, A. N., Levy, S., MacCracken, P. R. and Mastrandrea, L. L. (Eds), *Contribution of Working Group II to the Fifth Assessment Report of the Intergovernmental Panel on Climate Change*, Cambridge and New York: Cambridge University Press.

Mkonda, M.Y. (2021), "The underway to pragmatic implementations of sustainable and intensive agricultural systems in Tanzania", *Environmental and Sustainability Indicators*, Vol. 11, doi: 10.1016/j.indic.2021.100117.

Mkonda, M.Y. (2022), "Stakeholders' engagement in the process of adapting to climate change impacts. A case of central Tanzania", *Management of Environmental Quality, An International Journal*. https://www.emerald.com/insight/1477-7835.htm.

Mkonda, M.Y., He, X.H. and Festin, E.S. (2018), "Comparing smallholder farmers' perception of climate change with meteorological data: experiences from seven agro-ecological zones of Tanzania", *Weather Climate and Society*, Vol. 10, No. 3, doi: 10.1175/WCAS-D-17-0036.1.

Nicholson, S.E. (2018), "The ITCZ and the seasonal cycle over equatorial Africa", *Bulletin of the American Meteorological Society*, Vol. 99, pp. 337–348, doi: 10.1175/BAMS-D-16-0287.1.

Paavola, J. (2008), "Livelihoods, vulnerability and adaptation to climate change in Morogoro, Tanzania", CSERGE Working Paper EDM 04-12, Environmental Science and Policy, doi: 10.1016/j.envsci.2008.06.002.

Rainey, D.L. (2006), *Sustainable Business Development*, 1st ed., Cambridge: Cambridge University Press.

Rao, K., Ndegwa, W., Kizito, K. and Oyoo, A. (2011), "Climate variability and change: farmer perceptions and understanding of intra-seasonal variability in rainfall and associated risk in semi-arid Kenya", *Experimental Agriculture*, Vol. 47, pp. 267–291.

Tsufac, A., Awazi, N. and Yerima, B. (2021), "Characterization of agroforestry systems and their effectiveness in soil fertility enhancement in the south-west region of Cameroon", *Current Research in Environmental Sustainability*, Vol. 3, p. 100024.

UNPD Department of Economic and Social Affairs. (2019). World Population Prospects 2019. https://population.un.org/wpp/publications/files/wpp2019_highlights.pdf.

Vaidya, B., Shrestha, S. and Ghimire, A. (2021), "Water footprint assessment of food-water-energy systems at Kathmandu University, Nepal", *Current Research in Environmental Sustainability*, Vol. 3, p. 100044.

Vogel, B. and Henstra, D. (2015), "Studying local climate adaptation: heuristic research framework for comparative policy analysis", *Global Environmental Change*, Vol. 31. doi: 10.1016/j.gloenvcha.2015.01.001.

Wamsler, C. (2017), "Stakeholder involvement in strategic adaptation planning: transdisciplinary and co-production at stake?", *Environmental Science and Policy*, Vol. 75, pp. 148–157.

Index

Note: Bold page numbers refer to tables; *italic* page numbers refer to figures.

Printed in the United States
by Baker & Taylor Publisher Services